INTERNATIONAL UNION OF PURE AND APPLIED CHEMISTRY

ANALYTICAL CHEMISTRY DIVISION
COMMISSION ON SOLUBILITY DATA

SOLUBILITY DATA SERIES

Volume 32

HYDROGEN SULFIDE, DEUTERIUM SULFIDE AND HYDROGEN SELENIDE

SOLUBILITY DATA SERIES

Editor-in-Chief

A. S. KERTES
The Hebrew University
Jerusalem, Israel

EDITORIAL BOARD

A. F. M. Barton (Australia)

R. Battino (USA)

R. W. Cargill (UK)

H. L. Clever (USA)

Rosa Crovetto (Argentina)

W. Gerrard (UK)

L. H. Gevantman (USA)

W. Hayduk (Canada)

J. W. Lorimer (Canada)

A. E. Mather (Canada)

M. Salomon (USA)

D. G. Shaw (USA)

T. Tominaga (Japan)

R. P. T. Tomkins (USA)

Managing Editor

P. D. GUJRAL
IUPAC Secretariat, Oxford, UK

INTERNATIONAL UNION OF PURE AND APPLIED CHEMISTRY
IUPAC Secretariat: Bank Court Chambers, 2-3 Pound Way,
Cowley Centre, Oxford OX4 3YF, UK

NOTICE TO READERS

Dear Reader

If your library is not already a standing-order customer or subscriber to the Solubility Data Series, may we recommend that you place a standing order or subscription order to receive immediately upon publication all new volumes published in this valuable series. Should you find that these volumes no longer serve your needs, your order can be cancelled at any time without notice.

Robert Maxwell
Publisher at Pergamon Press

A complete list of volumes published in the Solubility Data Series will be found on p. 352.

SOLUBILITY DATA SERIES

Editor-in-Chief
A.S. KERTES

Volume 32

HYDROGEN SULFIDE, DEUTERIUM SULFIDE AND HYDROGEN SELENIDE

Volume Editors

PETER G. T. FOGG

*Polytechnic of North London
Holloway, London, UK*

COLIN L. YOUNG

*University of Melbourne
Parkville, Victoria, Australia*

Contributors

H. LAWRENCE CLEVER

*Emory University
Atlanta, Georgia, USA*

ELIZABETH L. BOOZER

*Emory University
Atlanta, Georgia, USA*

WALTER HAYDUK

*University of Ottawa
Ottawa, Ontario, Canada*

PERGAMON PRESS

OXFORD · NEW YORK · BEIJING · FRANKFURT
SÃO PAULO · SYDNEY · TOKYO · TORONTO

U.K.	Pergamon Press plc, Headington Hill Hall, Oxford OX3 0BW, England
U.S.A.	Pergamon Press, Inc., Maxwell House, Fairview Park, Elmsford, New York 10523, U.S.A.
PEOPLE'S REPUBLIC OF CHINA	Pergamon Press, Room 4037, Qianmen Hotel, Beijing, People's Republic of China
FEDERAL REPUBLIC OF GERMANY	Pergamon Press GmbH, Hammerweg 6, D-6242 Kronberg, Federal Republic of Germany
BRAZIL	Pergamon Editora Ltda, Rua Eça de Queiros, 346, CEP 04011, Paraiso, São Paulo, Brazil
AUSTRALIA	Pergamon Press Australia Pty Ltd., P.O. Box 544, Potts Point, N.S.W. 2011, Australia
JAPAN	Pergamon Press, 5th Floor, Matsuoka Central Building, 1-7-1 Nishishinjuku, Shinjuku-ku, Tokyo 160, Japan
CANADA	Pergamon Press Canada Ltd., Suite No. 271, 253 College Street, Toronto, Ontario, Canada M5T 1R5

First edition 1988

The Library of Congress has catalogued this serial title as follows:

Solubility data series. — Vol. 1 — Oxford; New York;
Pergamon, c 1979-
v.; 28 cm.
Separately cataloged and classified in LC before no. 18.
ISSN 0191-5622 = Solubility data series.
1. Solubility — Tables — Collected works.
QD543.S6629   541.3′42′05-dc19   85-641351
AACR 2   MARC-S

British Library Cataloguing in Publication Data

Hydrogen sulfide, deuterium sulfide and
hydrogen selenide.
1. Hydrogen compounds. Solubility
I. Fogg, P. G. T.   II. Young, Colin L.
III. Clever, H. Laurence   IV. Boozer,
Elizabeth L.   V. Haydak, Walter
IV. Series
546.2642

ISBN 0-08-032481-9

Printed in Great Britain by A. Wheaton & Co. Ltd., Exeter

CONTENTS

FOREWORD

*If the knowledge is
undigested or simply wrong,
more is not better*

How to communicate and disseminate numerical data effectively in chemical science and technology has been a problem of serious and growing concern to IUPAC, the International Union of Pure and Applied Chemistry, for the last two decades. The steadily expanding volume of numerical information, the formulation of new interdisciplinary areas in which chemistry is a partner, and the links between these and existing traditional subdisciplines in chemistry, along with an increasing number of users, have been considered as urgent aspects of the information problem in general, and of the numerical data problem in particular.

Among the several numerical data projects initiated and operated by various IUPAC commissions, the *Solubility Data Project* is probably one of the most ambitious ones. It is concerned with preparing a comprehensive critical compilation of data on solubilities in all physical systems, of gases, liquids and solids. Both the basic and applied branches of almost all scientific disciplines require a knowledge of solubilities as a function of solvent, temperature and pressure. Solubility data are basic to the fundamental understanding of processes relevant to agronomy, biology, chemistry, geology and oceanography, medicine and pharmacology, and metallurgy and materials science. Knowledge of solubility is very frequently of great importance to such diverse practical applications as drug dosage and drug solubility in biological fluids, anesthesiology, corrosion by dissolution of metals, properties of glasses, ceramics, concretes and coatings, phase relations in the formation of minerals and alloys, the deposits of minerals and radioactive fission products from ocean waters, the composition of ground waters, and the requirements of oxygen and other gases in life support systems.

The widespread relevance of solubility data to many branches and disciplines of science, medicine, technology and engineering, and the difficulty of recovering solubility data from the literature, lead to the proliferation of published data in an ever increasing number of scientific and technical primary sources. The sheer volume of data has overcome the capacity of the classical secondary and tertiary services to respond effectively.

While the proportion of secondary services of the review article type is generally increasing due to the rapid growth of all forms of primary literature, the review articles become more limited in scope, more specialized. The disturbing phenomenon is that in some disciplines, certainly in chemistry, authors are reluctant to treat even those limited-in-scope reviews exhaustively. There is a trend to preselect the literature, sometimes under the pretext of reducing it to manageable size. The crucial problem with such preselection - as far as numerical data are concerned - is that there is no indication as to whether the material was excluded by design or by a less than thorough literature search. We are equally concerned that most current secondary sources, critical in character as they may be, give scant attention to numerical data.

On the other hand, tertiary sources - handbooks, reference books and other tabulated and graphical compilations - as they exist today are comprehensive but, as a rule, uncritical. They usually attempt to cover whole disciplines, and thus obviously are superficial in treatment. Since they command a wide market, we believe that their service to the advancement of science is at least questionable. Additionally, the change which is taking place in the generation of new and diversified numerical data, and the rate at which this is done, is not reflected in an increased third-level service. The emergence of new tertiary literature sources does not parallel the shift that has occurred in the primary literature.

With the status of current secondary and tertiary services being as briefly stated above, the innovative approach of the *Solubility Data Project* is that its compilation and critical evaluation work involve consolidation and reprocessing services when both activities are based on intellectual and scholarly reworking of information from primary sources. It comprises compact compilation, rationalization and simplification, and the fitting of isolated numerical data into a critically evaluated general framework.

The *Solubility Data Project* has developed a mechanism which involves a number of innovations in exploiting the literature fully, and which contains new elements of a more imaginative approach for transfer of reliable information from primary to secondary/tertiary sources. *The fundamental trend of the Solubility Data Project is toward integration of secondary and tertiary services with the objective of producing in-depth critical analysis and evaluation which are characteristic to secondary services, in a scope as broad as conventional tertiary services.*

Fundamental to the philosophy of the project is the recognition that the basic element of strength is the active participation of career scientists in it. Consolidating primary data, producing a truly critically-evaluated set of numerical data, and synthesizing data in a meaningful relationship are demands considered worthy of the efforts of top scientists. Career scientists, who themselves contribute to science by their involvement in active scientific research, are the backbone of the project. The scholarly work is commissioned to recognized authorities, involving a process of careful selection in the best tradition of IUPAC. This selection in turn is the key to the quality of the output. These top experts are expected to view their specific topics dispassionately, paying equal attention to their own contributions and to those of their peers. They digest literature data into a coherent story by weeding out what is wrong from what is believed to be right. To fulfill this task, the evaluator must cover *all* relevant open literature. No reference is excluded by design and every effort is made to detect every bit of relevant primary source. Poor quality or wrong data are mentioned and explicitly disqualified as such. In fact, it is only when the reliable data are presented alongside the unreliable data that proper justice can be done. The user is bound to have incomparably more confidence in a succinct evaluative commentary and a comprehensive review with a complete bibliography to both good and poor data.

It is the standard practice that the treatment of any given solute-solvent system consists of two essential parts: I. Critical Evaluation and Recommended Values, and II. Compiled Data Sheets.

The Critical Evaluation part gives the following information:

(i)   a verbal text of evaluation which discusses the numerical solubility information appearing in the primary sources located in the literature. The evaluation text concerns primarily the quality of data after consideration of the purity of the materials and their characterization, the experimental method employed and the uncertainties in control of physical parameters, the reproducibility of the data, the agreement of the worker's results on accepted test systems with standard values, and finally, the fitting of data, with suitable statistical tests, to mathematical functions;

(ii) a set of recommended numerical data. Whenever possible, the set of recommended data includes weighted average and standard deviations, and a set of smoothing equations derived from the experimental data endorsed by the evaluator;

(iii) a graphical plot of recommended data.

The Compilation part consists of data sheets of the best experimental data in the primary literature. Generally speaking, such independent data sheets are given only to the best and endorsed data covering the known range of experimental parameters. Data sheets based on primary sources where the data are of a lower precision are given only when no better data are available. Experimental data with a precision poorer than considered acceptable are reproduced in the form of data sheets when they are the only known data for a particular system. Such data are considered to be still suitable for some applications, and their presence in the compilation should alert researchers to areas that need more work.

The typical data sheet carries the following information:

(i) components - definition of the system - their names, formulas and Chemical Abstracts registry numbers;
(ii) reference to the primary source where the numerical information is reported. In cases when the primary source is a less common periodical or a report document, published though of limited availability, abstract references are also given;
(iii) experimental variables;
(iv) identification of the compiler;
(v) experimental values as they appear in the primary source. Whenever available, the data may be given both in tabular and graphical form. If auxiliary information is available, the experimental data are converted also to SI units by the compiler.

Under the general heading of Auxiliary Information, the essential experimental details are summarized:

(vi) experimental method used for the generation of data;
(vii) type of apparatus and procedure employed;
(viii) source and purity of materials;
(ix) estimated error;
(x) references relevant to the generation of experimental data as cited in the primary source.

This new approach to numerical data presentation, formulated at the initiation of the project and perfected as experience has accumulated, has been strongly influenced by the diversity of background of those whom we are supposed to serve. We thus deemed it right to preface the evaluation/compilation sheets in each volume with a detailed discussion of the principles of the accurate determination of relevant solubility data and related thermodynamic information.

Finally, the role of education is more than corollary to the efforts we are seeking. The scientific standards advocated here are necessary to strengthen science and technology, and should be regarded as a major effort in the training and formation of the next generation of scientists and engineers. Specifically, we believe that there is going to be an impact of our project on scientific-communication practices. The quality of consolidation adopted by this program offers down-to-earth guidelines, concrete examples which are bound to make primary publication services more responsive than ever before to the needs of users. The self-regulatory message to scientists of the early 1970s to refrain from unnecessary publication has not achieved much. A good fraction of the literature is still cluttered with poor-quality articles. The Weinberg report (in 'Reader in Science Information', ed. J. Sherrod and A. Hodina, Microcard Editions Books, Indian Head, Inc., 1973, p. 292) states that 'admonition to authors to restrain themselves from premature, unnecessary publication can have little effect unless the climate of the entire technical and scholarly community encourages restraint...' We think that projects of this kind translate the climate into operational terms by exerting pressure on authors to avoid submitting low-grade material. The type of our output, we hope, will encourage attention to quality as authors will increasingly realize that their work will not be suited for permanent retrievability unless it meets the standards adopted in this project. It should help to dispel confusion in the minds of many authors of what represents a permanently useful bit of information of an archival value, and what does not.

If we succeed in that aim, even partially, we have then done our share in protecting the scientific community from unwanted and irrelevant, wrong numerical information.

A. S. Kertes

PREFACE

This volume of The Solubility Data Series is concerned with the solubility of hydrogen sulfide and of hydrogen selenide in aqueous and non-aqueous solvents. Data on the solubility of deuterium sulfide in deuterium oxide are also included. The editors believe that all relevant data published before January 1987 have been included but will be grateful for details of significant omissions.

Solubilities of hydrogen sulfide in a wide range of aqueous and non-aqueous solvents have been reported in the chemical literature. Much of this work is of a high standard. Reliance may be placed on data if there is consistency between measurements by different workers. In the case of some systems, however, there are wide divergences of data. As far as possible the evaluator has indicated which he considers to be reliable data where there are discrepancies but some systems merit re-examination using modern techniques. It is hoped that the publication of this, and similar volume will stimulate further experimental work on systems for which data is in doubt or is lacking.

In aqueous solutions of hydrogen sulfide equilibria involving hydrosulfide and sulfide ions play an important role in determining the solubility of the gas. If other weak acids and bases are also present there will be other equilibria established. Literature references to the theoretical treatment of some such complex systems are included but there has been no attempt to analyse the merits of different theoretical models for prediction of solubility which have been published. Similar equilibria exist in aqueous solutions of hydrogen selenide but these have received little study.

The editors recommend that all published reports of experimental measurements of the solubility of gases should include a complete record of primary experimental observations of temperature, pressure, composition of the phases etc. If such full publication is not possible then authors should endeavour to deposit such a record in an accessible data store.

The editors are grateful for support and encouragement from fellow members of the I.U.P.A.C. Commission on Solubility Data. In particular we wish to acknowledge help and advice given by Prof. A.L. Mather and we are grateful for the provision of data awaiting publication by Prof. W. Hayduk and by Prof. S. Lynn. In addition we would like to express our appreciation for the opportunity to use a daisy-wheel printer owned by Pergamon Press and to Lesley Flanagan for help in preparing the manuscript.

Peter Fogg   Colin Young

London   Melbourne

August 1987

THE SOLUBILITY OF GASES IN LIQUIDS

R. Battino, H. L. Clever and C. L. Young

INTRODUCTION

The Solubility Data Project aims to make a comprehensive search of the literature for data on the solubility of gases, liquids and solids in liquids. Data of suitable accuracy are compiled into data sheets set out in a uniform format. The data for each system are evaluated and where data of sufficient accuracy are available values recommended and in some cases a smoothing equation suggested to represent the variation of solubility with pressure and/or temperature. A text giving an evaluation and recommended values and the compiled data sheets are published on consecutive pages.

DEFINITION OF GAS SOLUBILITY

The distinction between vapor-liquid equilibria and the solubility of gases in liquids is arbitrary. It is generally accepted that the equilibrium set up at 300K between a typical gas such as argon and a liquid such as water is gas liquid solubility whereas the equilibrium set up between hexane and cyclohexane at 350K is an example of vapor-liquid equilibrium. However, the distinction between gas-liquid solubility and vapor-liquid equilibrium is often not so clear. The equilibria set up between methane and propane above the critical temperature of methane and below the critical temperature of propane may be classed as vapor-liquid equilibrium or as gas-liquid solubility depending on the particular range of pressure considered and the particular worker concerned.

The difficulty partly stems from our inability to rigorously distinguish between a gas, a vapor, and a liquid, which has been discussed in numerous textbooks. We have taken a fairly liberal view in these volumes and have included systems which may be regarded, by some workers, as vapor-liquid equilibria.

UNITS AND QUANTITIES

The solubility of gases in liquids is of interest to a wide range of scientific and technological disciplines and not solely to chemistry. Therefore a variety of ways for reporting gas solubility have been used in the primary literature and inevitably sometimes, because of insufficient available information, it has been necessary to use several quantities in the compiled tables. Where possible, the gas solubility has been quoted as a mole fraction of the gaseous component in the liquid phase. The units of pressure used are bar, pascal, millimeters of mercury and atmosphere. Temperatures are reported in Kelvin.

EVALUATION AND COMPILATION

The solubility of comparatively few systems is known with sufficient accuracy to enable a set of recommended values to be presented. This is true both of the measurement near atmospheric pressure and at high pressures. Although a considerable number of systems have been studied by at least two workers, the range of pressures and/or temperatures is often sufficiently different to make meaningful comparison impossible.

Occasionally, it is not clear why two groups of workers obtained very different sets of results at the same temperature and pressure, although both sets of results were obtained by reliable methods and are internally consistent. In such cases, sometimes an incorrect assessment has been given. There are several examples where two or more sets of data have been classified as tentative although the sets are mutually inconsistent.

Many high pressure solubility data have been published in a smoothed form. Such data are particularly difficult to evaluate, and unless specifically discussed by the authors, the estimated error on such values can only be regarded as an "informed guess".

Many of the high pressure solubility data have been obtained in a more general study of high pressure vapor-liquid equilibrium. In such cases a note is included to indicate that additional vapor-liquid equilibrium data are given in the source. Since the evaluation is for the compiled data, it is possible that the solubility data are given a classification which is better than that which would be given for the complete vapor-liquid data (or vice versa). For example, it is difficult to determine coexisting liquid and vapor compositions near the critical point of a mixture using some widely used experimental techniques which yield accurate high pressure solubility data. For example, conventional methods of analysis may give results with an expected error which would be regarded as sufficiently small for vapor-liquid equilibrium data but an order of magnitude too large for acceptable high pressure gas-liquid solubility.

It is occasionally possible to evaluate data on mixtures of a given substance with a member of a homologous series by considering all the available data for the given substance with other members of the homologous series. In this study the use of such a technique has been very limited.

The estimated error is often omitted in the original article and sometimes the errors quoted do not cover all the variables. In order to increase the usefulness of the compiled tables estimated errors have been included even when absent from the original article. If the error on *any* variable has been inserted by the compiler this has been noted.

PURITY OF MATERIALS

The purity of materials has been quoted in the compiled tables where given in the original publication. The solubility is usually more sensitive to impurities in the gaseous component than to liquid impurities in the liquid component. However, the most important impurities are traces of a gas dissolved in the liquid. Inadequate degassing of the absorbing liquid is probably the most often overlooked serious source of error in gas solubility measurements.

APPARATUS AND PROCEDURES

In the compiled tables brief mention is made of the apparatus and procedure. There are several reviews on experimental methods of determining gas solubilities and these are given in References 1-7.

METHODS OF EXPRESSING GAS SOLUBILITIES

Because gas solubilities are important for many different scientific and engineering problems, they have been expressed in a great many ways:

The Mole Fraction, x (g)

The mole fraction solubility for a binary system is given by:

$$x(g) = \frac{n(g)}{n(g) + n(l)}$$

$$= \frac{W(g)/M(g)}{[W(g)/M(g)] + [W(l)/M(l)]}$$

here n is the number of moles of a substance (an *amount* of substance), W is the mass of a substance, and M is the molecular mass. To be unambiguous, the partial pressure of the gas (or the total pressure) and the temperature of measurement must be specified.

The Weight Per Cent Solubility, wt%

For a binary system this is given by

$$wt\% = 100 \ W(g)/[W(g) + W(l)]$$

where W is the weight of substance. As in the case of mole fraction,
the pressure (partial or total) and the temperature must be specified.
The weight per cent solubility is related to the mole fraction solubility by

$$x(g) = \frac{[wt\%/M(g)]}{[wt\%/M(g)] + [(100 - wt\%)/M(1)]}$$

The Weight Solubility, C_w

The weight solubility is the number of moles of dissolved gas per
gram of solvent when the partial pressure of gas is 1 atmosphere.
The weight solubility is related to the mole fraction solubility at
one atmosphere partial pressure by

$$x(g) \text{ (partial pressure 1 atm)} = \frac{C_w M(1)}{1 + C_w M(1)}$$

where M(1) is the molecular weight of the solvent.

The Moles Per Unit Volume Solubility, n

Often for multicomponent systems the density of the liquid mixture is
not known and the solubility is quoted as moles of gas per unit volume of liquid mixture. This is related to the mole fraction solubility by

$$x = \frac{n \, v^o(1)}{1 + n \, v^o(1)}$$

where $v^o(1)$ is the molar volume of the liquid component.

The Bunsen Coefficient, α

The Bunsen coefficient is defined as the volume of gas reduced to
273.15K and 1 atmosphere pressure which is absorbed by unit volume of
solvent (at the temperature of measurement) under a partial pressure
of 1 atmosphere. If ideal gas behavior and Henry's law is assumed
to be obeyed,

$$\alpha = \frac{V(g)}{V(1)} \frac{273.15}{T}$$

where V(g) is the volume of gas absorbed and V(1) is the original
(starting) volume of absorbing solvent. The mole fraction solubility is related to the Bunsen coefficient by

$$x(g, 1 \text{ atm}) = \frac{\alpha}{\alpha + \dfrac{273.15}{T} \dfrac{v^o(g)}{v^o(1)}}$$

where $v^o(g)$ and $v^o(1)$ are the molar volumes of gas and solvent at a
pressure of one atmosphere. If the gas is ideal,

$$x(g) = \frac{\alpha}{\alpha + \dfrac{273.15R}{v^o(1)}}$$

Real gases do not follow the ideal gas law and it is important to es-
tablish the real gas law used for calculating α in the original publication and to make the necessary adjustments when calculating the
mole fraction solubility.

The Kuenen Coefficient, S

This is the volume of gas, reduced to 273.15K and 1 atmosphere pressure, dissolved at a partial pressure of gas of 1 atmosphere by 1
gram of solvent.

The Ostwald Coefficient, L

The Ostwald coefficient, L, is defined at the ratio of the volume of
gas absorbed to the volume of the absorbing liquid, all measured at
the same temperature:

$$L = \frac{V(g)}{V(l)}$$

If the gas is ideal and Henry's Law is applicable, the Ostwald coefficient is independent of the partial pressure of the gas. .It is
necessary, in practice, to state the temperature and total pressure
for which the Ostwald coefficient is measured. The mole fraction
solubility, x, is related to the Ostwald coefficient by

$$x(g) = \left[\frac{RT}{P(g)\ L\ v^{o}(l)} + 1 \right]^{-1}$$

where P is the partial pressure of gas. The mole fraction solubility
will be at a partial pressure of P(g).

The Absorption Coefficient, β

There are several "absorption coefficients", the most commonly used
one being defined as the volume of gas, reduced to 273.15K and 1 atmosphere, absorbed per unit volume of liquid when the total pressure is 1 atmosphere. β is related to the Bunsen coefficient by

$$\beta = \alpha(1-P(l))$$

where P(l) is the partial pressure of the liquid in atmosphere.

The Henry's Law Contant

A generally used formulation of Henry's Law may be expressed as

$$P(g) = K_H\ x(g)$$

where K_H is the Henry's Law constant and x the mole fraction solubility. Other formulations are

$$P(g) = K_2 C(l)$$

or

$$C(g) = K_c C(l)$$

where K_2 and K_c are constants, C the concentration, and (l) and (g)
refer to the liquid and gas phases. Unfortunately, K_H, K_2 and K_c
are all sometimes referred to as Henry's Law constants. Henry's
Law is a limiting law but can sometimes be used for converting solubility data from the experimental pressure to a partial gas pressure
of 1 atmosphere, provided the mole fraction of the gas in the liquid
is small, and that the difference in pressures is small. Great caution must be exercised in using Henry's Law.

The Mole Ratio, N

The mole ratio, N, is defined by

$$N = n(g)/n(l)$$

Table 1 contains a presentation of the most commonly used inter-con-
versions not already discussed.

For gas solubilities greater than about 0.01 mole fraction at a partial
pressure of 1 atmosphere there are several additional factors which must
be taken into account to unambiguously report gas solubilities. Solution
densities or the partial molar volume of gases must be known. Corrections
should be made for the possible non-ideality of the gas or the non-ap-
plicability of Henry's Law.

TABLE 1.   Interconversion of parameters used for reporting solubility.

$$L = \alpha(T/273.15)$$

$$C_w = \alpha/v_o \rho$$

$$K_H = \frac{17.033 \times 10^6 \rho(soln)}{\alpha\, M(l)} + 760$$

$$L = C_w\, v_{t,gas}\, \rho$$

where v_o is the molal volume of the gas in $cm^3\ mol^{-1}$ at 0°C, ρ the density of the solvent at the temperature of the measurement, ρ_{soln} the density of the solution at the temperature of the measurement, and $v_{t,gas}$ the molal volume of the gas ($cm^3\ mol^{-1}$) at the temperature of the measurement.

REFERENCES

1.   Battino, R.; Clever, H. L. *Chem. Rev.* 1966, *66*, 395.

2.   Clever, H. L.; Battino, R. in *Solutions and Solubilities*, Ed. M. R. J. Dack, J. Wiley & Sons, New York, 1975, Chapter 7.

3.   Hildebrand, J. H.; Prausnitz, J. M.; Scott, R. L. *Regular and Related Solutions*, Van Nostrand Reinhold, New York, 1970, Chapter 8.

4.   Markham, A. E.; Kobe, K. A. *Chem. Rev.* 1941, *63*, 449.

5.   Wilhelm, E.; Battino, R. *Chem. Rev.* 1973, *73*, 1.

6.   Wilhelm, E.; Battino, R.; Wilcock, R. J. *Chem. Rev.* 1977, *77*, 219.

7.   Kertes, A. S.; Levy, O.; Markovits, G. Y. in *Experimental Thermo-chemistry* Vol. II, Ed. B. Vodar and B. LeNaindre, Butterworth, London, 1974, Chapter 15.

Revised: December 1984 (CLY)

APPENDIX I.   Conversion Factors k and k^{-1}.

	k 1 (non-SI Unit) = k (SI Unit)	k^{-1} 1 (SI Unit) = k^{-1} (non-SI Unit)
LENGTH		SI Unit, m
Å (angstrom	1×10^{-10} (*)	1×10^{10} (*)
cm (centimeter)	1×10^{-2} (*)	1×10^{2} (*)
in (inch)	254×10^{-4} (*)	$3\ 937\ 008 \times 10^{-5}$
ft (foot)	$3\ 048 \times 10^{-4}$ (*)	$3\ 280\ 840 \times 10^{-6}$
AREA		SI Unit, m^2
cm^2	1×10^{-4} (*)	1×10^{4} (*)
in^2	$64\ 516 \times 10^{-8}$ (*)	$1\ 550\ 003 \times 10^{-3}$
ft^2	$9\ 290\ 304 \times 10^{-8}$ (*)	$1\ 076\ 391 \times 10^{-5}$
VOLUME		SI Unit, m^3
cm^3	1×10^{6} (*)	1×10^{6} (*)
in^3	$16\ 387\ 064 \times 10^{-12}$ (*)	$6\ 102\ 374 \times 10^{-2}$
ft^3	$2\ 831\ 685 \times 10^{-8}$	$3\ 531\ 467 \times 10^{-5}$
l (litre)	1×10^{-3} (*)	1×10^{3} (*)
UKgal (UK gallon)	$45\ 461 \times 10^{-7}$	$21\ 997 \times 10^{-2}$
USgal (US gallon)	$37\ 854 \times 10^{-7}$	$26\ 417 \times 10^{-2}$
MASS		SI Unit, kg
g (gram)	1×10^{-3} (*)	1×10^{3} (*)
t (tonne)	1×10^{3} (*)	1×10^{-3} (*)
lb (pound)	$45\ 359\ 237 \times 10^{-8}$ (*)	$2\ 204\ 623 \times 10^{-6}$
DENSITY		SI Unit, $kg\ m^{-3}$
$g\ cm^{-3}$	1×10^{3} (*)	1×10^{-3} (*)
$g\ l^{-1}$	1 (*)	1 (*)
$lb\ in^{-3}$	$2\ 767\ 991 \times 10^{-2}$	$3\ 612\ 728 \times 10^{-11}$
$lb\ ft^{-3}$	$1\ 601\ 847 \times 10^{-5}$	$6\ 242\ 795 \times 10^{-8}$
$lb\ UKgal^{-1}$	$99\ 776 \times 10^{-3}$	$100\ 224 \times 10^{-7}$
$lb\ USgal^{-1}$	$1\ 198\ 264 \times 10^{-4}$	$8\ 345\ 406 \times 10^{-9}$
PRESSURE		SI Unit, Pa (pascal, $kg\ m^{-1}\ s^{-2}$)
$dyn\ cm^{-2}$	1×10^{-1} (*)	1×10 (*)
at ($kgf\ cm^{-2}$)	$980\ 665 \times 10^{-1}$ (*)	$1\ 019\ 716 \times 10^{-11}$
atm (atmosphere)	$101\ 325$ (*)	$9\ 869\ 233 \times 10^{-12}$
bar	1×10^{5} (*)	1×10^{-5} (*)
$lbf\ in^{-2}$ (p.s.i.)	$6\ 894\ 757 \times 10^{-3}$	$1\ 450\ 377 \times 10^{-10}$
$lbf\ ft^{-2}$	$47\ 880 \times 10^{-3}$	$20\ 886 \times 10^{-6}$
inHg (inch of mercury)	$3\ 386\ 388 \times 10^{-3}$	$2\ 952\ 999 \times 10^{-10}$
mmHg (millimeter of mercury, torr)	$1\ 333\ 224 \times 10^{-4}$	$7\ 500\ 617 \times 10^{-9}$
ENERGY		SI Unit, J (joule, $kg\ m^2\ s^{-2}$)
erg	1×10^{-7} (*)	1×10^{7} (*)
cal_{IT} (I.T. calorie)	$41\ 868 \times 10^{-4}$ (*)	$2\ 388\ 459 \times 10^{-7}$
cal_{th} (thermochemical calorie)	$4\ 184 \times 10^{-3}$ (*)	$2\ 390\ 057 \times 10^{-7}$
kW h (kilowatt hour)	36×10^{5} (*)	$2\ 777\ 778 \times 10^{-13}$
l atm	$101\ 325 \times 10^{-3}$ (*)	$9\ 869\ 233 \times 10^{-9}$
ft lbf	$1\ 355\ 818 \times 10^{-6}$	$7\ 375\ 622 \times 10^{-7}$
hp h (horse power hour)	$2\ 684\ 519$	$3\ 725\ 062 \times 10^{-13}$
Btu (British thermal unit)	$1\ 055\ 056 \times 10^{-3}$	$9\ 478\ 172 \times 10^{-10}$

An asterisk (*) denotes an exact relationship.

COMPONENTS:	EVALUATOR:
1. Hydrogen sulfide; H₂S; [7783-06-4] 2. Water; H₂O; [7732-18-5]	Peter G.T. Fogg, Department of Applied Chemistry and Life Sciences, Polytechnic of North London, Holloway, London N7 8DB, U.K. February 1987

CRITICAL EVALUATION:

The solubility of hydrogen sulfide in water has been investigated by numerous workers since the nineteenth century. Some of this early work (1-3) is in good agreement with more recent work.

Hydrogen sulfide exists, in aqueous solution, in equilibrium with its ions. Solubilities which have been compiled correspond to bulk solubilities. For this purpose the number of moles of hydrogen sulfide in solution has been taken to be the sum of the number of moles of the uncharged H₂S plus those of HS- and those of S--.

Measurements have been made at pressures from 0.02 to 207 bar and temperatures from 273.2 to 603.2 K. The most extensive series of measurements over a pressure range were carried out by Lee & Mather (4) and by Selleck *et al.* (5). Above 344 K the mole fraction solubilities found by Lee & Mather tend to be higher than those found by Selleck *et al.* but there is better agreement at lower temperatures. However measurements indicate that, for many purposes, the mole fraction solubility can be considered to be close to a linear function of pressure to mole fraction solubilities of about 0.02. i.e. Henry's law is a useful approximation. The higher the concentration of hydrogen sulfide the greater the deviation from linear behaviour. The measurements of solubilities at partial pressures of hydrogen sulfide less then 1.013 bar made by Clarke & Glew (6) are consistent with the earlier measurements by Wright & Maass (7). When values for 298.2 K are extrapolated down to about 0.1 bar there is good agreement with solubilities measured by McLauchlan (1).

Many values of mole fraction solubilities for a partial pressure of hydrogen sulfide of 1.013 bar (1 atm) at temperatures in the range 273.2 K to 453.2 K have been published or may be obtained by extrapolation from published solubilities (1-13). Measurements published by Kozintseva (8) are the only set of measurements made in the range 453.2 K to 603.2 K. Available values indicate that mole fraction solubilities for a partial pressure of 1.013 bar decrease with increase in temperature to about 450 K but at higher temperatures values increase. Although there is no reason to doubt the reliability of the measurements reported by Kozintseva these high temperature values must be considered to be tentative until they are confirmed by other workers.

Mole fraction solubilities for a partial pressure of hydrogen sulfide of 1.013 bar, based upon measurements made since 1930, may be represented by the equation:

$$\ln x_{H_2S} = -24.912 + 3477.1/(T/K) + 0.3993 \ln (T/K) + 0.015700 \, T/K$$

The standard deviation for x_{H_2S} is given by:

$$\delta x_{H_2S} = \pm \, 0.000065$$

This equation is valid for temperatures from 283.2 K to 603.2 K. The variation of x_{H_2S} with T/K corresponding to this equation and the values of x_{H_2S} from experimental measurements made since 1930 are plotted in fig 1.

Values in the range 273.2 to 283.2 are better fitted by the equation:

$$\ln x_{H_2S} = -14.761 + 2507.2/(T/K)$$

$$\delta x_{H_2S} = \pm \, 0.000030$$

The average value of x_{H_2S} for a partial pressure of 1.013 bar at 298.15 K is 0.001830 ± 0.000007. This is based upon five sets of measurements (6,9, 11-13).

COMPONENTS:	EVALUATOR:
1. Hydrogen sulfide; H_2S; [7783-06-4] 2. Water; H_2O; [7732-18-5]	Peter G.T. Fogg, Department of Applied Chemistry and Life Sciences, Polytechnic of North London, Holloway, London N7 8DB, U.K. February 1987

CRITICAL EVALUATION:

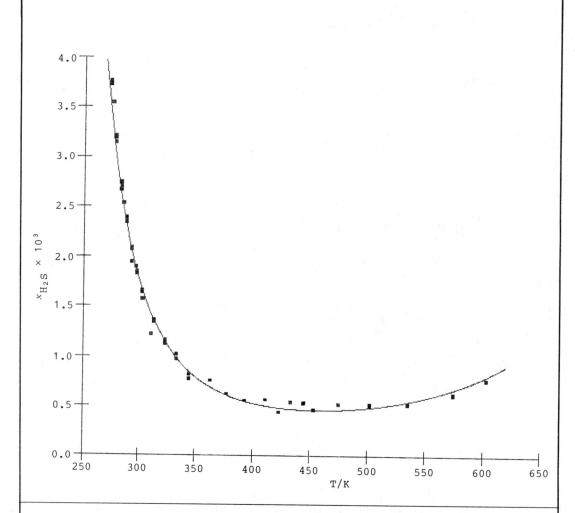

Fig. 1.   Variation with temperature of the mole fraction solubility of
hydrogen sulfide in water at a partial pressure of 1.013 bar from
experimental measurements published since 1930

Experimental points have been superimposed upon the curve corresponding to the
equation:

$$\ln x_{H_2S} = -24.912 + 3477.1/(T/K) + 0.3993 \ln (T/K) + 0.015700\ T/K$$

COMPONENTS:	EVALUATOR:
1. Hydrogen sulfide; H_2S; [7783-06-4] 2. Water; H_2O; [7732-18-5]	Peter G.T. Fogg, Department of Applied Chemistry and Life Sciences, Polytechnic of North London, Holloway, London N7 8DB, U.K. February 1987

CRITICAL EVALUATION:

References

1.   McLauchlan, W.H. *Zeit. physik. Chem.* 1903, *44*, 600-633.

2.   Winkler, L.W. *Zeit. physik. Chemie, (Leipzig)* 1906, *55*, 344-354.

3.   Pollitzer, F. *Zeit. anorg. Chem.* 1909, *64*, 121-148.

4.   Lee, J.I.; Mather, A.E. *Ber. Bunsenges. phys. Chem.* 1977, *81*, 1020-3.

5.   Selleck, F.T.; Carmichael, L.T; Sage, B.H.
     Ind. Eng. Chem. 1952, *44*, 2219-2226.

6.   Clarke, E.C.W.; Glew, D.N. *Can. J. Chem.* 1971, *49*, 691-698.

7.   Wright, R.H.; Maass, D. *Can. J. Research* 1932, *6*, 94-101.

8.   Kozintseva, T.N.; *Geokhimiya* 1964, *8*, 758-765.

9.   Kapustinsky, A.F.; Anvaer, B.I.
     Compt. Rend. Acad. Sci. URSS 1941, *30*, 625-628.

10. Byeseda, J.J. *Proc. Laurance Reid Gas. Cond. Conf.* 1985.

11. Gamsjäger, H.; Rainer, W.; Schindler, P.
     Monatsh. Chem. 1967, *98*, 1782-1802.

12. Douabul, A.A.; Riley, J.P. *Deep-Sea Res.* 1979, *26A*, 259-268.

13. Kiss, A. v.; Lajtai, I.; Thury, G.
     Zeit. anorg. allgem. Chem. 1937, *233*, 346-352.

COMPONENTS:	ORIGINAL MEASUREMENTS:
1. Hydrogen sulfide; H_2S; [7783-06-4] 2. Water; H_2O; [7732-18-5]	McLauchlan, W.H. Z. physik. Chem. <u>1903</u>, 44, 600-633.
VARIABLES: Pressure	PREPARED BY: P.G.T. Fogg

EXPERIMENTAL VALUES:

P_{H_2S}/mmHg	P_{H_2S}/bar [*]	Conc. of H_2S in solution / mol dm^{-3}
74.7	0.0996	0.0100
88.7	0.1183	0.01194
99.8	0.1331	0.0135
101.4	0.1352	0.01400
105.5	0.1407	0.0144

T/K = 298.2

   * calculated by compiler.

AUXILIARY INFORMATION

METHOD/APPARATUS/PROCEDURE:	SOURCE AND PURITY OF MATERIALS:
Solutions of hydrogen sulfide were prepared by dilution of a stock solution. The concentration of H_2S in solution was determined by iodimetry. Hydrogen, generated by electrolysis of water, was passed through a solution under test. The H_2S in the emergent gas stream was absorbed in copper sulfate solution and estimated from the change in electrical conductivity of this solution. The partial pressure of H_2S was calculated from the volume of hydrogen and the total pressure of the emergent gas stream.	1. prepared from FeS.
	ESTIMATED ERROR:
	REFERENCES:

COMPONENTS:	ORIGINAL MEASUREMENTS:
1. Hydrogen sulfide; H_2S; [7783-06-4] 2. Water; H_2O; [7732-18-5]	Winkler, L.W. *Zeit. phys. Chem. (Leipzig)* <u>1906</u>, *55*, 344-354.
VARIABLES: Temperature	PREPARED BY: P.G.T. Fogg

EXPERIMENTAL VALUES:

T/K	Bunsen absorption coeff. α	Mole fraction solubility, x_{H_2S}, (1.013 bar)*
273.2	4.621	0.003747
283.2	3.362	0.002729
293.2	2.554	0.002078
303.2	2.014	0.001643
313.2	1.642	0.001345
323.2	1.376	0.001132
333.2	1.176	0.000972

   *   calculated by the compiler.

AUXILIARY INFORMATION

METHOD/APPARATUS/PROCEDURE:	SOURCE AND PURITY OF MATERIALS:
No information	No information
	ESTIMATED ERROR:
	REFERENCES:

COMPONENTS:	ORIGINAL MEASUREMENTS:
1. Hydrogen sulfide; H_2S [7783-06-4] 2. Water; H_2O; [7732-18-5]	Wright, R.H.; Maass, O. *Can. J. Research*, <u>1932</u>, *6*, 94-101.

VARIABLES:	PREPARED BY:
Temperature, concentration	P.G.T. Fogg

EXPERIMENTAL VALUES:

T/K	Total pressure /mmHg	P_{H_2S}/mmHg	H_2S in vapor /mol dm^{-3}	H_2S in liquid /mol dm^{-3}	Mole fraction in liquid x_{H_2S}
278.15	274.5	268.0	0.0155	0.0635	0.00114
	560	553	0.0321	0.1302	0.00234
	838	831	0.0484	0.1910	0.00343
	1176	1169	0.0685	0.2682	0.00481
283.15	303.8	294.7	0.0168	0.0597	0.00107
	615	606	0.0346	0.1220	0.00219
	914	905	0.0518	0.1801	0.00324
	1279	1270	0.0731	0.2511	0.00450
	1567	1558	0.0900	0.3060	0.00548
	2112	2103	0.1221	0.4099	0.00733
288.15	333.3	320.6	0.0179	0.0560	0.00101
	670	657	0.0369	0.1144	0.00206
	991	978	0.0551	0.1693	0.00304
	1382	1369	0.0774	0.2346	0.00421
	1692	1679	0.0953	0.2877	0.00516
	2284	2271	0.1297	0.3866	0.00692

* calculated by compiler on the assumption that dissolution of H_2S caused negligible change in the volume of the liquid phase.

AUXILIARY INFORMATION

METHOD/APPARATUS/PROCEDURE:	SOURCE AND PURITY OF MATERIALS:
A measured quantity of water was run into the a cell of known volume attached to a glass diaphragm manometer. Dissolved air was removed by repeated freezing and melting *in vacuo*. A measured amount of H_2S was condensed in the cell by liquid air and the cell sealed. The temperature of the cell and manometer were then controlled to 0.1 °C. P_{H_2S} was taken to be $P_{total} - P_{H_2O}$ where P_{H_2O} is the vapor pressure of pure water. The weight of H_2S in the gas phase was calculated from P_{H_2S} and the volume of the gas phase. The difference between this weight and the total weight of H_2S used gave the weight of H_2S in solution.	1. From a cylinder; purified by fractional distillation as described in ref. (1). 2. Laboratory distilled and degassed.

	ESTIMATED ERROR: $\delta T = \pm 0.1$ K (authors)

REFERENCES:

1. Wright, R.H.; Maass, O.; *Can. J. Research*, <u>1931</u>, *5*, 436.

COMPONENTS:	ORIGINAL MEASUREMENTS:
1. Hydrogen sulfide; H₂S ; [7783-06-4] 2. Water; H₂O; [7732-18-5]	Wright, R.H.; Maass, O. Can. J. Research, 1932, 6, 94-101.

EXPERIMENTAL VALUES:

T/K	Total pressure /mmHg	P_{H_2S}/mmHg	H₂S in vapor /mol dm⁻³	H₂S in liquid /mol dm⁻³	Mole fraction * in liquid x_{H_2S}
293.15	362.8	345.4	0.0190	0.0528	0.00095
	724	707	0.0390	0.1074	0.00193
	1067	1050	0.0581	0.1594	0.00287
	1483	1466	0.0816	0.2188	0.00393
	1817	1800	0.1005	0.2696	0.00484
	2454	2437	0.1371	0.3642	0.00653
298.15	392.6	369.1	0.0199	0.0497	0.00090
	778	754	0.0409	0.1010	0.00182
	1144	1120	0.0610	0.1499	0.00270
	1581	1557	0.0851	0.2050	0.00369
	1935	1911	0.1049	0.2544	0.00458
	2622	2598	0.1437	0.3437	0.00617
303.15	422.8	391.3	0.0208	0.0470	0.00085
	830	798	0.0425	0.0955	0.00173
	1219	1187	0.0636	0.1413	0.00255
	1672	1640	0.0882	0.1932	0.00348
	2052	2020	0.1091	0.2398	0.00432
	2785	2753	0.1498	0.3247	0.00584
313.15	486.5	431.6	0.0222	0.0426	0.00077
	934	879	0.0454	0.0858	0.00156
	1370	1315	0.0682	0.1260	0.00229
	1853	1798	0.0937	0.1722	0.00312
	2278	2223	0.1162	0.2149	0.00389
	3095	3040	0.1603	0.2921	0.00529
323.15	562.4	470.4	0.0235	0.0387	0.00071
	1040	948	0.0474	0.0789	0.00144
	1522	1430	0.0719	0.1139	0.00207
	2033	1941	0.0979	0.1560	0.00284
	2505	2413	0.1223	0.1937	0.00352
	3402	3310	0.1690	0.2647	0.00480
333.15	652.2	503.3	0.0243	0.0359	0.00066
	1162	1013	0.0492	0.0730	0.00134
	1681	1532	0.0747	0.1045	0.00191
	2213	2064	0.1010	0.1440	0.00263
	2731	2582	0.1269	0.1777	0.00325
	3707	3558	0.1762	0.2424	0.00442

* calculated by compiler on the assumption that dissolution of H₂S caused negligible change in the volume of the liquid phase.

Values of x_{H_2S} from interpolated values of x_{H_2S}/P_{H_2S} supplied by authors:

T/K	P_{H_2S}=1 atm.	P_{H_2S}=2 atm.	P_{H_2S}=3 atm.
278.15	0.00321	0.00626	0.00921
283.15	0.00275	0.00543	0.00807
288.15	0.00239	0.00473	0.00707
293.15	0.00209	0.00416	0.00620
298.15	0.00185	0.00368	0.00549
303.15	0.00166	0.00330	0.00492
313.15	0.00136	0.00271	0.00404
323.15	0.00116	0.00228	0.00340
333.15	0.00102	0.00199	0.00296

760 mmHg = 1 atm = 1.01325 bar

COMPONENTS:	ORIGINAL MEASUREMENTS:
1. Hydrogen sulfide; H_2S [7783-06-4] 2. Water; H_2O; [7732-18-5]	Kapustinsky, A.F.; Anvaer, B.I. *Compt.Rend.Acad.Sci.URSS* <u>1941</u>, *30*, 625-8.
VARIABLES:	PREPARED BY: P.G.T. Fogg

EXPERIMENTAL VALUES:

T/K	P_{H_2S}/mmHg	Concn. of H_2S in liquid / mol dm^{-3}	Mole fraction in liquid* x_{H_2S}
298.2	760	0.1013	0.001826

* calculated by the compiler

$$760 \text{ mmHg} = 1.01325 \text{ bar}$$

AUXILIARY INFORMATION

METHOD/APPARATUS/PROCEDURE:	SOURCE AND PURITY OF MATERIALS:
Hydrogen sulfide, saturated with water vapor, was bubbled through a 300 cm³ thermostatted absorption vessel for 3 - 4 hours. Samples of solution were then withdrawn for analysis by iodimetry. Allowance was made for the vapor pressure of water.	1. from chemically pure hydrochloric acid and NaHS; CO_2 removed by $Ba(OH)_2$.

ESTIMATED ERROR:

$\delta T/K = \pm 0.1$ (authors)

REFERENCES:

COMPONENTS:	ORIGINAL MEASUREMENTS:
1. Hydrogen sulfide; H_2S; [7783-06-4] 2. Water; H_2O; [7732-18-5]	Selleck, F.T.; Carmichael, L.T.; Sage, B.H. _Ind. Eng. Chem._ 1952, 44, 2219-2226
VARIABLES: Temperature, pressure	PREPARED BY: P.G.T. Fogg

EXPERIMENTAL VALUES:

T/°F	T/K	Total pressure /(Lb/Sq.Inch Absolute)	Total pressure /bar*	Mole fraction of H_2S Gas phase	Liquid phase
100	310.93	100	6.89	0.9894	0.0082
		150	10.34	0.9925	0.0123
		200	13.79	0.9940	0.0165
		250	17.24	0.9949	0.0207
		300	20.68	0.9954	0.0250
		390.1†	26.90	0.9960	0.0333
160	344.26	100	6.89	0.9493	0.0050
		150	10.34	0.9643	0.0076
		200	13.79	0.9726	0.0102
		250	17.24	0.9771	0.0128
		300	20.68	0.9801	0.0154
		400	27.58	0.9837	0.0206
		500	34.47	0.9856	0.0258
		600	41.37	0.9865	0.0310
		700	48.26	0.9868	0.0364
		759.4†	52.36	0.9869	0.0395

* calculated by the compiler;
† three-phase equilibria involving aqueous liquid, H_2S-rich liquid, & gas.

AUXILIARY INFORMATION

METHOD/APPARATUS/PROCEDURE:	SOURCE AND PURITY OF MATERIALS:
The measurements were obtained as part of a study of the complete phase behaviour of the H_2S/H_2O system. The apparatus consisted of a spherical pressure vessel from which samples of gas or liquid phases could be withdrawn for analysis. The vessel was fitted with a metal diaphragm which responded to pressure differentials. The pressure within the system was measured by a pressure balance calibrated against the vapor pressure of CO_2 at the ice point (1).	1. prepared by hydrolysis of Al_2S_3; dried and fractionated. 2. distilled; deaerated at low pressure.

ESTIMATED ERROR:

$\delta P = \pm 0.1\%$ or 0.07 bar, whichever the greater.
$\delta T = \pm 0.02$ K; $\delta x_{H_2S}/x_{H_2S} = \pm 0.0015$ (authors)

REFERENCES:

1. Sage, B.H.; Lacey, W.N. _Trans.Am.Inst.Mining Met.Engrs._ 1948, _174_, 102.

COMPONENTS:	ORIGINAL MEAASUREMENTS:
1. Hydrogen sulfide; H_2S; [7783-06-4]	Selleck, F.T.; Carmichael, L.T.; Sage, B.H.
2. Water; H_2O; [7732-18-5]	*Ind. Eng. Chem.* <u>1952</u>, *44*, 2219-2226.

T/°F	T/K	Total pressure /(Lb/Sq.Inch Absolute)	Total pressure /bar*	Mole fraction of H_2S Gas phase	Liquid phase
220	377.59	200	13.79	0.9046	0.0077
		400	27.58	0.9477	0.0156
		600	41.37	0.9597	0.0230
		800	55.16	0.9647	0.0301
		1000	68.95	0.9664	0.0371
		1250	86.18	0.9665	0.0463
		1500	103.4	0.9651	0.0577
		1750	120.7	0.9630	0.0690[§]
		2000	137.9	0.9602	0.0823
		2250	155.1	0.9568	0.0973
		2500	172.4	0.9531	0.1145
		2750	189.6	0.9491	0.1346
		3000	206.8	0.9451	0.1586
280	410.93	200	13.79	0.7375	0.0057
		400	27.58	0.8589	0.0127
		600	41.37	0.8984	0.0191
		800	55.16	0.9155	0.0250
		1000	68.95	0.9248	0.0308
		1250	86.18	0.9307	0.0382
		1500	103.4	0.9330	0.0463
		1750	120.7	0.9327	0.0550
		2000	137.9	0.9303	0.0647[§]
		2250	155.1	0.9263	0.0750
		2500	172.4	0.9213	0.0860
		2750	189.6	0.9159	0.0978
		3000	206.8	0.9104	0.1106
340	444.26	200	13.79	0.3981	0.0029
		400	27.58	0.6828	0.0094
		600	41.37	0.7772	0.0155
		800	55.16	0.8224	0.0214
		1000	68.95	0.8466	0.0273
		1250	86.18	0.8646	0.0351
		1500	103.4	0.8742	0.0435
		1750	120.7	0.8788	0.0520
		2000	137.9	0.8797	0.0610[§]
		2250	155.1	0.8779	0.0705
		2500	172.4	0.8740	0.0805
		2750	189.6	0.8681	0.0910
		3000	206.8	0.8606	0.1024

* calculated by the compiler;

§ at this temperature the authors obtained the composition of aqueous layers at all higher pressures by extrapolation.

COMPONENTS:	ORIGINAL MEASUREMENTS:
1. Hydrogen sulfide; H_2S; [7783-06-4] 2. Water; H_2O; [7732-18-5]	Kozintseva, T.N. *Geokhimiya* <u>1964</u>, *8*, 758-765.
VARIABLES: Temperature, concentration	PREPARED BY: P.G.T. Fogg

EXPERIMENTAL VALUES:

T/K	Composition of gas phase			Composition of liquid phase			Henry's constant[†] /atm
	Conc.of H_2S /mol dm^{-3}	Mole ratio H_2S/H_2O	P_{H_2S} /atm	Conc.of H_2S /mol dm^{-3}	Mole ratio H_2S/H_2O	Mole fraction[*] x_{H_2S}	
433.2	0.00831	0.224	1.44	0.0440	0.00078	0.00078	1850
475.2	0.00618	0.0663	1.17	0.0339	0.00061	0.00061	1920
502.2	0.00507	0.0325	1.02	0.0296	0.000532	0.000532	1920
502.2	0.00419	0.0269	0.84	0.0234	0.000422	0.000422	2000
502.2	0.00725	0.0465	1.46	0.0423	0.000762	0.000761	1910
535.2	0.00792	0.0283	1.69	0.0494	0.000887	0.000886	1910
535.2	0.00507	0.0181	1.08	0.0319	0.000575	0.000575	1890
535.2	0.00583	0.0208	1.25	0.0368	0.000660	0.000660	1890
535.2	0.00677	0.0242	1.45	0.0432	0.000778	0.000777	1860
535.2	0.00675	0.0241	1.44	0.0428	0.000770	0.000769	1880
574.2	0.00740	0.0138	1.70	0.0585	0.00105	0.00105	1620
574.2	0.00531	0.0099	1.22	0.0428	0.000767	0.000766	1590
603.2	0.00860	0.0098	2.07	0.0884	0.00159	0.00159	1300
603.2	0.00746	0.0085	1.80	0.0756	0.00136	0.00136	1320

* calculated by compiler.

† Henry's constant = P_{H_2S}/x_{H_2S}

  1 atm = 760 mmHg = 1.01325 bar

AUXILIARY INFORMATION

METHOD/APPARATUS/PROCEDURE:	SOURCE AND PURITY OF MATERIALS:
A special bomb, from which samples of either gas phase or liquid phase could be withdrawn, was used. The bomb was flushed with oxygen-free nitrogen, water (50 cm^3) added and the apparatus saturated with H_2S (pressure not stated). The bomb and contents were then heated in a 'stirrer-furnace' to the required temperature for 25 to 50 hrs. A sample of the liquid phase was then analysed iodimetrically. A sample of the hot gas phase was passed through traps to collect the condensed water and through iodine solution. The H_2S in the hot gas phase was determined iodimetrically and the water content found from the weight of condensate.	No information
	ESTIMATED ERROR: $\delta T/K = \pm 2$ (author)
	REFERENCES:

COMPONENTS:	ORIGINAL MEASUREMENTS:
1. Hydrogen sulfide; H_2S; [7783-06-4] 2. Water; H_2O; [7732-18-5]	Clarke, E.C.W.; Glew, D.N. *Can. J. Chem.* 1971, 49, 691-698
VARIABLES: Temperature, pressure	PREPARED BY: C.L. Young

EXPERIMENTAL VALUES:

T/K	Total Pressure P/atm	P/kPa	H_2S in gas y_{H_2S}	H_2S in liquid $10^3 x_{H_2S}$	HS^- in liquid $10^6 x_{HS^-}$
273.151	0.53985	54.700	0.98964	2.0189	1.14
273.151	0.64663	65.520	0.99048	2.4198	1.25
273.150	0.81496	82.576	0.99240	3.0497	1.40
278.106	0.49205	49.857	0.98228	1.5515	1.11
278.108	0.57457	58.218	0.98478	1.8146	1.20
278.105	0.69533	69.533	0.98738	2.1988	1.32
278.105	0.89032	90.212	0.99008	2.8171	1.50
283.169	0.46154	46.766	0.97334	1.2365	1.09
283.168	0.54041	54.757	0.97717	1.4522	1.18
283.169	0.65667	66.537	0.98115	1.7697	1.31
283.169	0.84654	85.776	0.98529	2.2864	1.49
288.158	0.57265	58.024	0.97010	1.3234	1.24
288.158	0.70044	70.972	0.97546	1.6255	1.37
288.156	0.91275	92.484	0.98105	2.1258	1.57
293.160	0.49456	50.111	0.95265	0.9847	1.16
293.161	0.58325	59.098	0.95975	1.1689	1.27
293.161	0.71673	72.622	0.96712	1.4456	1.41
293.159	0.94115	95.362	0.97479	1.9091	1.62
298.168	0.55244	55.976	0.94250	0.9616	1.25
298.166	0.66766	67.651	0.95227	1.1729	1.38
298.165	0.85319	86.449	0.96245	1.5122	1.56
298.168	0.58192	58.963	0.94536	1.0162	1.28
298.167	0.71834	72.786	0.95557	1.2663	1.43
298.167	0.95039	96.298	0.96620	1.6904	1.65
298.180	0.55377	56.111	0.94259	0.9645	1.25
298.179	0.68442	69.349	0.95338	1.2042	1.39

AUXILIARY INFORMATION

METHOD/APPARATUS/PROCEDURE:	SOURCE AND PURITY OF MATERIALS:
Static equilibrium cell described in source. Vapor pressures over saturated aqueous solutions were measured with a fused quartz precision Bourdon gauge. Great care was taken to make the necessary correction for water in the vapor phase.	1. Prepared by the action of distilled water on aluminium sulfide (pure grade K and K Lab). 2. Distilled.
	ESTIMATED ERROR: $\delta T/K = 0.0015$; $\delta P/P = \pm 0.0002$; $\delta x_{H_2S}/x_{H_2S} = \pm 0.002$
	REFERENCES:

COMPONENTS:	ORIGINAL MEASUREMENTS:
1. Hydrogen sulfide; H_2S; [7783-06-4]	Clarke, E.C.W.; Glew, D.N.
2. Water; H_2O; [7732-18-5]	*Can. J. Chem.* 1971, *49*, 691-698

EXPERIMENTAL VALUES:

	Total Pressures		Mole fractions		
T/K	P/atm	P/kPa	H_2S in gas y_{H_2S}	H_2S in liquid $10^3 x_{H_2S}$	HS^- in liquid $10^6 x_{HS^-}$
298.181	0.90833	92.036	0.96465	1.6137	1.61
303.145	0.56822	57.575	0.92521	0.8680	1.27
303.142	0.70403	71.336	0.93942	1.0907	1.43
303.145	0.93731	94.973	0.95421	1.4719	1.66
313.141	0.49072	49.722	0.84998	0.5616	1.17
313.141	0.61249	62.061	0.87943	0.7245	1.33
313.142	0.82651	83.746	0.91016	1.0100	1.57
323.137	0.54699	55.424	0.77499	0.4801	1.22
323.134	0.68605	69.154	0.82000	0.6364	1.40
323.135	0.93305	94.541	0.86687	0.9132	1.68

Values of x_{HS^-} were calculated from the equation

$$\log_{10} K_{H_2S} = 55.06 - 3760 \, (T/K) - 20 \log (T/K)$$

COMPONENTS:	ORIGINAL MEASUREMENTS:
1. Hydrogen sulfide; H_2S; [7783-06-4] 2. Water; H_2O; [7732-18-5]	Lee, J.I.; Mather, A.E. *Ber. Bunsenges. phys. Chem.* <u>1977</u>, *81*, 1020-1023. (Numerical data deposited in the National Depository of Unpublished Data, Ottawa, Canada)*
VARIABLES: Temperature, pressure	PREPARED BY: P.G.T. Fogg

EXPERIMENTAL VALUES:

T/K	Total pressure /kPa	Mole fraction of H_2S in liquid	T/K	Total pressure /kPa	Mole fraction of H_2S in liquid
283.2	154.8	0.00405	293.2	1073.1	0.02058
	182.4	0.00500		1086.9	0.02068
	200.3	0.00556		1183.5	0.02212
	265.8	0.00692		1273.1	0.02409
	271.8	0.00716		1355.8	0.02519
	283.0	0.00697	303.2	214.5	0.00326
	292.4	0.00762		352.4	0.00556
	292.8	0.00747		507.5	0.00792
	361.7	0.00894		614.4	0.00936
293.2	173.4	0.00328		724.6	0.01124
	323.7	0.00649		934.8	0.01433
	452.6	0.00916		1024.6	0.01590
	569.8	0.01137		1196.8	0.01790
	684.7	0.01340		1273.4	0.01923
	783.3	0.01500		1278.9	0.01945
	824.7	0.01577		1466.4	0.02203
	835.0	0.01605		1569.8	0.02337
	955.9	0.01817		1590.5	0.02355
	976.6	0.01821		1623.4	0.02430
	1045.6	0.01986		1645.7	0.02426

* National Depository of Unpublished Data, National Science Library,
  National Research Council, Ottawa, Ontario, K1A OS2, Canada

AUXILIARY INFORMATION

METHOD/APPARATUS/PROCEDURE:	SOURCE AND PURITY OF MATERIALS:
A recirculating equilibrium cell was used. It consisted of a high pressure liquid level gauge and vapor reservoir. The gas phase was circulated by a magnetic pump. Samples of the liquid phase could be withdrawn. The total quantity of H_2S in a sample was found from the volume of gas evolved when the pressure over the sample dropped to atmospheric and the residual gas in solution, as measured by gas-liquid chromatography.	1. Matheson C.P. grade. 2. Laboratory distilled. ESTIMATED ERROR: $\delta T/K = \pm 0.5$ below 403 K; $\delta T/K = \pm 0.7$ above 403 K (authors) REFERENCES:

COMPONENTS:	ORIGINAL MEASUREMENTS:
1. Hydrogen sulfide; H_2S; [7783-06-4] 2. Water; H_2O; [7732-18-5]	Lee, J.I.; Mather, A.E. *Ber. Bunsenges. phys. Chem.* 1977, *81*, 1020-1023. (Numerical data deposited in the National Depository of Unpublished Data, Ottawa, Canada)*

EXPERIMENTAL VALUES:

T/K	Total pressure /kPa	Mole fraction of H_2S in liquid	T/K	Total pressure /kPa	Mole fraction of H_2S in liquid
303.2	1659.5	0.02453	313.2	2452.9	0.02822
	1758.9	0.02582		2458.9	0.02926
	1900.5	0.02806		2479.4	0.02873
	1945.3	0.02873		2561.6	0.02983
	1955.9	0.02833	323.1	718.3	0.00783
	1997.0	0.02956		892.9	0.00963
	2034.3	0.02993		1005.5	0.01090
	2059.1	0.03054		1165.5	0.01231
	2079.1	0.03063		1510.9	0.01542
	2106.7	0.03077		1510.9	0.01525
	2169.0	0.03195		1683.2	0.01750
	2262.1	0.03275		1848.4	0.01878
	2265.5	0.03292		1928.1	0.01997
313.2	472.0	0.00620		1940.1	0.01996
	730.4	0.00915		1976.3	0.01975
	970.8	0.01249		2010.1	0.02022
	1070.6	0.01373		2048.1	0.02048
	1133.0	0.01434		2133.8	0.02114
	1235.8	0.01565		2178.0	0.02210
	1469.9	0.01810		2265.8	0.02327
	1534.6	0.01950		2383.1	0.02383
	1583.1	0.01979		2482.5	0.02458
	1609.1	0.01941		2518.3	0.02437
	1636.8	0.01975		2637.1	0.02588
	1671.4	0.02013		2637.1	0.02594
	1671.5	0.02061		2658.1	0.02564
	1707.2	0.02077		2711.2	0.02640
	1788.3	0.02126		2737.2	0.02663
	1892.8	0.02282		2744.5	0.02743
	1934.1	0.02334		2755.4	0.02710
	1981.9	0.02347		2806.1	0.02779
	1983.1	0.02398		2808.4	0.02742
	2052.2	0.02408		2816.5	0.02722
	2059.9	0.02498		2893.7	0.02822
	2061.3	0.02517	333.2	240.6	0.00211
	2100.7	0.02526		406.1	0.00365
	2100.8	0.02546		575.0	0.00530
	2128.3	0.02529		836.9	0.00769
	2227.0	0.02612		1195.4	0.01100
	2417.4	0.02826		1541.7	0.01425

* National Depository of Unpublished Data, National Science Library, National Research Council, Ottawa, Ontario, K1A OS2, Canada.

COMPONENTS:	ORIGINAL MEASUREMENTS:
1. Hydrogen sulfide; H_2S; [7783-06-4]	Lee, J.I.; Mather, A.E.
2. Water; H_2O; [7732-18-5]	*Ber. Bunsenges. phys. Chem.* 1977, *81*, 1020-1023. (Numerical data deposited in the National Depository of Unpublished Data, Ottawa, Canada)*

EXPERIMENTAL VALUES:

T/K	Total pressure /kPa	Mole fraction of H_2S in liquid	T/K	Total pressure /kPa	Mole fraction of H_2S in liquid
333.2	1650.4	0.01546	333.2	3961.9	0.03402
	1762.3	0.01626		4003.6	0.03425
	1914.0	0.01779		4111.8	0.03487
	2015.6	0.01858		4160.6	0.03497
	2077.6	0.01913		4160.7	0.03516
	2105.2	0.01951		4217.2	0.03557
	2148.4	0.01967		4223.6	0.03553
	2284.5	0.02118		4224.1	0.03525
	2325.8	0.02141	344.2	374.8	0.00277
	2346.5	0.02168		546.3	0.00421
	2348.4	0.02127		760.9	0.00583
	2444.0	0.02221		802.2	0.00622
	2546.5	0.02322		829.0	0.00647
	2554.3	0.02304		1022.9	0.00783
	2560.3	0.02292		1050.5	0.00814
	2567.4	0.02333		1132.4	0.00914
	2569.0	0.02328		1450.3	0.01147
	2679.6	0.02436		1492.0	0.01184
	2692.0	0.02454		1518.5	0.01210
	2693.1	0.02438		1780.5	0.01429
	2843.2	0.02562		1905.7	0.01478
	2871.3	0.02615		2181.4	0.01720
	3038.1	0.02717		2214.9	0.01735
	3098.6	0.02765		2511.3	0.01981
	3174.7	0.02821		2636.5	0.02012
	3210.5	0.02863		2877.5	0.02230
	3250.0	0.02922		2919.2	0.02280
	3257.1	0.02885		3091.5	0.02406
	3326.1	0.02974		3215.4	0.02515
	3415.7	0.03046		3381.1	0.02622
	3457.1	0.03052		3546.3	0.02646
	3498.7	0.03102		3911.7	0.03029
	3601.6	0.03184		4194.4	0.03196
	3651.5	0.03175		4711.5	0.03546
	3652.0	0.03204		4757.3	0.03518
	3734.2	0.03281		5091.7	0.03662
	3844.7	0.03358		5112.3	0.03694
	3877.9	0.03407		5126.1	0.03715
	3892.9	0.03397	363.2	238.2	0.00126
	3899.7	0.03380		514.0	0.00317

* National Depository of Unpublished Data, National Science Library, National Research Council, Ottawa, Ontario, K1A OS2, Canada.

COMPONENTS:	ORIGINAL MEASUREMENTS:
1. Hydrogen sulfide; H_2S; [7783-06-4]	Lee, J.I.; Mather, A.E.
2. Water; H_2O; [7732-18-5]	*Ber. Bunsenges. phys. Chem.* 1977, *81*, 1020-1023. (Numerical data deposited in the National Depository of Unpublished Data, Ottawa, Canada)*

EXPERIMENTAL VALUES:

T/K	Total pressure /kPa	Mole fraction of H_2S in liquid	T/K	Total pressure /kPa	Mole fraction of H_2S in liquid
363.2	896.5	0.00577	363.2	6574.0	0.03943
	1217.1	0.00793	393.2	496.2	0.00164
	1530.8	0.01012		858.2	0.00375
	1834.2	0.01217		1154.7	0.00547
	2120.3	0.01370		1499.4	0.00707
	2341.0	0.01522		1762.1	0.00860
	2616.6	0.01729		2072.3	0.01050
	2733.6	0.01786		2427.4	0.01236
	2774.9	0.01823		2617.0	0.01367
	2850.8	0.01855		2941.2	0.01536
	2988.9	0.01959		3196.3	0.01666
	3216.4	0.02092		3527.2	0.01831
	3264.4	0.02142		3610.0	0.01887
	3436.4	0.02231		3891.9	0.02043
	3532.7	0.02299		3954.7	0.02073
	3533.3	0.02300		4023.6	0.02116
	3698.4	0.02403		4205.6	0.02199
	3843.2	0.02481		4643.4	0.02426
	4029.3	0.02620		4849.8	0.02545
	4222.4	0.02749		4919.2	0.02571
	4256.9	0.02756		4953.2	0.02590
	4470.6	0.02841		5518.6	0.02879
	4574.0	0.02925		5746.5	0.02984
	4733.5	0.03006		5863.3	0.03050
	4939.4	0.03132		5987.4	0.03099
	4954.2	0.03129		6049.5	0.03168
	5091.1	0.03205		6194.7	0.03206
	5126.5	0.03223		6208.5	0.03177
	5229.0	0.03311		6228.7	0.03223
	5264.4	0.03323		6394.6	0.03317
	5277.3	0.03317		6394.7	0.03315
	5526.0	0.03438		6491.1	0.03332
	5608.8	0.03456		6573.9	0.03359
	5609.2	0.03469		6587.3	0.03388
	5622.5	0.03500		6670.4	0.03416
	5642.7	0.03500	423.2	692.5	0.00095
	5712.6	0.03533		933.8	0.00216
	5828.8	0.03610		1154.4	0.00333
	6043.1	0.03709		1388.8	0.00442
	6332.7	0.03804		1471.4	0.00499

* National Depository of Unpublished Data, National Science Library, National Research Council, Ottawa, Ontario, K1A OS2, Canada.

COMPONENTS:	ORIGINAL MEASUREMENTS:
1. Hydrogen sulfide; H_2S; [7783-06-4] 2. Water; H_2O; [7732-18-5]	Lee, J.I.; Mather, A.E. *Ber. Bunsenges. phys. Chem.* 1977, *81*, 1020-1023. (Numerical data deposited in the National Depository of Unpublished Data, Ottawa, Canada)*

EXPERIMENTAL VALUES:

T/K	Total pressure /kPa	Mole fraction of H_2S in liquid	T/K	Total pressure /kPa	Mole fraction of H_2S in liquid
423.2	1568.9	0.00552	423.2	5560.2	0.02520
	1741.0	0.00635		5863.6	0.02650
	1768.2	0.00626		6277.3	0.02822
	1898.9	0.00749		6339.3	0.02896
	2031.8	0.00787		6401.4	0.02897
	2078.2	0.00816		6601.3	0.02993
	2223.6	0.00863	453.2	1071.3	0.00032
	2230.5	0.00877		1422.9	0.00199
	2376.5	0.00957		1865.1	0.00397
	2575.2	0.01036		2154.7	0.00531
	2775.2	0.01147		2354.6	0.00621
	2980.0	0.01268		2451.1	0.00657
	3051.0	0.01283		2540.8	0.00700
	3202.0	0.01364		2629.3	0.00742
	3450.2	0.01498		3229.2	0.01033
	3739.8	0.01644		3484.3	0.01158
	4105.2	0.01846		3911.7	0.01356
	4256.9	0.01893		4242.7	0.01497
	4403.4	0.01978		4615.0	0.01650
	4623.0	0.02056		5076.9	0.01866
	4691.9	0.02109		5421.7	0.02029
	5133.2	0.02324		5918.1	0.02257
	5222.8	0.02347			

T/K	Total pressure /kPa	Mole fraction of H_2S in vapour	T/K	Total pressure /kPa	Mole fraction of H_2S in vapour
363.2	1921	0.9614	423.2	1942	0.7409
	2011	0.9606		2196	0.7812
	2389	0.9709		2624	0.8199
393.2	1486	0.8865		2699	0.8373
	2500	0.9298		3003	0.8292
	3161	0.9415		3079	0.8356
	3174	0.9529			
	3230	0.9314			
	3401	0.9526			

* National Depository of Unpublished Data, National Science Library,
  National Research Council, Ottawa, Ontario, K1A OS2, Canada.

COMPONENTS:	ORIGINAL MEASUREMENTS:
1. Hydrogen sulfide; H_2S; [7783-06-4] 2. Water; H_2O; [7732-18-5]	Byeseda, J.J.; Deetz, J.A.; Manning, W.P. *Proc.Laurance Reid Gas Cond.Conf.* 1985.

VARIABLES:	PREPARED BY:
	P.G.T. Fogg

EXPERIMENTAL VALUES:

T/K	P_{H_2S}/psia	P_{H_2S}/bar*	Ostwald coeff. L	Mole fraction in liquid* x_{H_2S}
297.1	14.73	1.016	2.5	0.0019

* calculated by compiler

AUXILIARY INFORMATION	

METHOD/APPARATUS/PROCEDURE:	SOURCE AND PURITY OF MATERIALS:
The H_2S was contained in a thermostatted metal cylinder connected to a pressure gage, vacuum pump and supply of gas. A tight fitting internal piston sealed with an O-ring fitted into the cylinder so that the volume of gas could be changed by controlled movement of the piston. A measured volume of solvent was injected into the cylinder by a syringe. The absorption of gas was found from the movement of the piston which was necessary to maintain constant pressure.	No information.
	ESTIMATED ERROR:
	REFERENCES:

COMPONENTS:	EVALUATOR:
1. Hydrogen sulfide; H₂S; [7783-06-4] 2. Aqueous solutions of weak or non-electrolytes	Peter G.T. Fogg, Department of Applied Chemistry and Life Sciences, Polytechnic of North London, Holloway, London N7 8DB, U.K. July 1987

CRITICAL EVALUATION:

Carbon dioxide; CO_2; [124-38-9]
Phenol; C_6H_6O; [108-95-2]
1,2,3-Propanetriol, (*glycerol*); $C_3H_8O_3$; [56-81-5]
Ethanol; C_2H_6O; [64-17-5]
Urea; CH_4N_2O; [57-13-6]

The solubility of hydrogen sulfide in water containing carbon dioxide was measured by Golutvin, Malysheva & Skorobogatova (1). It was part of a study of the ratios of concentrations of H_2S/CO_2 in the gas phase to that in the liquid phase, in the presence of aqueous solutions. Water at 293.2 K was saturated with various mixtures of hydrogen sulfide and carbon dioxide. The total pressure of gas (CO_2 + H_2S + H_2O) was equal to barometric pressure. Within the limits of the experimental error the molar concentration of hydrogen sulfide in the saturated solution was proportional to the partial pressure of the hydrogen sulfide. The varying partial pressure of carbon dioxide did not cause any deviation from this linear relationship between solubility and partial pressure of hydrogen sulfide. Extrapolation of the solubility to a partial pressure of 1.013 bar gives a solubility of 0.098 mol dm⁻³ at 293.2 K. This may be compared with the solubility of 0.112 mol dm⁻³ calculated from the recommended value of the mole fraction solubility for H_2S in pure water (see Critical Evaluation - pure water).

In addition, Golutvin *et al.*(1) measured solubilities of mixtures of carbon dioxide and hydrogen sulfide in aqueous solutions of ammonia and of ammonia and phenol. There is no obvious inconsistency between these measurements of solubility in ammonia solutions and measurements made by other workers (see Critical Evaluation - aqueous solutions of ammonia). No other measurements of solubilities in the presence of phenol are available for comparison.

Kiss, Lajtai & Thury (2) measured solubilities of hydrogen sulfide in pure water and aqueous solutions of ethanol, glycerol and of urea in the temperature range 273.2 K to 298.15 K. The total pressure was barometric but the reported solubilities were corrected to a partial pressure of hydrogen sulfide of 1.013 bar. Solubilities in pure water are within a standard deviation of the recommended solubilities (see Critical Evaluation - pure water) and are therefore consistent with measurements by other workers.

Concentrations of ethanol ranged from 2 mol dm⁻³ to 16 mol dm⁻³. Concentrations of glycerol and of urea ranged from 2 mol dm⁻³ to 8 mol dm⁻³. The solubilities in ethanol solutions show a self-consistent variation with temperature and concentration of ethanol. They may be extrapolated to give estimated solubilities in pure ethanol. i.e, ethanol at a concentration of approximately 17.6 mol dm⁻³. The extrapolated value for 273.2 K is about 6% lower than the solubility calculated from direct measurements of the solubility in pure ethanol made by Gerrard (3). The extrapolated solubility for 285.5 K is about 40% greater and the extrapolated value for 298.15 K about 100% greater than solubilities estimated from Gerrard's data.

Solubility measurements for aqueous solutions of glycerol are self-consistent but, in this case, the concentrations of glycerol do not extend high enough to enable extrapolation to pure glycerol (approx. 13.7 mol dm⁻³). Comparison with other systems is not possible for this system or for the urea-water-hydrogen sulfide system.

References

1.   Golutvin, Yu. M.; Malysheva, T.V.; Skorobogatova, V.I.
       Izvest. Sibir. Otdel. Akad. Nauk. SSSR 1958, *No.8*, 83-87.

2.   Kiss, A.; Lajtai, I.; Thury, G.
       Z. anorg. allgem. Chem. 1937, *233*, 346-352.

3.   Gerrard, W. *J. Appl. Chem. Biotechnol.* 1972, *22*, 623-650.

COMPONENTS:	ORIGINAL MEASUREMENTS:
1. Hydrogen sulfide; H_2S; [7783-06-4] 2. Carbon dioxide; CO_2; [124-38-9] 3. Water; H_2O; [7732-18-5]	Golutvin, Yu. M.; Malysheva, T.V.; Skorobogatova, V.I. *Izvest. Sibir. Otdel. Akad. Nauk. S.S.S.R.* 1958, *No.8, 83-7.*

VARIABLES:	PREPARED BY:
Concentration of ammonia	P.G.T. Fogg

EXPERIMENTAL VALUES:

	Liquid phase		Gas phase	
Conc. of H_2S /mol dm^{-3}	Conc. of CO_2 /mol dm^{-3}	P_{H_2S} / P_{CO_2}	P_{H_2S}/bar[*]	P_{CO_2}/bar[*]
0.000000	0.030170	0.0000	0.0000	0.9899
0.001425	0.025979	0.0338	0.0324	0.9575
0.003377	0.020680	0.0378	0.0361	0.9538
0.011310	0.018775	0.1263	0.1110	0.8789
0.012310	0.018070	0.2208	0.1790	0.8109
0.015380	0.021050	0.1927	0.1599	0.8300
0.017690	0.019175	0.3044	0.2310	0.7589
0.018500	0.017340	0.4000	0.2828	0.7071
0.020400	0.017342	0.3871	0.2762	0.7137
0.021100	0.019970	0.3973	0.2815	0.7084
0.038720	0.014550	0.7084	0.4105	0.5794
0.040200	0.014540	0.6881	0.4035	0.5864
0.043380	0.011980	0.8133	0.4440	0.5459
0.048630	0.011810	0.7740	0.4319	0.5580
0.052400	0.006827	1.6650	0.6184	0.3715
0.056920	0.010088	1.2220	0.5444	0.4455
0.072746	0.004295	3.7350	0.7808	0.2091
0.082180	0.002274	3.9930	0.7916	0.1983
0.091180	0.001491	16.4510	0.9331	0.0568

T/K = 293.2;   [*]calculated by the compiler on the assumption that the sum of the partial pressure of H_2S, CO_2 & H_2O is 1.013 bar. Concentrations of H_2S and of CO_2 include all ionic species derived from these compounds.

AUXILIARY INFORMATION

METHOD/APPARATUS/PROCEDURE:	SOURCE AND PURITY OF MATERIALS:
Premixed hydrogen sulfide and carbon dioxide were bubbled through water in a thermostatted saturator for 4.5 to 5 hrs. The gas and the liquid phases were analysed by chemical methods. The solubility of H_2S, in the absence of CO_2, was measured by the authors but there is clearly a mistake in the figures which were published.	1. from a Kipp's apparatus. 2. by action of hydrochloric acid on marble chips.
	ESTIMATED ERROR: $\delta T/K$ = ± 0.1 (authors)
	REFERENCES:

COMPONENTS:	ORIGINAL MEASUREMENTS:
1. Hydrogen sulfide; H_2S; [7783-06-4] 2. Water; H_2O; [7732-18-5] 3. Urea; CH_4N_2O; [57-13-6]	Kiss, A. v.; Lajtai, I.; Thury, G. *Zeit. anorg. allgem. Chem.* 1937, *233*, 346-352.

VARIABLES:	PREPARED BY:
Temperature, concentration of urea.	P.G.T. Fogg

EXPERIMENTAL VALUES:

Conc. of urea /mol dm^{-3}	Conc. of H_2S in liquid /mol dm^{-3}			Ostwald coefficient, L		
	273.20 K	285.65 K	298.15 K	273.20 K	285.65 K	298.15 K
0	0.2079	0.1410	0.1014	4.662	3.307	2.482
2	0.2095	0.1445	0.1055	4.698	3.389	2.582
4	0.2110	0.1462	0.1081	4.732	3.424	2.646
6	0.2129	0.1487	0.1103	4.774	3.487	2.698
8		0.1497	0.1117		3.511	2.734

P_{H_2S} = 1.013 bar

AUXILIARY INFORMATION

METHOD/APPARATUS/PROCEDURE:	SOURCE AND PURITY OF MATERIALS:
H_2S was bubbled through about 100 cm^3 of water & urea in an absorption vessel at barometric pressure for 1 to 1.5 h. Dissolved H_2S was estimated iodimetrically. Allowance was made for the partial pressure of water and the head of liquid in the absorption vessel in the calculation of the partial pressure of H_2S. Solubilities were corrected to a partial pressure of 1.013 bar by use of Henry's law.	1. From iron sulfide (ex Merck) and dilute H_2SO_4; washed with H_2O and solvent mixture. 2. Distilled. 3. *p.a.* or *puriss.* from Merck or Kahlbaum-Schering;

ESTIMATED ERROR:
Quoted values are averages from several measurements with a maximum deviation from the mean of ± 0.2% (authors)

REFERENCES:

COMPONENTS:	ORIGINAL MEASUREMENTS:
1. Hydrogen sulfide; H_2S; [7783-06-4] 2. Water; H_2O; [7732-18-5] 3. Ethanol; C_2H_6O; [64-17-5]	Kiss, A. v.; Lajtai, I.; Thury, G. *Zeit. anorg. allgem. Chem.* <u>1937</u>, *233*, 346-352.

VARIABLES:	PREPARED BY:
Temperature, concentration of ethanol.	P.G.T. Fogg

EXPERIMENTAL VALUES:

Conc. of ethanol /mol dm^{-3}	Conc. of H_2S in liquid /mol dm^{-3}			Ostwald coefficient, L		
	273.20 K	285.65 K	298.15 K	273.20 K	285.65 K	298.15 K
0	0.2079	0.1410	0.1014	4.662	3.307	2.482
2	0.1881	0.1320	0.0994	4.218	3.096	2.433
4	0.1680	0.1276	0.1034	3.767	2.988	2.531
5	0.1590	0.1285	0.1074	3.566	3.013	2.629
6	0.1632	0.1345	0.1155	3.660	3.154	2.827
7	0.1761	0.1513	0.1324	3.949	3.631	3.240
8	0.2030	0.1709	0.1470	4.552	3.989	3.598
10	0.2832	0.2350	0.1970	6.351	5.500	4.281
12	0.3982	0.3182	0.2621	8.029	7.462	6.415
14	0.5435	0.4255	0.3431	12.19	9.978	8.397
16	0.7115	0.5650	0.4469	15.95	13.25	10.94

P_{H_2S} = 1.013 bar

AUXILIARY INFORMATION

METHOD/APPARATUS/PROCEDURE:	SOURCE AND PURITY OF MATERIALS:
H_2S was bubbled through about 100 cm^3 of water & ethanol in an absorption vessel at barometric pressure for 1 to 1.5 h. Dissolved H_2S was estimated iodimetrically. Allowances were made for the partial pressure of water, the partial pressure of ethanol and the head of liquid in the absorption vessel when the partial pressure of H_2S was calculated. Solubilities were corrected to 1.013 bar by use of Henry's law.	1. From iron sulfide (ex Merck) and dilute H_2SO_4; washed with H_2O and solvent mixture. 2. Distilled. 3. *p.a.* or *puriss.* from Merck or Kahlbaum-Schering; distilled

ESTIMATED ERROR:
Quoted values are averages from several measurements with a maximum deviation from the mean of ± 0.2% (authors)

REFERENCES:

COMPONENTS:	ORIGINAL MEASUREMENTS:
1. Hydrogen sulfide; H_2S; [7783-06-4] 2. Water; H_2O; [7732-18-5] 3. 1,2,3-Propanetriol (*glycerol*); $C_3H_8O_3$; [56-81-5]	Kiss, A. v.; Lajtai, I.; Thury, G. *Zeit. anorg. allgem. Chem.* 1937, *233*, 346-352.

VARIABLES:	PREPARED BY:
Temperature, concentration of glycerol	P.G.T. Fogg

EXPERIMENTAL VALUES:

Conc. of glycerol	Conc. of H_2S in liquid /mol dm^{-3}			Ostwald coefficient, L		
/mol dm^{-3}	273.20 K	285.65 K	298.15 K	273.20 K	285.65 K	298.15 K
0	0.2079	0.1410	0.1014	4.662	3.307	2.482
2	0.1752	0.1180	0.0930	3.929	2.768	2.276
4	0.1514	0.1066	0.0850	3.396	2.500	2.080
6	0.1311	0.0967	0.0793	2.940	2.267	1.938
8	0.1165	0.0913	0.0770	2.613	2.140	1.885

P_{H_2S} = 1.013 bar

AUXILIARY INFORMATION

METHOD/APPARATUS/PROCEDURE:	SOURCE AND PURITY OF MATERIALS:
H_2S was bubbled through about 100 cm^3 of water & glycerol in an absorption vessel at barometric pressure for 1 to 1.5 h. Dissolved H_2S was estimated iodimetrically. Allowance was made for the partial pressure of water and the head of liquid in the absorption vessel in the calculation of the partial pressure of H_2S. Solubilities were corrected to a partial pressure of 1.013 bar by use of Henry's law.	1. From iron sulfide (ex Merck) and dilute H_2SO_4; washed with H_2O and solvent mixture. 2. Distilled. 3. *p.a.* or *puriss.* from Merck or Kahlbaum-Schering; distilled.

	ESTIMATED ERROR: Quoted values are averages from several measurements with a maximum deviation from the mean of ± 0.2% (authors)
	REFERENCES:

COMPONENTS:	EVALUATOR:
1. Hydrogen sulfide; H$_2$S; [7783-06-4] 2. Strong electrolytes	Peter G.T. Fogg, Department of Applied Chemistry and Life Sciences, Polytechnic of North London, Holloway, London N7 8DB, U.K. July 1987

CRITICAL EVALUATION:

Several workers have reported the solubilities of hydrogen sulfide in solutions of salts and of acids and, as part of the same work, the solubility of the gas in pure water. If such measurements using pure water are in accord with values given by other workers then measurements of solubilities in the presence of electrolytes which have been carried out as part of the same research are likely to be reliable.

The published solubilities indicate that addition of an acid to water usually increases the solubility of hydrogen sulfide whereas addition of a salt often has the opposite effect. Experimental values for low electrolyte concentrations sometimes approximately obey empirical equations based upon the Sechenov relation (1). i.e.

$$s = s' \exp(-kc)$$

where s is the solubility of the gas in solution, s' the solubility in pure water, c the concentration of the electrolyte and k an empirical constant specific for the electrolyte. Concentrations may be expressed as molarities, molalities or mole fractions. When the equation is expressed in a form such as:

$$\ln(s/\text{mol dm}^{-3}) = \ln(s'/\text{mol dm}^{-3}) - kc$$

the effects of different electrolytes are approximately additive i.e.

$$\ln(s/\text{mol dm}^{-3}) = \ln(s'/\text{mol dm}^{-3}) - (k_1 c_1 + k_2 c_2 + \ldots)$$

where k_1, k_2 etc. are appropriate constants for electrolyte 1,2, etc. at concentrations c_1, c_2, etc. Better agreement with experimental data is obtained if additional terms depending on concentrations raised to a higher power, are introduced.

Solubilities in hydrochloric acid solutions at 298.15 K and a partial pressure of 1.013 bar have been reported by Kendall & Andrews (2), Kapustinsky & Anvaer (3) and by Gamsjäger & Schindler (4). Solubilities in pure water at the same temperature and pressure, which were reported by each group, are within 1% of the recommended value.

Gamsjäger & Schindler investigated the solubilities in the concentration range of hydrochloric acid of 0 to 3 mol dm^{-3}. Values are within about 2% of solubilities reported by Kendall & Andrews who made measurements at acid concentrations to 4.9 mol dm^{-3}. Kapustinsky & Anvaer measured solubilities in hydrochloric acid of concentrations from 5 to 11.4 mol dm^{-3}. Extrapolation of these measurements to lower concentrations of hydrochloric acid confirms that Gamsjäger's experimental measurements are likely to be reliable. However the empirical equation for the ratio of solubility in acid to that in pure water as a function of acid concentration, reported by Gamsjäger, does not appear to be applicable to the acid concentration range investigated by Kapustinsky.

Measurements reported by Gamsjäger & Schindler (4) for solutions containing sodium chloride are self-consistent. The solubility at 298.15 K for a partial pressure of 1.013 bar in sodium chloride of concentration 3.0 mol dm^{-3} is 0.062 mol dm^{-3} which is greater than the value of 0.0548 reported by Kapustinsky & Anvaer (3). The evaluator considers that data given by Gamsjäger & Schindler are likely to be the more reliable and these should be accepted on a tentative basis.

Kapustinsky and Anvaer (3) also reported the solubility at 298.15K and 1.013 bar in a solution of potassium chloride (1.0 mol dm^{-3}) to be 0.0877 mol dm^{-3} which is close to the value of the value of 0.089 estimated from measurements by McLauchlin (5). In view of the uncertainty over the solubility in sodium chloride solution this value for potassium chloride solutions can only be accepted on a tentative basis.

Kozintseva (6) measured solubilities in solutions of sodium chloride, calcium chloride and sodium sulfate at temperatures of 475 K and 535 K. These solubilities appear to be reliable and may be accepted as tentative values.

COMPONENTS:	EVALUATOR:
1. Hydrogen sulfide; H_2S; [7783-06-4] 2. Strong electrolytes	Peter G.T. Fogg, Department of Applied Chemistry and Life Sciences, Polytechnic of North London, Holloway, London N7 8DB, U.K. July 1987

CRITICAL EVALUATION:

Comparison with measurements by other workers at lower temperatures is not possible.

Measurements by Gamsjäger and Schindler (4) of solubilities in solutions containing both hydrochloric acid and sodium chloride are consistent with their measurements on solutions containing either one or other of these two compounds.

Solubilities in solutions of hydrochloric acid and zinc chloride and in solutions of hydrochloric acid and ferrous chloride were investigated by Kapustinsky and Anvaer. Their results show that, over the concentration ranges studied, addition of hydrochloric acid and of zinc chloride increased the solubility of hydrogen sulfide but addition of ferrous chloride decreased solubility. Further work on these systems is required before quantitative relationships can be reliably established.

Dede & Becker (7) measured solubilities, in solutions of calcium chloride, of sodium perchlorate and of sodium sulfate at 293.15 K and a total pressure equal to barometric. The value of the solubility in pure water, measured as part of the same research, is about 7% less than the recommended value based upon more modern measurements. There is a linear variation of log(concentration of H_2S in solution) with concentration of sodium perchlorate over the concentration range studied (0 to 4 mol dm^{-3}) showing that a Sechenov equation is applicable over this range. The other two salts deviate from linearity over the range 0 to 2 mol dm^{-3}.

Solubilities in sodium perchlorate solutions, perchloric acid solutions and in mixtures of the two have been reported by Gamsjäger, Rainer & Schindler (8) at 298.15 K and a partial pressure of 1.013 bar. The measurements are self consistent. An empirical equation relating solubility of hydrogen sulfide to concentration of sodium perchlorate and of perchloric acid was derived for total concentrations of perchlorates to 3 mol dm^{-3}. In the absence of acid the variation of solubility with concentration of sodium perchlorate found by Gamjager at 298.15 K is consistent with the similar behaviour at 293.15 K found by Dede & Becker (7).

Pollitzer (9) measured the solubility of hydrogen sulfide at 298.15 K and a total pressure of 1.013 bar in solutions of hydrogen iodide of concentrations 1.01 to 9.21 mol dm^{-3}. His value for the solubility in pure water corresponds to a mole fraction solubility at a partial pressure of 1.013 bar of 1.864×10^{-3} compared with the recommended value of $(1.830 \pm 0.014) \times 10^{-3}$ based upon more modern determinations (see Critical Evaluation - pure water). The solubilities over the hydrogen iodide concentration range follow a Sechenov relation fairly closely and can be fitted to the equation:

$$\log_{10}(\text{concn. } H_2S/\text{mol dm}^{-3}) = -0.9979 + 0.04494 \text{ (concn. HI/mol dm}^{-3})$$
$$\text{Standard deviation of } \log_{10}(\text{concn. of } H_2S/\text{mol dm}^{-3}) = \pm 0.0122$$

Aleksandrova and Yaroshchuk (10) measured the solubility in pure water, in dilute sulfuric acid and in concentrated sulfuric acid at 323.15 K, each at a different partial pressure of H_2S in the range 0.02 to 0.05 bar. Measurements indicate that, if the pressure were kept constant, solubility would increase with concentration of acid. The value of the solubility in pure water is about double the value expected on the assumption that the variation of mole fraction solubility would vary linearly with pressure to 1.013 bar. The solubilities reported for sulfuric acid are therefore of uncertain reliability.

Litvinenko (11) investigated the reversible equilibria between hydrogen sulfide gas and solutions of potassium carbonate and of sodium carbonate at temperatures from 298.15 K to 333.15 K.   Values of the ratio:

$$\frac{[HCO_3^-] \, [HS^-]}{[CO_3^-] \, P_{H_2S}}$$

depend upon the concentration of alkali metal ions and change with temperature. Measurements made by Dryden (12) for the the equilibrium involving potassium salts indicate that the ratio increases with decrease in concentration of

COMPONENTS:	EVALUATOR:
1. Hydrogen sulfide; H_2S; [7783-06-4] 2. Strong electrolytes	Peter G.T. Fogg, Department of Applied Chemistry and Life Sciences, Polytechnic of North London, Holloway, London N7 8DB, U.K. July 1987

CRITICAL·EVALUATION:

potassium ions. Litvinenko made measurements at a concentration of potassium
ions of about 2 mol dm^{-3} and obtained values of the ratio about 20% lower than
values presented by Dryden. In the opinion of the evaluator, the data presented
by Dryden are the more reliable.

Equilibria in the system hydrogen sulfide-carbon dioxide-potassium
carbonate-water were also investigated by Dryden (12) and by Litvinenko (11).
Both workers presented values of the the ratio:

$$\frac{[KHCO_3] \; P_{H_2S}}{[KHS] \; P_{CO_2}}$$

Dryden's measurements indicate that the ratio changes with total concentration
of potassium ions. Dryden made measurements at 293.2 K and 323.2 K. A value of
about 2.5 for a concentration of potassium ions of 2 mol dm^{-3} at 298.2 K may be
obtained from these measurements by interpolation. This is close to the average
value of 2.6 under these conditions from Litvinenko's measurements. If Dryden's
measurements are extrapolated to give the ratio at 333.2 K for this
concentration of potassium ions a value of about 1.0 is obtained. The average
of the values for 333.2 K obtained by Litvinenko is 1.5 so there is poor
agreement in this case.

The corresponding system containing sodium ions instead of potassium ions was
investigated by Berl & Rittener (13). They gave values of the ratio

$$\frac{\text{(molar concentration of carbonate + bicarbonate)} \; P_{H_2S}}{\text{(molar concentration of sulfide + bisulfide)} \; P_{CO_2}}$$

Under the conditions of the experiments the concentrations of carbonate and
sulfide ions are small compared with the concentrations of bicarbonate and
bisulfide ions and the ratio is equivalent to

$$\frac{[NaHCO_3] \; P_{H_2S}}{[NaHS] \; P_{CO_2}}$$

Values of this ratio for a concentration of sodium ions of about 0.5 mol dm^{-3}
are very close to values reported by Dryden for the similar ratio for the
potassium compounds at this concentration. This is shown in the table below :

	T/K	$[M^+]$/mol dm^{-3}	Ratio
Berl & Rittener (13)	291.2	0.544	2.01
	291.2	0.544	1.97
	328.2	0.532	0.90
Dryden (12)	293.2	0.49	1.9
	323.2	0.50	0.93

On the basis of the evidence which is available, the general relationships for
the two systems which have been presented by Dryden and by Berl & Rittener,
appear to be reliable.

References

1.   Sechenov,   Z. physik. Chem. 1889, 4,·121.

2.   Kendall, J.; Andrews, J.C. J. Amer. Chem. Soc. 1921, 43, 1545-1560.

COMPONENTS:	EVALUATOR:
1. Hydrogen·sulfide; H_2S; [7783-06-4] 2. Strong electrolytes	Peter G.T. Fogg, Department of Applied Chemistry and Life Sciences, Polytechnic of North London, Holloway, London N7 8DB, U.K. July 1987

CRITICAL EVALUATION:

3.   Kapustinsky, A.F.; Anvaer, B.I.

       Compt. Rend. Acad. Sci. URSS <u>1941</u>, *30*, 625-628.

4.   Gamsjäger, H.; Schindler, P. *Helv. Chim. Acta* <u>1969</u>, *52*, 1395-1402.

5.   McLauchlin, W.H. *Z. physik. Chem.* <u>1903</u>, *44*, 600-633.

6.   Kozintseva, T.N.
       Geokhim. Issled. v Obl. Povyshennykh Davlenii i Temperatur., Akad. Nauk
       SSSR, Inst. Geokhim. i Analit. Khim., Sb. Statei <u>1965</u>, 121-134.

7.   Dede, L.; Becker, Th. *Z. anorg. allgem. Chem.* <u>1926</u>, *152*, 185-196.

8.   Gamsjäger, H.; Rainer, W.; Schindler, P.
       Monatsh. Chem. <u>1967</u>, *98*, 1782-1802.

9.   Pollitzer, F. *Z. anorg. Chem.* <u>1909</u>, *64*, 121-148.

10.   Aleksandrova, M.V.; Yaroshchuk, E.G.
       Zh. Prikl. Khim. (Leningrad) <u>1978</u>, *51*, 1273-1276.
       J. Applied Chem. (USSR) <u>1978</u>, *51*, 1221-1223.

11.   Litvinenko, M.S. *Zh. Prikl. Khim.* <u>1952</u>, *25*, 516-532.
       J. Applied Chem. (USSR) <u>1952</u>, *25*, 579-595.

12.   Dryden, I.G.C. *J. Soc. Chem. Ind.* <u>1947</u>, *66*, 59-64.

13.   Berl, A.; Rittener, A. *Z. angew. Chem.* <u>1907</u>, *20*, 1637-1642.

COMPONENTS:	ORIGINAL MEASUREMENTS:
1. Hydrogen sulfide; H_2S; [7783-06-4] 2. Water; H_2O; [7732-18-5] 3. Hydrogen chloride, HCl; [7647-01-0]	Kendall, J.; Andrews, J.C. *J. Amer. Chem. Soc.* <u>1921</u>, *43*, 1545-1560.

VARIABLES:	PREPARED BY:
Concentration of HCl	P.G.T. Fogg

EXPERIMENTAL VALUES:

Concentration of HCl /mol dm^{-3}	Absorption coefficient*	Concentration of H_2S /mol dm^{-3} (P_{H_2S} = 1.013 bar)
0.0	2.266	0.1023
0.1348	2.253	0.1018
0.2828	2.247	0.1015
0.6308	2.250	0.1016
1.180	2.260	0.1020
1.848	2.272	0.1026
2.498	2.281	0.1030
3.040	2.291	0.1034
3.308	2.301	0.1039
4.410	2.384	0.1076
4.874	2.413	0.1090

T/K = 298.15

Mole fraction solubility in pure water, x_{H_2S} (P_{H_2S} = 1.013 bar) = 0.001845

This has been estimated by the compiler assuming no change of volume on dissolution of gas.

* Volume of gas, measured at 273.15 K and 1.013 bar, absorbed by one volume of solution at the temperature specified, when P_{H_2S} = 1.013 bar.

AUXILIARY INFORMATION

METHOD/APPARATUS/PROCEDURE:	SOURCE AND PURITY OF MATERIALS:
Hydrogen sulfide, produced from aqueous ammonium sulfide and concentrated hydrochloric acid, was washed with dilute sulfuric acid and with concentrated sodium sulfide solution. It then passed through hydrochloric acid at the same concentration as the solution in which solubility was to be determined. The gas then passed into standardised hydrochloric acid in a calibrated absorption burette held in a thermostat controlled to ± 0.01 K. Absorption was complete in 2-3 h. Hydrogen sulfide in the solution was estimated by iodimetry. Corrections were made for the barometric pressure, the vapor pressure of the solution and the head of liquid over the solution.	1. from ammonium sulfide solution and concentrated hydrochloric acid.
	ESTIMATED ERROR: $\delta T/K$ = ± 0.01 (authors)
	REFERENCES:

COMPONENTS:	ORIGINAL MEASUREMENTS:
1. Hydrogen sulfide; H_2S; [7783-06-4] 2. Water; H_2O; [7732-18-5] 3. Hydrogen chloride; HCl; [7647-01-0]	Kapustinsky, A.F.; Anvaer, B.I. *Compt.Rend.Acad.Sci.URSS* 1941, *30*, 625-8.
VARIABLES: Concentration of HCl	PREPARED BY: P.G.T. Fogg

EXPERIMENTAL VALUES:

Concentration of HCl in liquid			Conc. of H_2S in liquid		
Wt%	/mol dm^{-3}	/mol kg^{-1} of H_2O	/mol dm^{-3} *	/mol kg^{-1} * of H_2O	/mol kg^{-1} † of H_2O
16.8	5.0	5.55	0.1098	0.1227	0.1200
20.2	6.1	7.0	0.1124	0.1291	0.1266
23.7	7.3	8.4	0.1165	0.1375	0.1351
28.1	8.8	10.7	0.1192	0.1462	0.1437
34.9	11.4	14.7	0.1169	0.1542	0.1319

T/K = 298.2

* correspond to a partial pressure of H_2S of 760 mmHg (1.013 bar)

† correspond to a total pressure of 760 mmHg (1.013 bar)

AUXILIARY INFORMATION

METHOD/APPARATUS/PROCEDURE:	SOURCE AND PURITY OF MATERIALS:
Hydrogen sulfide, saturated with vapor from a sample of the solution under test, was bubbled through further solution in a thermostatted absorption vessel for 3 - 4 h. Samples were then withdrawn for analysis by iodimetry. Allowance was made for the vapor pressure of water and hydrogen chloride and for the density of the solutions in the calculation of solubilities.	1. from chemically pure hydrochoric acid and NaHS; CO_2 removed by Ba(OH)$_2$.
	ESTIMATED ERROR: $\delta T/K = \pm 0.1$ (authors)
	REFERENCES:

COMPONENTS:	ORIGINAL MEASUREMENTS:
1. Hydrogen sulfide; H_2S; [7783-06-4] 2. Hydrogen iodide; HI; [10034-85-2] 3. Water; H_2O; [7732-18-5]	Pollitzer, F. *Zeit. anorg. Chem.* <u>1909</u>, *64*, 121-148.

VARIABLES:
 	
  Concentration
	PREPARED BY:
	
	  P.G.T. Fogg

EXPERIMENTAL VALUES:

Conc. of HI /mol dm^{-3}	Conc. of H_2S /mol dm^{-3}
0.00	0.1004
1.01	0.111
1.51	0.113
1.93	0.125
2.45	0.130
2.64	0.138
3.42	0.142
4.38	0.163
5.005	0.165
5.695	0.181
6.935	0.197
9.21	0.267

T/K = 298.2   Total pressure = 760 mmHg = 1.01325 bar

AUXILIARY INFORMATION

METHOD/APPARATUS/PROCEDURE:	SOURCE AND PURITY OF MATERIALS:
H_2S was passed into solutions of HI which were then analysed iodimetrically.	1. from CaS and dilute HCl.
	ESTIMATED ERROR:
	REFERENCES:

COMPONENTS:	ORIGINAL MEASUREMENTS:
1. Hydrogen sulfide; H_2S; [7783-06-4] 2. Water; H_2O; [7732-18-5] 3. Sulfuric acid; H_2SO_4; [7664-93-9]	Aleksandrova, M.V.; Yaroshchuk, E.G. *Zh. Prikl. Khim. (Leningrad)* 1978, *51*, 1273-76. *J. Appl. Chem. (USSR)* 1978, *51*, 1221-3.

VARIABLES:	PREPARED BY:
Concentration of acid	P.G.T. Fogg

EXPERIMENTAL VALUES:

Solvent	P_{H_2S}/mmHg	P_{H_2S}/bar[*]	Total pressure[**] /mmHg	Mole fraction of H_2S in liquid x_{H_2S}
Water	16.5	0.0220	786.0	0.000055
Sulfuric acid 7.25 mol dm^{-3}	5.0	0.0067	780.5	0.00009
Sulfuric acid (concentrated)	36.6	0.0488	769.0	0.00129

   Temperature = 323.2 K

[*] calculated by compiler

[**] the compiler considers that an inert gas must also have been present to give the reported total pressure although this was not stated by the authors.

AUXILIARY INFORMATION

METHOD/APPARATUS/PROCEDURE:	SOURCE AND PURITY OF MATERIALS:
A flow method described in ref.(1) was used. A gas mixture was passed into the apparatus for 1 - 1.5 h. Samples of liquid and of gas were then analysed.	No information
	ESTIMATED ERROR:
	REFERENCES:
	1. Hals, E.; Pick, J.; Fried, V.; Vilim, O. *Vapor-Liquid Equilibrium*, Pergamon Press, New York

COMPONENTS:	ORIGINAL MEASUREMENTS:
1. Hydrogen sulfide; H_2S; [7783-06-4] 2. Water; H_2O; [7732-18-5] 3. Sulfuric acid; H_2SO_4; [7664-93-9]	Douabul, A.A.; Riley, J.P. *Deep-Sea Res.* <u>1979</u>, *26A*, 259 - 268.

VARIABLES:	PREPARED BY:
Temperature	P.G.T. Fogg

EXPERIMENTAL VALUES:

Solubility of H_2S at a H_2S fugacity of 1 atm (1.01325 bar).

T/K	Unacidified distilled water		Acidified[†] distilled water
	/mol dm^{-3}	mole fraction[*] x_{H_2S}	/mol dm^{-3}
275.25	0.1944	0.003490	0.1944
278.20	0.1769	0.003177	0.1769
283.34	0.1518	0.002728	0.1517
288.19	0.1314	0.002364	0.1314
293.25	0.1156	0.002082	0.1155
297.87	0.1022	0.001843	0.1023
302.97	0.0915	0.001653	0.0913

[*] Calculated by compiler

[†] Dilute sulfuric acid was prepared by adding 4 cm^3 of sulfuric acid (5 mol dm^{-3}) to 5 dm^3 of distilled water.

AUXILIARY INFORMATION

METHOD/APPARATUS/PROCEDURE:	SOURCE AND PURITY OF MATERIALS:
Distilled water or dilute acid was placed in a glass vessel. The liquid under test was deoxygenated by a stream of nitrogen. Hydrogen sulfide was then bubbled through the liquid for about 15 h. with continuous magnetic stirring. Gas flow rate was then reduced and use made of a shorter inlet tube so as to ensure that the pressure of the gas was within 0.5 mmHg of barometric pressure. After a further 2 h a sample of the liquid was analysed by iodimetry.	1. from Air Products Ltd; purity > 99.6%

	ESTIMATED ERROR: δ(solubility) = ± 0.2%; $\delta P/Pa$ = ± 70; $\delta T/K$ = ± 0.02 (authors)
	REFERENCES:

COMPONENTS:	ORIGINAL MEASUREMENTS:
1. Hydrogen sulfide; H_2S; [7783-06-4] 2. Water; H_2O; [7732-18-5] 3. Hydrogen chloride; HCl; [7647-01-0] 4. Iron(II) chloride; $FeCl_2$; [7758-94-3]	Kapustinsky, A.F.; Anvaer, B.I. *Compt.Rend.Acad.Sci.URSS* <u>1941</u>, *30*, 625-8.

VARIABLES:	PREPARED BY:
Concentration	P.G.T. Fogg

EXPERIMENTAL VALUES:

T/K	Total pressure /mmHg	Conc. of HCl /mol kg^{-1} of water	Conc. of $FeCl_2$ /mol kg^{-1} of water	Conc. of H_2S /mol. kg^{-1} of water
298.15	760	1.6	0.9	0.085
		3.4	1.0	0.089
		7.8	1.1	0.108
		10.6	1.2	0.120
		9	saturated	0.101

AUXILIARY INFORMATION

METHOD/APPARATUS/PROCEDURE:	SOURCE AND PURITY OF MATERIALS:
Hydrogen sulfide, saturated with vapor from a sample of the solution under test, was bubbled through further solution in a 300 cm^3 thermostatted absorption vessel for 3 - 4 hours. Samples of this solution were then withdrawn for analysis by iodimetry.	1. from chemically pure hydrochoric acid and NaHS; CO_2 removed by $Ba(OH)_2$. 4. from FeS and hydrochloric acid.

	ESTIMATED ERROR:
	$\delta T/K$ = ± 0.1 (authors)

	REFERENCES:

COMPONENTS:	ORIGINAL MEASUREMENTS:
1. Hydrogen sulfide; H₂S; [7783-06-4] 2. Water; H₂O; [7732-18-5] 3. Hydrogen chloride; HCl; [7647-01-0] 4. Zinc chloride; ZnCl₂; [7646-85-7]	Kapustinsky, A.F.; Anvaer, B.I. *Compt.Rend.Acad.Sci.URSS* 1941, *30*, 625-8.

VARIABLES:	PREPARED BY:
Concentration	P.G.T. Fogg

EXPERIMENTAL VALUES:

T/K	Total pressure /mmHg	Conc. of HCl /mol kg^{-1} of water	Conc. of ZnCl₂ /mol kg^{-1} of water	Conc. of H₂S /mol. kg^{-1} of water
298.15	760	3.9	2.1	0.1157
	758	5.0	4.7	0.1501
	755	7.8	1.0	0.1345
	746	9.1	1.5	0.1433
	755	10.6	1.1	0.1495
	750	13.5	0.8	0.1416

AUXILIARY INFORMATION

METHOD/APPARATUS/PROCEDURE:	SOURCE AND PURITY OF MATERIALS:
Hydrogen sulfide, saturated with vapor from a sample of the solution under test, was bubbled through further solution in a 300 cm³ thermostatted absorption vessel for 3 - 4 hours. Samples of this solution were then withdrawn for analysis by iodimetry.	1. from chemically pure hydrochoric acid and NaHS; CO₂ removed by Ba(OH)₂. 4. from pure zinc and hydrochloric acid.

	ESTIMATED ERROR:
	δT/K = ± 0.1 (authors)

	REFERENCES:

COMPONENTS:	ORIGINAL MEASUREMENTS:
1. Hydrogen sulfide; H_2S; [7783-06-4] 2. Water; H_2O; [7732-18-5] 3. Hydrogen chloride; HCl; [7647-01-0] 4. Sodium chloride; NaCl; [7647-14-5]	Gamsjäger, H.; Schindler, P. *Helv. Chim. Acta* <u>1969</u>, *52*, 1395-1402.

VARIABLES:	PREPARED BY:
Concentrations of HCl and NaCl	P.G.T. Fogg

EXPERIMENTAL VALUES:

Ionic strength /mol dm^{-3}	Conc. of HCl /mol dm^{-3}	Conc. of NaCl[†] /mol dm^{-3}	Absorption constant[*] /mol dm^{-3} atm^{-1}
0.500	0.000	0.500	0.0925
0.500	0.250	0.250	0.0970
0.500	0.500	0.000	0.1005
1.000	0.000	1.000	0.0856
1.000	0.000	1.000	0.0858
1.000	0.400	0.600	0.0913
1.000	0.780	0.220	0.0976
1.000	1.000	0.000	0.1008
1.500	0.000	1.500	0.0784
1.500	0.500	1.000	0.0856
1.500	0.750	0.750	0.0899
1.500	1.000	0.500	0.0933
1.500	1.400	0.100	0.0997
1.500	1.500	0.000	0.1004
1.500	1.500	0.000	0.1010

T/K = 298.15

[*] Mole of gas absorbed by one dm^3 of solution per atm pressure of H_2S. It was assumed that the following form of Henry's law was applicable:

$$\text{molar concentration of } H_2S = \text{constant} \times P_{H_2S}$$

[†] calculated by compiler; ionic strength - conc. of HCl.

AUXILIARY INFORMATION

METHOD/APPARATUS/PROCEDURE:	SOURCE AND PURITY OF MATERIALS:
The method was described in ref.(1). P_{H_2S} was taken to be equal to barometric pressure minus P_{H_2O}. Allowance was made for vapor pressure lowering resulting from the presence of electrolytes.	1. From Fa. Gerling, Holz & Co. (99.0 ± 0.5%) or Matheson & Co. (99.5 ± 0.5%) 3. Merck *p.a.* grade. 4. Merck *p.a.* grade; dried at 150 °C.
	ESTIMATED ERROR: δ(absorption constant) = ± 0.5% (authors)
	REFERENCES: 1. Gamsjäger, H.; Rainer, W.; Schindler, P. *Mh. Chem.* <u>1967</u>, *98*, 1793.

COMPONENTS:	ORIGINAL MEASUREMENTS:
1. Hydrogen sulfide; H_2S; [7783-06-4] 2. Water; H_2O; [7732-18-5] 3. Hydrogen chloride; HCl; [7647-01-0] 4. Sodium chloride; NaCl; [7647-14-5]	Gamsjäger, H.; Schindler, P. *Helv. Chim. Acta* <u>1969</u>, *52*, 1395-1402.

EXPERIMENTAL VALUES:

Ionic strength /mol dm^{-3}	Conc. of HCl /mol dm^{-3}	Conc. of NaCl[†] /mol dm^{-3}	Absorption constant[*] /mol dm^{-3} atm^{-1}
2.000	0.000	2.000	0.0724
2.000	0.000	2.000	0.0727
2.000	0.200	1.800	0.0753
2.000	0.500	1.500	0.0794
2.000	1.000	1.000	0.0867
2.000	1.500	0.500	0.0940
2.000	1.800	0.200	0.0989
2.000	2.000	0.000	0.1014
2.500	0.000	2.500	0.0668
2.500	0.250	2.250	0.0705
2.500	0.600	1.900	0.0749
2.500	1.250	1.250	0.0839
2.500	1.875	0.625	0.0933
2.520	2.270	0.250	0.0983
2.500	2.500	0.000	0.1018
3.000	0.000	3.000	0.0619
3.000	0.000	3.000	0.0621
3.000	0.300	2.700	0.0660
3.000	0.750	2.250	0.0712
3.000	1.600	1.400	0.0824
3.000	1.600	1.400	0.0825
3.000	2.250	0.750	0.0919
3.000	2.700	0.300	0.0986
3.000	2.900	0.100	0.1010
3.000	3.000	0.000	0.1031

T/K = 298.15

[*] Mole of gas absorbed by one dm^3 of solution per atm pressure of H_2S. It was assumed that the following form of Henry's law was applicable:

$$\text{molar concentration of } H_2S = \text{constant} \times P_{H_2S}$$

[†] calculated by compiler; ionic strength - conc. of HCl.

COMPONENTS:	ORIGINAL MEASUREMENTS:
1. Hydrogen sulfide; H_2S; [7783-06-4] 2. Water; H_2O; [7732-18-5] 3. Perchloric acid; $HClO_4$; [7601-90-3] 4. Sodium perchlorate; $NaClO_4$; [7601-89-0]	Gamsjäger, H.; Rainer, W.; Schindler, P. *Monatsh. Chem.* <u>1967</u>, *98*, 1782-1802.

VARIABLES:	PREPARED BY:
Concentrations of 3 & 4	P.G.T. Fogg

EXPERIMENTAL VALUES:

Concn. of solution / mol dm^{-3}			Concn. of solution / mol dm^{-3}		
$NaClO_4$	$HClO_4$	H_2S	$NaClO_4$	$HClO_4$	H_2S
0	0.000	0.1021	1.20	0.800	0.0881
0	0.100	0.1022	0.80	1.200	0.0936
0	0.200	0.1022	0.40	1.600	0.0994
0	0.400	0.1028	3.00	0.000	0.0677
0	0.601	0.1031	2.996	0.004	0.0678
0	1.000	0.1040	2.99	0.010	0.0678
0	2.015	0.1055	2.90	0.100	0.0685
0	2.300	0.1058	2.80	0.200	0.0699
0	2.714	0.1059	2.70	0.300	0.0710
0	3.021	0.1059	2.60	0.400	0.0724
1.00	0.000	0.0895	2.50	0.500	0.0735
0.90	0.100	0.0902	2.40	0.600	0.0743
0.80	0.200	0.0915	2.80	1.200	0.0818
0.50	0.500	0.0959	1.20	1.800	0.0898
0.20	0.800	0.1004	1.76	2.240	0.0952
2.00	0.000	0.0777	1.60	2.400	0.0971
1.80	0.200	0.0798	1.40	2.600	0.1001
1.60	0.400	0.0826	1.20	2.800	0.1031

Partial pressure of H_2S / bar = 1.013;   Temperature/K = 298.2

AUXILIARY INFORMATION

METHOD/APPARATUS/PROCEDURE	SOURCE AND PURITY OF MATERIALS
Hydrogen sulfide was passed into the perchlorate solution until the concentration of H_2S, as estimated by iodimetry, was constant. Allowance was made for the partial vapor pressure of water over the solution in the calculation of the solubility for a partial pressure of H_2S of 1.013 bar.	1. from Fa. Gerling & Holz, Hanau; 98.5 - 99.5 % pure. 2. & 3. prepared as in ref.(1)
	ESTIMATED ERROR $\delta T/K$ = ± 0.1 ; concentrations accurate to ± 1% (authors)
	REFERENCES 1. Gämsjager, H.; Kraft, W.; Rainer, W. *Monatsh. Chem.* <u>1966</u>, *97*, 833.

COMPONENTS:	ORIGINAL MEASUREMENTS:
1. Hydrogen sulfide; H_2S; [7783-06-4] 2. Water; H_2O; [7732-18-5] 3. Salts	McLauchlan, W.H. Z. physik. Chem. 1903, 44, 600-633.
VARIABLES:	PREPARED BY:
Concentration of salt	P.G.T. Fogg

EXPERIMENTAL VALUES:

Salt	Conc. of salt /mol dm^{-3}	P_{H_2S}/mmHg	P_{H_2S}/bar[*]	Conc. of H_2S in solution / mol dm^{-3}
Sodium sulfate;	0.5	124	0.165	0.0143
Na_2SO_4; [7757-82-6]	0.25	156.5	0.209	0.0187
Potassium sulfate;	0.5	126.7	0.169	0.0136
K_2SO_4; [7778-80-5]	0.25	110.8	0.148	0.01388
Ammonium sulfate;	0.5	149	0.199	0.0167
$(NH_4)_2SO_4$; [7783-20-2]	0.25	111.5	0.149	0.0139
Sodium chloride;	1	161	0.215	0.0188
NaCl; [7647-14-5]	0.5	186.5	0.249	0.0236
Potassium chloride;	1	156.2	0.208	0.0183
KCl; [7447-40-7]				
Ammonium chloride;	1	148	0.197	0.0193
NH_4Cl; [12125-02-9]				
Sodium nitrate;	1	100	0.133	0.0122
$NaNO_3$; [7631-99-4]				
Potassium nitrate;	1	133.7	0.178	0.0164
KNO_3; [7757-79-1]				
Ammonium nitrate;	1	94	0.125	0.0126
NH_4NO_3; [6484-52-2]				

T/K = 298.2   [*] calculated by compiler.

AUXILIARY INFORMATION

METHOD/APPARATUS/PROCEDURE:	SOURCE AND PURITY OF MATERIALS:
Solutions of hydrogen sulfide were prepared by dilution of a stock solution. The concentration of H_2S in solution was determined by iodimetry. Hydrogen, generated by electrolysis of water, was passed through a solution under test. The H_2S in the emergent gas stream was absorbed in copper sulfate solution and estimated from the change in electrical conductivity of this solution. The partial pressure of H_2S was calculated from the volume of hydrogen and the total pressure of the emergent gas stream.	1. prepared from FeS.
	ESTIMATED ERROR:
	REFERENCES:

COMPONENTS:	ORIGINAL MEASUREMENTS:
1. Hydrogen sulfide; H_2S [7783-06-4] 2. Water; H_2O; [7732-18-5] 3. Salts	Dede, L; Becker, Th. Z. anorg. allgem. Chem. 1926, 152, 185-196.
VARIABLES: Concentration of salt.	PREPARED BY: P.G.T. Fogg

EXPERIMENTAL VALUES:

Salt	Concn. of salt / normality	Concn. of salt* / mol dm^{-3}	Concn. of H_2S /g per 100 g of solution
Calcium chloride;	0	0	0.392
$CaCl_2$; [10043-52-4]	1	0.5	0.350
	2	1	0.313
	4	2	0.270
Sodium perchlorate;	1	1	0.340
$NaClO_4$; [7601-89-0]	2	2	0.293
	4	4	0.220
Sodium sulfate;	1	0.5	0.348
Na_2SO_4; [7757-82-6]	2	1	0.306
	4	2	0.257

Temperature = 293.2 K   Pressure = barometric (unspecified)

* the compiler has assumed that the equivalent of $CaCl_2$ and of Na_2SO_4 is half a mole in each each case.

AUXILIARY INFORMATION

METHOD/APPARATUS/PROCEDURE:	SOURCE AND PURITY OF MATERIALS:
Hydrogen sulfide from a Kipp's apparatus was washed with water and passed through the solution under test for about 15 min. Passing the gas for a longer period did not increase the amount dissolved. The temperature was controlled by a thermostat.	No information.
	ESTIMATED ERROR:
	REFERENCES:

COMPONENTS:	ORIGINAL MEASUREMENTS:
1. Hydrogen sulfide; H_2S; [7783-06-4] 2. Water; H_2O; [7732-18-5] 3. Alkali chlorides	Kapustinsky, A.F.; Anvaer, B.I. *Compt.Rend.Acad.Sci.URSS* *1941*, *30*, 625-8.

VARIABLES:	PREPARED BY:
	P.G.T. Fogg

EXPERIMENTAL VALUES:

Halide solution	T/K	P_{H_2S} /mmHg	conc. of halide /mol dm^{-3}	concn. of H_2S in liquid /mol dm^{-3}
Sodium chloride; NaCl; [7647-14-5]	298.2	760	3.0	0.0548
Potassium chloride; KCl; [7447-40-7]	298.2	760	1.0	0.0877

AUXILIARY INFORMATION

METHOD/APPARATUS/PROCEDURE:	SOURCE AND PURITY OF MATERIALS:
Hydrogen sulfide, saturated with water vapor, was bubbled through a 300 cm^3 thermostatted absorption vessel for 3 - 4 hours. Samples of solution were then withdrawn for analysis by iodimetry. Allowance was made for the vapor pressure of water.	1. from chemically pure hydrochloric acid and NaHS; CO_2 removed by $Ba(OH)_2$.

ESTIMATED ERROR:
$\delta T/K = \pm 0.1$ (authors)

REFERENCES:

COMPONENTS:	ORIGINAL MEASUREMENTS:
1. Hydrogen sulfide; H_2S; [7783-06-4] 2. Water; H_2O; [7732-18-5] 3. Sodium Chloride; NaCl; [7647-14-5]	Kozintseva, T. N. *Geokhim. Issled. v Obl. Povyshennykh Davlenii i Temperatur., Akad. Nauk SSSR, Inst. Geokhim. i Analit. Khim., Sb. Statei* 1965, 121-134

VARIABLES:	PREPARED BY:
Temperature, concentration of salt	P.G.T. Fogg

EXPERIMENTAL VALUES:

T/K	Conc. of NaCl /mol. dm^{-3}	P_{H_2S}/atm	P_{H_2S}/bar[+]	Distribution ratio, A[*]	Henry's constant /atm
475.2	0.53	1.1	1.1	0.0087	2190
475.2	1.11	1.1	1.1	0.0071	2480
475.2	1.17	1.3	1.3	0.0070	2510
535.2	0.56	0.67	0.68	0.0279	2155
535.2	1.13	0.24	0.24	0.0253	2360

[+] calculated by the compiler.

[*] A = mole ratio H_2S to H_2O in liquid phase / corresponding mole ratio in the gas phase.

The definition of Henry's constant was not stated clearly. Kozintseva compared values with those published by Ellis & Golding for CO_2-salt-H_2O systems (1). The compiler has assumed that Henry's constants given by Kozintseva were calculated in an analogous way to those given by Ellis & Golding.

$$\text{i.e.} \qquad \text{Henry's constant} = P_{H_2S} / x$$

$$x = \frac{\text{number of moles of } H_2S \text{ in the liquid phase}}{\text{number of moles of } H_2S + H_2O \text{ in the liquid phase}}$$

AUXILIARY INFORMATION

METHOD/APPARATUS/PROCEDURE:	SOURCE AND PURITY OF MATERIALS:
A special bomb, from which samples of either gas phase or liquid phase could be withdrawn, was used. The bomb was flushed with oxygen-free nitrogen, solution of salt (50cm^3) added and the apparatus saturated with H_2S (pressure not stated). The bomb and contents were heated and agitated for 25 - 50 hrs. A sample of the liquid phase was analysed iodimetrically to determine the H_2S content and evaporated to dryness to determine NaCl. A sample of the hot gas phase was passed through traps to collect the condensed water and through iodine solution. The H_2S in the gas phase was determined iodimetrically and the water content found from the weight of condensate.	No information
	ESTIMATED ERROR:
	$\delta T/K = \pm 2$
	REFERENCES:
	1. Ellis, A.J.; Golding, R.M. *Amer. J. Sci.* 1963, 261, 47.

COMPONENTS:	ORIGINAL MEASUREMENTS:
1. Hydrogen sulfide; H_2S; [7783-06-4] 2. Water; H_2O; [7732-18-5] 3. Sodium Sulfate; Na_2SO_4; [7757-82-6]	Kozintseva, T. N. *Geokhim. Issled. v Obl. Povyshennykh Davlenii i Temperatur., Akad. Nauk SSSR, Inst. Geokhim. i Analit. Khim., Sb. Statei* <u>1965</u>, 121-134
VARIABLES: Temperature, concentration of salt	PREPARED BY: P.G.T. Fogg

EXPERIMENTAL VALUES:

T/K	Conc. of Na_2SO_4 /mol. dm^{-3}	P_{H_2S}/atm	P_{H_2S}/bar[†]	Distribution ratio, A[*]	Henry's constant /atm
475.2	0.24	1.00	1.01	0.0086	2050
475.2	0.26	0.71	0.72	0.0086	2050
475.2	0.50	0.86	0.87	0.0075	2355
475.2	0.52	1.13	1.14	0.0076	2330

[†] calculated by the compiler.

[*] A = mole ratio H_2S to H_2O in liquid phase / corresponding mole ratio in the gas phase.

The definition of Henry's constant was not stated clearly. Kozintseva compared values with those published by Ellis & Golding for CO_2-salt-H_2O systems (1). The compiler has assumed that Henry's constants given by Kozintseva were calculated in an analogous way to those given by Ellis & Golding.

i.e.   Henry's constant = P_{H_2S} / x

$$x = \frac{\text{number of moles of } H_2S \text{ in the liquid phase}}{\text{number of moles of } H_2S + H_2O \text{ in the liquid phase}}$$

AUXILIARY INFORMATION

METHOD/APPARATUS/PROCEDURE:	SOURCE AND PURITY OF MATERIALS:
A special bomb, from which samples of either gas phase or liquid phase could be withdrawn, was used. The bomb was flushed with oxygen-free nitrogen, solution of salt ($50cm^3$) added and the apparatus saturated with H_2S (pressure not stated). The bomb and contents were heated and agitated for 25 - 50 hrs. A sample of the liquid phase was analysed iodimetrically to determine the H_2S content. The sulfate content was determined as barium sulfate. A sample of the gas phase was passed through traps to collect the condensed water and through iodine solution. The H_2S in the gas phase was determined iodimetrically and the water content found from the weight of condensate.	No information
	ESTIMATED ERROR: $\delta T/K$ = ± 2
	REFERENCES: 1. Ellis, A.J.; Golding, R.M. *Amer. J. Sci.* <u>1963</u>, *261*, 47.

COMPONENTS:	ORIGINAL MEASUREMENTS:
1. Hydrogen sulfide; H_2S; [7783-06-4] 2. Water; H_2O; [7732-18-5] 3. Calcium Chloride; $CaCl_2$; [10043-52-4]	Kozintseva, T. N. *Geokhim. Issled. v Obl. Povyshennykh Davlenii i Temperatur., Akad. Nauk SSSR, Inst. Geokhim. i Analit. Khim., Sb. Statei* <u>1965</u>, 121-134
VARIABLES: Temperature, concentration of salt	PREPARED BY: P.G.T. Fogg

EXPERIMENTAL VALUES:

T/K	Conc. of $CaCl_2$ /mol. dm^{-3}	P_{H_2S}/atm	P_{H_2S}/bar[†]	Distribution ratio, A[*]	Henry's constant /atm
475.2	0.23	1.07	1.08	0.0083	2130
475.2	0.57	1.06	1.07	0.0070	2530
475.2	0.54	0.99	1.00	0.0071	2496

[†] calculated by the compiler.

[*] A = mole ratio H_2S to H_2O in liquid phase / corresponding mole ratio in the gas phase.

The definition of Henry's constant was not stated clearly. Kozintseva compared values with those published by Ellis & Golding for CO_2-salt-H_2O systems (1). The compiler has assumed that Henry's constants given by Kozintseva were calculated in an analogous way to those given by Ellis & Golding.

i.e. Henry's constant = P_{H_2S} / x

$$x = \frac{\text{number of moles of } H_2S \text{ in the liquid phase}}{\text{number of moles of } H_2S + H_2O \text{ in the liquid phase}}$$

AUXILIARY INFORMATION

METHOD/APPARATUS/PROCEDURE:	SOURCE AND PURITY OF MATERIALS:
A special bomb, from which samples of either gas phase or liquid phase could be withdrawn, was used. The bomb was flushed with oxygen-free nitrogen, solution of salt ($50cm^3$) added and the apparatus saturated with H_2S (pressure not stated). The bomb and contents were heated and agitated for 25 - 50 hrs. A sample of the liquid phase was analysed iodimetrically to determine the H_2S content and titrated with oxalate to determine $CaCl_2$. A sample of the hot gas phase was passed through traps to collect the condensed water and through iodine solution. The H_2S in the gas phase was determined iodimetrically and the water content found from the weight of condensate.	No information
	ESTIMATED ERROR: $\delta T/K = \pm 2$
	REFERENCES: 1. Ellis, A.J.; Golding, R.M. *Amer. J. Sci.* <u>1963</u>, *261*, 47.

COMPONENTS:	ORIGINAL MEASUREMENTS:
1. Hydrogen sulfide; H_2S; [7783-06-4] 2. Water; H_2O; [7732-18-5] 3. Sodium carbonate; Na_2CO_3; [497-19-8]	Litvinenko, M.S.; Zh. Prikl. Khim. 1952, 25, 516-532. J. Appl. Chem. (USSR) 1952, 25, 579-595.
VARIABLES: Temperature, pressure	PREPARED BY: P.G.T. Fogg

EXPERIMENTAL VALUES:

Conc. of Na_2CO_3* before addition of H_2S/mol dm^{-3}	T/K	Conc. of NaHS in liquid /mol dm^{-3}	Conc. of H_2S in gas /g m^{-3}	P_{H_2S} /mmHg	Equilib. constant K /mol dm^{-3} mmHg^{-1}
0.491	293.2	0.170	3.5	1.7	0.0529
0.491		0.223	7.9	3.8	0.0488
0.491		0.261	13.0	6.3	0.0470
0.491		0.294	18.5	9.0	0.0487
0.491		0.319	24.0	11.6	0.0510
0.491		0.340	28.6	13.9	0.0552
0.492	333.2	0.135	4.6	1.8	0.0284
0.496		0.185	8.5	3.6	0.0306
0.496		0.242	19.5	8.0	0.0288
0.496		0.281	28.0	11.8	0.0311

* stated in the version in J.Appl.Chem. (USSR) to be the concentration of sodium in the solution. This statement is inconsistent with calculations carried out by the author and appears to be erroneous. (compiler)

The author evaluated the constants for the equilibrium:

$$Na_2CO_3(aq) + H_2S(gas) \rightleftharpoons NaHCO_3(aq) + NaHS(aq)$$

i.e
$$K = \frac{[NaHCO_3] \, [NaHS]}{[Na_2CO_3] \, P_{H_2S}}$$

Values of K for individual pressures of H_2S are given in the final column of the table.

AUXILIARY INFORMATION

METHOD/APPARATUS/PROCEDURE	SOURCE AND PURITY OF MATERIALS:
'Coke oven gas', freed from oxygen and acidic constituents, was mixed with H_2S, monitored by a flowmeter. The mixed gases were saturated with water vapor and then bubbled through sodium carbonate solution which had first been charged with H_2S. Samples of liquid were withdrawn every 30 min until two or three consecutive samples had identical sulfide content. The temperature of the absorption vessel was thermostatically controlled. Whether the initial sulfide was above or below the equilibrium value did not affect the consistency of the results.	No information
	ESTIMATED ERROR:
	REFERENCES:

COMPONENTS:	ORIGINAL MEASUREMENTS:
1. Hydrogen sulfide; H_2S; [7783-06-4] 2. Water; H_2O; [7732-18-5] 3. Potassium carbonate; K_2CO_3; [584-08-7]	Dryden, I.G.C J.Soc. Chem. Ind. 1947, 66. 59-64.
VARIABLES: Temperature	PREPARED BY: P.G.T. Fogg

EXPERIMENTAL VALUES

$$T/K \qquad K_2 \ / \ mol \ dm^{-3} \ mmHg^{-1}$$

273.7	$0.147 \ \pm \ 0.004$
283.1	$0.120 \ \pm \ 0.001$
293.9	$0.098 \ \pm \ 0.0015$
302.7	$0.086 \ \pm \ 0.0015$
312.7	$0.073 \ \pm \ 0.002$
322.8	$0.0635 \pm \ 0.0015$
332.8	$0.057 \ \pm \ 0.003$

Concentration of potassium ions = 2 mol dm^{-3}

$$K_2 \quad = \quad \frac{[KHCO_3] \ [KHS]}{[K_2CO_3] \ P_{H_2S}}$$

The authors stated that the value of K_2, extrapolated to zero concentration, at 298.2 K was 0.15 mol dm^{-3} $mmHg^{-1}$.

AUXILIARY INFORMATION

METHOD/APPARATUS/PROCEDURE:	SOURCE AND PURITY OF MATERIALS:
Nitrogen, purified by passage through soda-lime, was bubbled through solutions of potassium carbonate and bicarbonate previously saturated with H_2S and contained in a series of three Drechsel bottles in a thermostat. The proportion of H_2S in the emergent gas stream, and hence the partial pressure of H_2S, was determined by iodimetry. The concentration of KHS in solutions under test was also found by iodimetry.	1. by warming a suspension of MgHS in water.
	ESTIMATED ERROR: Average standard error is ± 2% (author).
	REFERENCES:

COMPONENTS:	ORIGINAL MEASUREMENTS:
1. Hydrogen sulfide; H_2S; [7783-06-4] 2. Water; H_2O; [7732-18-5] 3. Potassium carbonate; K_2CO_3; [584-08-7]	Litvinenko, M.S.; *Zh. Prikl. Khim.* 1952, *25*, 516-532. *J. Appl. Chem. (USSR)* 1952, *25*, 579-595.

VARIABLES:	PREPARED BY:
Temperature, pressure	P.G.T. Fogg

EXPERIMENTAL VALUES:

Conc. of K_2CO_3* before addition of H_2S/mol dm^{-3}	T/K	Conc. of KHS in liquid /mol dm^{-3}	Conc. of H_2S in gas† /g m^{-3}	P_{H_2S} /mmHg	Equilib. constant K /mol dm^{-3} mmHg^{-1}
1.032	298.2	0.356	4.9	2.4	0.0780
1.032		0.466	11.6	5.2	0.0710
1.046		0.587	21.5	10.1	0.0734
1.046		0.685	36.0	17.2	0.0756
1.046		0.670	37.0	17.4	0.0690
1.082	313.2	0.330	6.3	2.8	0.0517
1.082		0.457	14.3	6.5	0.0514
1.040		0.553	22.8	10.4	0.0609
1.113		0.606	27.7	12.8	0.0578
1.046		0.622	35.5	16.0	0.0572
1.032		0.618	36.0	16.3	0.0566
1.110	333.2	0.352	9.3	3.7	0.0447
1.113		0.470	20.42	8.2	0.0418
1.085		0.468	21.20	8.3	0.0427
1.113		0.564	32.71	12.9	0.0456

* stated in the version in J.Appl.Chem. (USSR) to be the concentration of potassium in the solution.
† stated in the translated paper to be the concentration in solution. These statements are inconsistent with calculations carried out by the author and appear to be errors. (compiler).

The author evaluated the constants for the equilibrium:

$$K_2CO_3(aq) + H_2S(gas) \rightleftharpoons KHCO_3(aq) + KHS(aq)$$

i.e

$$K = \frac{[KHCO_3] [KHS]}{[K_2CO_3] \ P_{H_2S}}$$

Values of K for individual pressures of H_2S are given above.

AUXILIARY INFORMATION

METHOD/APPARATUS/PROCEDURE	SOURCE AND PURITY OF MATERIALS:
'Coke oven gas', freed from oxygen and acidic constituents, was mixed with H_2S, monitored by a flowmeter. The mixed gases were saturated with water vapor and then bubbled through potassium carbonate solution which had first been charged with H_2S. Samples of liquid were withdrawn every 30 min until two or three consecutive samples had identical sulfide content. The temperature of the absorption vessel was thermostatically controlled. Whether the initial sulfide was above or below the equilibrium value did not affect the consistency of the results.	No information
	ESTIMATED ERROR:
	REFERENCES:

COMPONENTS:	ORIGINAL MEASUREMENTS:
1. Hydrogen sulfide; H_2S; [7783-06-4] 2. Carbon dioxide; CO_2; [124-38-9] 3. Water; H_2O; [7732-18-5] 4. Sodium carbonate, bicarbonate or sulfide	Berl, A.; Rittener, A. *Z. angew. Chem.* 1907, *20*, 1637-42.

VARIABLES:	PREPARED BY:
	P.G.T. Fogg

EXPERIMENTAL VALUES:

T/K	Salt	[salt][*] /g dm^{-3}	[Na$^+$][†] /mol dm^{-3}	Equilib.[§] [CO_2]/[H_2S]	Equilib. P_{CO_2}/P_{H_2S}	k
287.2	NaHCO$_3$ [144-55-8]	34.3	0.408	4.82	2.23	2.16
291.2	NaHCO$_3$	45.74	0.544	3.99	1.98	2.01
291.2	Na$_2$CO$_3$ [497-19-8]	29.8	0.562	0.712	0.362	1.97
291.2	Na$_2$S [1313-82-2]	35.5	0.910	1.61	0.864	1.86
291.2	Na$_2$S	29.1	0.746	1.76	0.889	1.98
328.2	Na$_2$CO$_3$	28.2	0.532	0.774	0.856	0.90
333.2	Na$_2$S	28.5	0.730	1.354	1.493	0.90
363.2	Na$_2$S	29.5	0.756	1.01	1.790	0.56

$$k = \frac{[CO_2] \; P_{H_2S}}{[H_2S] \; P_{CO_2}}$$

[*] Initial concentration of salt before addition of CO_2 and H_2S

[†] Calculated by the compiler

[§] [CO_2] is taken to be the sum of the concentrations of carbonate and bicarbonate ions in solution at equilibrium. [H_2S] is the corresponding sum of concentrations of sulfide and hydrosulfide ions.

AUXILIARY INFORMATION

METHOD/APPARATUS/PROCEDURE:	SOURCE AND PURITY OF MATERIALS:
The solution under test was shaken under either pure CO_2 or pure H_2S at barometric pressure. When equilibrium had been reached the gas phase was analysed for H_2S by iodimetry. The composition of the liquid phase was found by treating it with excess hydrochloric acid and analysing the mixture of CO_2 and H_2S evolved.	
	ESTIMATED ERROR:
	REFERENCES:

COMPONENTS:	ORIGINAL MEASUREMENTS:
1. Hydrogen sulfide; H_2S; [7783-06-4] 2. Carbon dioxide; CO_2; [124-38-9] 3. Water; H_2O; [7732-18-5] 4. Potassium carbonate; K_2CO_3; [584-08-7]	Dryden, I.G.C. *J.Soc. Chem. Ind.* 1947, *66*, 59-64.
VARIABLES: Temperature	PREPARED BY: P.G.T. Fogg

EXPERIMENTAL VALUES:

T/K	Concentration of K^+ / mol dm^{-3}	K_3
293.2	0.49	1.9 ± 0.07
293.2	0.98	2.05 ± 0.07
293.2	1.92	2.9 ± 0.15
323.2	0.50	0.93 ± 0.03
323.2	1.93	1.35 ± 0.07

$$K_3 = \frac{[KHCO_3]\, P_{H_2S}}{[KHS]\, P_{CO_2}}$$

The authors also determined the following values of the equilibrium constant K_1 given by :

$$K_1 = \frac{[KHCO_3]^2}{[K_2CO_3]\, P_{CO_2}}$$

The total concentration of potassium ion was 2 mol dm^{-3}

T/K	K_1 / mol dm^{-3} $mmHg^{-1}$
273.2	0.70 ± 0.01
283.2	0.44 ± 0.01
293.2	0.282 ± 0.004
303.2	0.182 ± 0.002
313.2	0.132 ± 0.002
323.2	0.093 ± 0.0015
333.6	0.072 ± 0.0015

AUXILIARY INFORMATION

METHOD/APPARAUS/PROCEDURE:	SOURCE AND PURITY OF MATERIALS:
A Winchester quart bottle, fitted with a rubber stopper and stopcock, was three times evacuated and filled with CO_2. It was then immersed in a warm water bath, excess pressure released and cooled to room temperature. Solutions were prepared by saturating solutions of $KHCO_3$ + K_2CO_3 with H_2S and introduced into the bottle without access of air. For measurements at 293.2 K the bottle was left overnight, shaken for 3 h and immersed in a thermostat for 1 h with frequent shaking. For measurements at 323.2 K, the vessel was rotated in a thermostat. Solutions were then analysed and compositions of the gas phase found by material balance.	1. from a Kipp's apparatus. ESTIMATED ERROR: Average value of δK_3 = ± 4% (author) REFERENCES:

COMPONENTS:	ORIGINAL MEASUREMENTS:
1. Hydrogen sulfide; H_2S; [7783-06-4] 2. Carbon dioxide; CO_2; [124-38-9] 3. Water; H_2O; [7732-18-5] 4. Potassium carbonate; K_2CO_3; [584-08-7]	Litvinenko, M.S.; *Zh. Prikl. Khim.* 1952, 25, 516-532. (*J. Appl. Chem.* (*USSR*) 1952, 25, 579-595)

VARIABLES:	PREPARED BY:
Temperature, pressure	P.G.T. Fogg

EXPERIMENTAL VALUES:

T/K	Conc. of H_2S in gas phase /mol m^{-3}	Conc. of CO_2 in gas phase /mol m^{-3}	Conc. of KHS in solution /mol dm^{-3}	Conc. of $KHCO_3$ in solution /mol dm^{-3}	Equilib. const. K
298.2	0.990	0.615	0.558	0.863	2.48
	1.310	0.669	0.610	0.865	2.78
	1.370	0.669	0.612	0.764	2.56
	0.680	0.464	0.473	0.784	2.46
333.2	0.870	0.803	0.500	0.572	1.24
	1.240	0.853	0.542	0.662	1.76
	1.094	0.781	0.567	0.638	1.58

The concentration of potassium carbonate before adsorption of gases was 1.0 mol dm^{-3}.

The equilibrium constant was defined as: $\quad K = \dfrac{[KHCO_3]\ P_{H_2S}}{[KHS]\ P_{CO_2}}$

AUXILIARY INFORMATION

METHOD/APPARATUS/PROCEDURE:	SOURCE AND PURITY OF MATERIALS:
'Coke oven gas', freed from oxygen and acidic constituents, was mixed with H_2S & CO_2, monitored by a flowmeter. The mixed gases were saturated with water vapor and then bubbled through potassium carbonate solution. Samples of liquid were withdrawn every 30 min until two or three consecutive samples had identical sulfide content. The temperature of the adsorption vessel was thermostatically controlled. Whether the initial sulfide was above or below the equilibrium value did not affect the consistency of the results.	No information
	ESTIMATED ERROR:
	REFERENCES:

COMPONENTS:	EVALUATOR:
1. Hydrogen sulfide; H_2S; [7783-06-4] 2. Seawater	Peter G.T. Fogg, Department of Applied Chemistry and Life Sciences, Polytechnic of North London, Holloway, London N7 8DB, U.K. February 1987

CRITICAL EVALUATION:

This system has been investigated at barometric pressure by Douabul and Riley (1) but no other set of measurements are available for direct comparison.

The seawater was made slightly acidic by adding sulfuric acid to give a concentration of 0.004 mol dm^{-3}. Ancillary measurements of the solubility of hydrogen sulfide in pure water and in sulfuric acid of this concentration demonstrated that the effect of addition of acid was almost imperceptible. The solubilities in pure water are in close agreement with measurements by other modern workers.

The authors have shown that their measurements are consistent with an equation for the solubility of gases in seawater developed by Weiss (2). They have also shown that extrapolation of solubilities of hydrogen sulfide in perchlorate solutions, as measured by Gamsjäger et al. (3), to an ionic strength of 0.7 gives a value which differs by about 3% from the solubility in sea-water of the same ionic strength. They suggested that this difference could be accounted for by differences in degrees in hydration of ions in the two media.

References

1.   Douabul, A.A.; Riley, J.P.   *Deep-Sea Res.* <u>1979</u>, *26A*, 259-268.

2.   Weiss, R.F. *Deep-Sea Res.* <u>1970</u>, *17*, 721-735.

3.   Gamsjäger, H.; Rainer, W.; Schindler, P.
      Monatsh. Chem. <u>1967</u>, *98*, 1793-1802.

COMPONENTS:	ORIGINAL MEASUREMENTS:
1. Hydrogen sulfide; H_2S; [7783-06-4] 2. Sea-water 3. Sulfuric acid; H_2SO_4; [7664-93-9]	Douabul, A.A.; Riley, J.P. *Deep-Sea Res.* 1979, *26A*, 259 - 268.
VARIABLES: Temperature	PREPARED BY: P.G.T. Fogg

EXPERIMENTAL VALUES:

T/K	Salinity /g kg^{-1}	Solubility at H_2S fugacity of 1 atm /mol dm^{-3}	T/K	Salinity /g kg^{-1}	Solubility at H_2S fugacity of 1 atm /mol dm^{-3}
275.25	9.972	0.1910	283.34	9.972	0.1496
	20.014	0.1878		20.014	0.1476
	24.958	0.1862		24.958	0.1466
	29.993	0.1846		29.993	0.1456
	34.994	0.1831		34.994	0.1447
	40.028	0.1816		40.028	0.1437
278.20	9.972	0.1741	288.19	9.972	0.1299
	20.014	0.1714		20.014	0.1284
	24.958	0.1701		24.958	0.1276
	29.993	0.1688		29.993	0.1269
	34.994	0.1675		34.994	0.1262
	40.028	0.1662		40.028	0.1255

Samples of seawater were acidified by adding 4 cm^3 of sulfuric acid (5 mol dm^{-3}) to 5 dm^3 of seawater before absorption of H_2S.

AUXILIARY INFORMATION

METHOD/APPARATUS/PROCEDURE:	SOURCE AND PURITY OF MATERIALS:
Acidified seawater was placed in a glass vessel. The liquid under test was deoxygenated by a stream of nitrogen. Hydrogen sulfide was then bubbled through the liquid for about 15 h. with continuous magnetic stirring. Gas flow rate was then reduced and use made of a shorter inlet tube so as to ensure that the pressure of the gas was within 0.5 mmHg of barometric pressure. After a further 2 h a sample of the liquid was analysed by iodimetry.	1. from Air Products Ltd; purity > 99.6% 2. surface water from Irish Sea, salinity approx. 33°/$_{oo}$; filtered; diluted or evaporated.
	ESTIMATED ERROR: δ(solubility) = ± 0.2%; δP/Pa = ± 70; δT/K = ± 0.02 (authors)
	REFERENCES:

COMPONENTS:	ORIGINAL MEASUREMENTS:

1. Hydrogen sulfide; H_2S;
     [7783-06-4]
 2. Sea-water
 	
 3. Sulfuric acid; H_2SO_4;
     [7664-93-9]
	Douabul, A.A.; Riley, J.P.
	
	 Deep-Sea Res. <u>1979</u>, *26A*, 259 - 268.

EXPERIMENTAL VALUES:

T/K	Salinity /g kg^{-1}	Solubility at H_2S fugacity of 1 atm /mol dm^{-3}	T/K	Salinity /g kg^{-1}	Solubility at H_2S fugacity of 1 atm /mol dm^{-3}
293.25	9.972	0.1143	297.87	29.993	0.0995
	20.014	0.1131		34.994	0.0990
	24.958	0.1126		40.028	0.0986
	29.993	0.1120	302.97	9.972	0.0907
	34.994	0.1114		20.014	0.0898
	40.028	0.1108		24.958	0.0894
297.87	9.972	0.1013		29.993	0.0891
	20.014	0.1004		34.994	0.0887
	24.958	0.0999		40.028	0.0884

The authors found that solubilities were fitted to better than ± 0.5% by the following equation developed by Weiss:

$$\ln C = A_1 + A_2(100/T) + A_3\ln(T/100) + S°/_{oo}[B_1 + B_2(T/100) + B_3(T/100)^2]$$

where C is the solubility in units of mol dm^{-3}; T is the temperature in units of K; $A_1 = -41.0563$; $A_2 = 66.4005$; $A_3 = 15.1060$; $B_1 = -0.60583$ $B_2 = 0.379753$; $B_3 = -0.602340$ and $S°/_{oo}$ is the salinity (g kg^{-1}).

COMPONENTS:	EVALUATOR:
1. Hydrogen sulfide; H_2S; [7783-06-4] 2. Ammonia; NH_3; [7664-4-17] 3. Water; H_2O; [7732-18-5]	Peter G.T. Fogg, Department of Applied Chemistry and Life Sciences, Polytechnic of North London, Holloway, London N7 8DB, U.K. February 1987

CRITICAL EVALUATION:

The original measurements on this system were published by van Krevelen, Hoftijzer, and Huntjens (1). Other work has since been published (2-7) and data cover temperatures from 293.15 K to 373.15 K and a wide range of concentrations of ammonia and of hydrogen sulfide. Direct comparison of different sets of data is only possible to a limited extent because of differing ranges of conditions studied by different authors but the general pattern of behaviour is the same. The concentration of hydrogen sulfide in solution increases with partial pressure of the gas and with concentration of ammonia in solution and decreases as the temperature is raised.

One apparent discrepancy between two sets of data has been found by the evaluator:

Authors	T/K	Wt.% in liquid H_2S	NH$_3$	Wt.% in gas H_2S	NH$_3$
Oratovskii *et al.* (2)	353.2	3.0	5.45	33.9	33.6
Ginzburg *et al.* (3)	353.8	3.20	5.04	26.02	36.02

Models for the behaviour of the system have been published by van Krevelen *et al.* (1), by Edwards *et al.* (8) and by Beutier & Renon (7). These can only be partially correlated with experimental data.

Van Krevelen *et al.* (1) investigated the systems H_2S-NH_3-NH_4Cl-H_2O and H_2S-NH_3-$(NH_4)_2SO_4$ and showed that data are consistent with those for the H_2S-NH_3-H_2O system.

The system H_2S-NH_3-CO_2-H_2O was investigated by Badger and Silver (9), by Dryden (10), by van Krevelen *et al.*(1) and by Wilson, Gillespie and Owens (11). Data span a range of concentrations of components and temperatures from 298.2 K to 533.2 K. There appear to be no serious discrepancies between the sets of data but detailed comparison is not possible because of the differences in conditions investigated by the different groups. Van Krevelen *et al.* have shown that experimental data can be partially correlated with a simple model for the system. Correlation models have also been published by Wilson *et al.*.

Golutvin *et al.* have studied the system H_2S-NH_3-CO_2-H_2O-phenol (4).

References

1.   van Krevelen, D.W.; Hoftijzer, P.J.; Huntjens, F.J.
      Recl. Trav. Chim. Pays-Bas <u>1949</u>, *68*, 191-216.

2.   Oratovskii, V.I.; Gamol'skii, A.M.; Klimenko, N.N.
      Zhur. Prik. Khim. <u>1964</u>, *37*, 2392-8.;
      J. Appl. Chem. USSR <u>1964</u>, *37*, 2363-7.

3.   Ginzburg, D.M.; Pikulina, N.S.; Litvin, V.P.
      Zhur. Prikl. Khim. <u>1965</u>, *38*, 2117-3.;
      J. Appl. Chem. USSR <u>1965</u>, *38*, 2071-3.

4.   Golutvin, Yu.M.; Malysheva, T.V.; Skorobogatova, V.I.
      Izvest. Sibir. Otdel. Akad. Nauk. S.S.S.R. <u>1958</u>, *No.8*, 83-87.

5.   Leyko, J.
      Bull. Acad. Polon. Sci. Ser. sci. chim. <u>1959</u>, *7*, 675-679.

COMPONENTS:	EVALUATOR:
1. Hydrogen sulfide; H_2S; [7783-06-4] 2. Ammonia; NH_3; [7664-4-17] 3. Water; H_2O; [7732-18-5]	Peter G.T. Fogg, Department of Applied Chemistry and Life Sciences, Polytechnic of North London, Holloway, London N7 8DB, U.K. February 1987

CRITICAL EVALUATION:

6. Leyko, J.; Piatkiewicz, J.
    Bull. Acad. Polon. Sci. Ser. sci. chim. <u>1964</u>, *12*, 445-446.

7. Beutier, D.; Renon, H.
    Ind. Eng. Chem. Proc. Des. Dev. <u>1978</u>, *17*, 220-230.

8. Edwards, T.J.; Newman, J.; Prausnitz, J.M.
    Amer. Inst. Chem. Eng. J. <u>1975</u>, *21(2)*, 248.

9. Badger, E.H.M.; Silver, L.
    J. Soc. Chem. Ind. <u>1938</u>, *57*, 110-112.

10. Dryden, I.G.C.
    J. Soc. Chem. Ind. <u>1947</u>, *66*, 59-64.

11. Wilson, G.M.; Gillespie, P.C.; Owens, J.L.
    Proc. 64th Ann. Conv. Gas Processors Association, <u>1955</u>, 282-288.

COMPONENTS:	ORIGINAL MEASUREMENTS:
1. Hydrogen sulfide; H_2S; [7783-06-4] 2. Ammonia; NH_3; [7664-41-7] 3. Water; H_2O; [7732-18-5]	van Krevelen, D.W.; Hoftijzer, P.J. Huntjens, F.J. *Recl. Trav. Chim. Pays-Bas* <u>1949</u>, *68*, 191–216.
VARIABLES: Temperature, concentration of liquid phase.	PREPARED BY: P.G.T. Fogg

EXPERIMENTAL VALUES:

Conc. of H_2S in liquid[*] /mol dm^{-3}	Conc. of NH_3 in liquid[*] /mol dm^{-3}	Partial pressures of H_2S / mmHg		
		293.2 K	313.2 K	333.2 K
0.007	0.280	0.3	1.3	
0.060	0.280			2.6
0.110	0.280	0.8	3.7	11.8
0.120	0.585	0.3	1.5	4.9
0.140	0.295			15.6
0.145	0.575	0.6		8.7
0.185	0.300	4.2	13.3	39.6
0.220	1.060	0.9	2.6	9.4
0.220	0.580	1.7	6.9	20.5
0.285	0.585	3.0	12.7	35.5
0.290	1.150	1.1	4.6	14.5
0.355	1.720	1.1	4.0	12.9
0.375	0.595	8.2	26.9	

[*] total H_2S & NH_3 including that equivalent to ionic species.

AUXILIARY INFORMATION

METHOD/APPARATUS/PROCEDURE:	SOURCE AND PURITY OF MATERIALS:
Solutions were prepared by passing H_2S, NH_3 and steam through water with exclusion of air. Samples of vapor in equilibrium with the solution under test were absorbed by solutions of cadmium acetate in acetic acid which were then analysed. These analyses together with P-V-T data were used to calculate partial pressures of NH_3 & H_2S. Another method consisted of passing a measured volume of N_2 at a measured total pressure through a series of vessels containing solution. H_2S in the effluent gas was determined iodimetrically. The two methods gave similar results.	No information
	ESTIMATED ERROR:
	REFERENCES:

COMPONENTS:	ORIGINAL MEASUREMENTS:
1. Hydrogen sulfide; H_2S; [7783-06-4] 2. Ammonia; NH_3; [7664-41-7] 3. Water; H_2O; [7732-18-5]	van Krevelen, D.W.; Hoftijzer, P.J. Huntjens, F.J. *Recl. Trav. Chim. Pays-Bas* <u>1949</u>, *68*, 191-216.

EXPERIMENTAL VALUES:

Conc. of H_2S in liquid[*] /mol dm^{-3}	Conc. of NH_3 in liquid[*] /mol dm^{-3}	Partial pressures of H_2S / mmHg		
		293.2 K	313.2 K	333.2 K
0.398	0.795	4.4		
0.420	1.637	1.7		
0.435	1.730	1.6	6.4	19.4
0.440	2.270	1.4	5.0	16.0
0.455	1.175	3.4	12.5	38.6
0.550	2.870	1.6	5.9	18.5
0.580	1.190	6.7	23.6	66.9
0.585	2.310	2.3	8.1	26.0
0.590	3.460	1.3	4.7	17.6
0.690	1.780	4.9	18.0	54.7
0.730	2.920	2.8	10.0	31.8
0.860	1.780	9.4	32.6	98.6
0.875	2.260	6.0	22.3	68.0
1.110	2.860	6.7	31.4	94.2
1.150	2.400	12.2	43.2	135.9
1.165	1.790	18.4		182.3
1.440	2.930	15.8	54.4	151.9
1.540	2.350	27.1		
1.890	2.920	38.5	125.4	310.5

[*] total H_2S & NH_3 including that equivalent to ionic species.

COMPONENTS:	ORIGINAL MEASUREMENTS:
1. Hydrogen sulfide; H_2S; [7783-06-4] 2. Ammonia; NH_3; [7664-41-7] 3. Water; H_2O; [7732-18-5]	Golutvin, Yu. M.; Malysheva, T.V.; Skorobogatova, V.I. *Izvest. Sibir. Otdel. Akad. Nauk. S.S.S.R.* 1958, *No.8, 83-7*
VARIABLES: Concentration of ammonia	PREPARED BY: P.G.T. Fogg

EXPERIMENTAL VALUES:

T/K	Concentration of NH_3 /mol dm^{-3}	Concentration of H_2S /mol dm^{-3}
293.2	0.70	0.78
	0.88	1.10
	1.19	1.23
	1.82	2.23
	2.24	2.87
	3.80	5.93
	4.98	7.40
	5.26	7.65
	6.16	8.57

Total pressure = barometric

Concentrations of NH_3 and of H_2S include all ionic species derived from these compounds.

AUXILIARY INFORMATION

METHOD/APPARATUS/PROCEDURE:	SOURCE AND PURITY OF MATERIALS:
Hydrogen sulfide from a Kipp's apparatus was bubbled through aqueous solutions of ammonia of various concentrations for six hours. The total ammonia in the resulting solution was analysed by titration with hydrochloric acid and the total hydrogen sulfide by iodimetry.	1. from a Kipp's apparatus.
	ESTIMATED ERROR: $\delta T/K = \pm 0.1$
	REFERENCES:

COMPONENTS:	ORIGINAL MEASUREMENTS:
1. Hydrogen sulfide; H_2S; [7783-06-4] 2. Ammonia; NH_3; [7664-41-7] 3. Water; H_2O; [7732-18-5]	Leyko, J. *Bull.Acad.Polon.Sci.Ser.sci.chim.,* <u>1959</u>, *7*, 675 - 679.
VARIABLES: Temperature, concentration of H_2S and NH_3 in the liquid phase.	PREPARED BY: P.G.T. Fogg

EXPERIMENTAL VALUES:

Conc. of H_2S in liquid /mol dm^{-3}	Conc. of NH_3 in liquid* /mol dm^{-3}	T/K	P_{H_2S}/mmHg	P_{NH_3}/mmHg	P_{H_2O}/mmHg
1.01	2.12	293.2	7.5	10.8	15.0
		313.2	31.1	27.5	42.0
		323.2	60.0	45.7	69.0
2.06	4.33	293.2	22.6	21.0	14.0
		313.2	73.5	58.3	41.0
		323.2	130.6	89.2	69.5
3.54	7.43	293.2	38.1	44.2	13.0
		303.2	74.5	79.0	20.0
		313.2	136.9	119.5	37.2
		323.2	235.0	190.0	55.1
5.20	10.92	293.2	62.2	80.0	10.0
		303.2	124.3	129.3	18.0
		313.2	238.4	215.0	30.0
		318.2	290.9	255.4	42.0

760 mmHg = 1 atm = 1.01325 bar

* calculated by compiler on the basis of the statement that molar conc.of NH_3 /molar conc. of H_2S = 2.1

AUXILIARY INFORMATION

METHOD/APPARATUS/PROCEDURE:	SOURCE AND PURITY OF MATERIALS:
Nitrogen was bubbled through the thermostatted samples of the solution under test. The content of NH_3, H_2S and H_2O in the gas stream was determined by analysis.	No information
	ESTIMATED ERROR:
	REFERENCES:

COMPONENTS:	ORIGINAL MEASUREMENTS:
1. Hydrogen sulfide; H_2S; [7783-06-4] 2. Ammonia; NH_3; [7664-41-7] 3. Water; H_2O; [7732-18-5]	Oratovskii, V.I.; Gamol'skii, A.M.; Klimenko, N.N. Zhur.Prik.Khim. 1964, 37, 2392-2398. J.Appl.Chem. USSR 1964, 37, 2363-7.
VARIABLES: Temperature, concentration of H_2S and NH_3 in the liquid phase.	PREPARED BY: P.G.T. Fogg

EXPERIMENTAL VALUES:

T/K	Wt% in liquid**		Mole fraction in liquid*		Wt% in gas		Mole fraction in gas*		Partial press./bar*	
	H_2S	NH_3	H_2S	NH_3	H_2S	NH_3	H_2S	NH_3	H_2S	NH_3
343.2	2.0	4.0	0.011	0.043	24.1	33.1	0.141	0.387	0.088	0.241
	3.0	5.5	0.016	0.059	31.9	36.0	0.194	0.437	0.152	0.341
	4.0	6.3	0.021	0.068	40.4	34.2	0.258	0.436	0.239	0.404
	5.0	7.1	0.027	0.077	48.1	31.2	0.321	0.417	0.342	0.444
	6.0	7.5	0.032	0.081	55.4	27.6	0.388	0.387	0.476	0.475
	7.0	8.0	0.038	0.087	61.0	24.2	0.444	0.352	0.593	0.470
	8.0	8.3	0.044	0.091	66.1	21.2	0.499	0.320	0.743	0.476
	9.0	8.5	0.049	0.093	70.9	18.1	0.554	0.283	0.908	0.464
	10.0	8.7	0.055	0.096	75.4	15.1	0.610	0.244	1.113	0.445

* estimated by the compiler. Partial pressures were assumed to be proportional to mole fractions in the gas phase. The partial pressure of water was taken as the product of mole fraction of water in the liquid phase and the vapor pressure of pure water.

** total H_2S & NH_3 including that equivalent to ionic species.

AUXILIARY INFORMATION

METHOD/APPARATUS/PROCEDURE:	SOURCE AND PURITY OF MATERIALS:
A solution of ammonium sulfide ($140-200$ g dm^{-3}) was evaporated at a rate of $0.8-1.0$ g min^{-1} from a thermostatted steel vessel of volume 100 cm^3. At various times the evaporation was discontinued, the vessel and contents weighed and samples analysed for NH_3 and H_2S. Compositions of the vapor phase were calculated from the rates of change of composition of the liquid phase as described in in ref. (1).	1 & 2 analytical grade ammonium sulfide used.
	ESTIMATED ERROR: $\delta T/K = \pm 0.05$ (authors)

	REFERENCES:
	1. Yarym-Agaev, N.L.; Kogan, E.A.
	    Zh. Fiz. Khim. 1956, 30 (11),
	    2150.

COMPONENTS:	ORIGINAL MEASUREMENTS:
1. Hydrogen sulfide; H_2S; [7783-06-4] 2. Ammonia; NH_3; [7664-41-7] 3. Water; H_2O; [7732-18-5]	Leyko, J.; Piatkiewicz, J. *Bull.Acad.Polon.Sci.Ser.sci.chim.*, <u>1964</u>, *12*, 445-446.
VARIABLES: Temperature	PREPARED BY: P.G.T. Fogg

EXPERIMENTAL VALUES:

Conc. of H_2S in liquid /mol dm^{-3}	Conc. of NH_2 in liquid /mol dm^{-3}	T/K	P_{H_2S}/atm	P_{NH_3}/atm	P_{H_2O}/atm
1.91	6.15	353.2	0.35	0.95	0.35
		363.2	0.51	1.31	0.53
		373.2	0.75	1.66	0.79
		383.2	1.17	2.05	1.18

AUXILIARY INFORMATION

METHOD/APPARATUS/PROCEDURE:	SOURCE AND PURITY OF MATERIALS:
Liquid under test was contained in a thermostatted pressure vessel connected to a manometer. Samples of the gas phase were taken through a valve into an evacuated gas sampling pipette and analysed by chemical methods.	No information
	ESTIMATED ERROR: $\delta P/P$ = ± 5% (authors)
	REFERENCES:

COMPONENTS:	ORIGINAL MEASUREMENTS:
1. Hydrogen sulfide; H_2S; [7783-06-4]	Oratovskii, V.I.; Gamol'skii, A.M.; Klimenko, N.N.
2. Ammonia; NH_3; [7664-41-7]	*Zhur.Prik.Khim.* 1964, *37*, 2392-2398.
3. Water; H_2O; [7732-18-5]	*J.Appl.Chem. USSR* 1964, *37*, 2363-7.

EXPERIMENTAL VALUES:

T/K	Wt% in liquid**		Mole fraction in liquid*		Wt% in gas		Mole fraction in gas*		Partial press./bar*	
	H_2S	NH_3	H_2S	NH_3	H_2S	NH_3	H_2S	NH_3	H_2S	NH_3
353.2	2.0	4.2	0.011	0.045	25.4	31.4	0.149	0.370	0.139	0.344
	3.0	5.45	0.016	0.058	33.9	33.6	0.208	0.413	0.241	0.479
	4.0	6.45	0.021	0.069	41.8	32.8	0.269	0.422	0.375	0.588
	5.0	7.1	0.027	0.077	49.5	29.7	0.334	0.401	0.545	0.642
	6.0	7.6	0.032	0.082	56.4	26.1	0.398	0.368	0.713	0.659
	7.0	7.95	0.038	0.087	62.8	22.5	0.463	0.332	0.936	0.671
	8.0	8.15	0.044	0.089	68.4	19.0	0.525	0.292	1.178	0.655
	9.0	8.35	0.049	0.092	73.9	15.5	0.591	0.248	1.502	0.630
	10.0	8.45	0.055	0.093	78.8	12.0	0.656	0.200	1.823	0.556
363.2	1.0	3.0	0.005	0.032	12.2	22.4	0.068	0.248	0.067	0.245
	2.0	4.45	0.011	0.047	25.1	32.2	0.147	0.378	0.205	0.527
	3.0	5.75	0.016	0.061	32.3	35.8	0.197	0.436	0.347	0.768
	4.0	6.8	0.021	0.073	38.4	35.8	0.242	0.451	0.501	0.933
	5.0	7.6	0.027	0.082	44.7	34.0	0.292	0.445	0.693	1.057
	6.0	8.15	0.032	0.088	50.3	31.4	0.340	0.425	0.895	1.119
	7.0	8.6	0.038	0.094	55.4	28.5	0.388	0.399	1.108	1.140
	8.0	9.4	0.044	0.103	60.4	25.3	0.437	0.367	1.335	1.121
	9.0	9.35	0.049	0.103	65.4	22.1	0.491	0.332	1.649	1.115
	10.0	9.5	0.055	0.105	70.3	18.7	0.547	0.291	1.988	1.058

* estimated by the compiler. Partial pressures were assumed to be proportional to mole fractions in the gas phase. The partial pressure of water was taken as the product of mole fraction of water in the liquid phase and the vapor pressure of pure water.

** total H_2S & NH_3 including that equivalent to ionic species.

COMPONENTS:	ORIGINAL MEASUREMENTS:
1. Hydrogen sulfide; H_2S; [7783-06-4] 2. Ammonia; NH_3; [7664-41-7] 3. Water; H_2O; [7732-18-5]	Ginzburg, D.M.; Pikulina, N.S.; Litvin, V.P. *Zhur.Prik.Khim.* 1965, *38*, 2117-9. *J.Appl.Chem. USSR* 1965, *38*, 2071-3.
VARIABLES: Temperature, concentration in liquid.	PREPARED BY: P.G.T. Fogg

EXPERIMENTAL VALUES:

T/K	Wt% in liquid**		Mole fraction in liquid*		Wt% in gas		Mole fraction in gas*		Partial press./bar*	
	H_2S	NH_3	H_2S	NH_3	H_2S	NH_3	H_2S	NH_3	H_2S	NH_3
335.7	9.24	8.80	0.0508	0.0968	61.84	23.51	0.4529	0.3445	0.459	0.349
337.7	8.90	10.03	0.0488	0.1101	51.46	32.35	0.3507	0.4411	0.355	0.447
340.2	7.28	7.00	0.0397	0.0764	64.67	18.31	0.4847	0.2746	0.491	0.278
341.5	7.10	7.22	0.0387	0.0787	63.98	19.67	0.4768	0.2933	0.483	0.297
342.4	7.00	9.01	0.0381	0.0980	40.15	40.06	0.2547	0.5084	0.258	0.515
342.6	7.04	8.90	0.0383	0.0969	41.10	39.10	0.2623	0.4992	0.266	0.506
345.8	5.21	4.99	0.0282	0.0540	61.33	16.11	0.4505	0.2368	0.456	0.240
347.6	5.00	7.16	0.0270	0.0772	27.10	43.80	0.1597	0.5165	0.162	0.523
347.7	5.00	5.11	0.0270	0.0552	65.10	11.70	0.4921	0.1769	0.499	0.179
347.8	5.01	7.20	0.0270	0.0777	27.40	43.90	0.1617	0.5185	0.164	0.525

Total pressure = 1 atm = 1.01325 bar

* estimated by the compiler. Partial pressures were assumed to be proportional to mole fractions in the gas phase.

** total H_2S & NH_3 including that equivalent to ionic species.

AUXILIARY INFORMATION

METHOD/APPARATUS/PROCEDURE:	SOURCE AND PURITY OF MATERIALS:
Partial pressures of NH_3, H_2S and H_2O over solutions of NH_3 and H_2S were measured by a static method described in ref. (1). The NH_3 content of the solutions were determined by titration against HCl solution (0.1 mol dm^{-3}) and that of H_2S by iodimetry.	No information
	ESTIMATED ERROR:

	REFERENCES:
	1. Ginzburg, D.M.
	   Zh. Fiz. Khim. 1965, *38 (10)*,
	   2197.

COMPONENTS:	ORIGINAL MEASUREMENTS:
1. Hydrogen sulfide; H_2S; [7783-06-4]	Ginzburg, D.M.; Pikulina, N.S.; Litvin, V.P.
2. Ammonia; NH_3; [7664-41-7]	
3. Water; H_2O; [7732-18-5]	*Zhur.Prik.Khim.* 1965, *38*, 2117-9. *J.Appl.Chem.* USSR 1965, *38*, 2071-3.

EXPERIMENTAL VALUES:

T/K	Wt% in liquid**		Mole fraction in liquid*		Wt% in gas		Mole fraction in gas*		Partial press./bar*	
	H_2S	NH_3	H_2S	NH_3	H_2S	NH_3	H_2S	NH_3	H_2S	NH_3
348.5	4.54	6.98	0.0244	0.0751	30.70	37.94	0.1852	0.4578	0.188	0.464
348.7	4.67	6.86	0.0251	0.0739	32.17	38.30	0.1955	0.4657	0.198	0.472
349.2	4.39	6.78	0.0236	0.0729	29.70	41.10	0.1778	0.4923	0.180	0.499
350.3	3.08	7.04	0.0165	0.0753	4.09	58.34	0.0213	0.6088	0.022	0.617
353.2	3.08	3.03	0.0165	0.0325	52.10	12.90	0.3619	0.1793	0.367	0.182
353.8	3.20	5.04	0.0171	0.0540	26.02	36.02	0.1533	0.4246	0.155	0.430
354.1	3.00	3.10	0.0161	0.0332	49.81	13.91	0.3409	0.1905	0.345	0.193
354.4	3.06	5.00	0.0164	0.0535	21.60	36.24	0.1244	0.4175	0.126	0.423
354.7	2.80	5.04	0.0150	0.0539	18.50	36.30	0.1048	0.4116	0.106	0.417
355.6	1.00	5.19	0.0053	0.0550	0.10	49.90	0.0005	0.5138	0.001	0.521
359.3	2.08	3.12	0.0111	0.0333	25.80	23.50	0.1531	0.2790	0.155	0.283
360.4	1.25	3.06	0.0066	0.0325	8.63	32.02	0.0467	0.3468	0.047	0.351
360.7	1.80	2.94	0.0096	0.0313	17.10	26.50	0.0968	0.3003	0.098	0.304
362.4	1.13	1.12	0.0060	0.0119	33.56	7.57	0.2100	0.0948	0.213	0.096
363.6	0.94	1.19	0.0050	0.0126	27.53	11.88	0.1662	0.1435	0.168	0.145
363.6	1.00	1.15	0.0053	0.0122	30.20	9.80	0.1852	0.1203	0.188	0.122
363.8	0.91	1.15	0.0048	0.0122	26.71	11.66	0.1606	0.1402	0.163	0.142

Total pressure = 1 atm = 1.01325 bar

* estimated by the compiler. Partial pressures were assumed to be proportional to mole fractions in the gas phase.

** total H_2S & NH_3 including that equivalent to ionic species.

COMPONENTS:	ORIGINAL MEASUREMENTS:
1. Hydrogen sulfide; H_2S; [7783-06-4] 2. Ammonia; NH_3; [7664-41-7] 3. Water; H_2O; [7732-18-5]	Sponsored by API (Brigham Young University, 1975) ; quoted by: Beutier, D.; Renon, H. *Ind.Eng.Chem.Proc.Des.Dev.* <u>1978</u>, *17*, 220-230.
VARIABLES: Concentrations	PREPARED BY: P.G.T. Fogg

EXPERIMENTAL VALUES:

T/K = 353.15

Conc. of NH_3 in liquid /mol dm^{-3}	Conc. of H_2S in liquid /mol dm^{-3}	Partial pressures P_{NH_3}/mmHg	P_{H_2S}/mmHg
0.960	0.971	12.0	2389
1.063	1.452	4.36	8556
2.332	1.151	136.0	319
5.112	1.143	574	94.8
5.538	5.305	72.6	12140
9.245	7.935	208	9506
10.201	5.983	597	1916
22.627	5.561	2285	241

760 mmHg = 1 atm = 1.01325 bar

AUXILIARY INFORMATION

METHOD/APPARATUS/PROCEDURE:	SOURCE AND PURITY OF MATERIALS:
No information	No information
	ESTIMATED ERROR:
	REFERENCES:

COMPONENTS:	ORIGINAL MEASUREMENTS:
1. Hydrogen sulfide; H_2S; [7783-06-4] 2. Ammonia; NH_3; [7664-41-7] 3. Water; H_2O; [7732-18-5] 4. Ammonium chloride NH_4Cl; [12125-02-9]	van Krevelen, D.W.; Hoftijzer, P.J. Huntjens, F.J. *Recl. Trav. Chim. Pays-Bas* <u>1949</u>, *68*, 191-216.

VARIABLES:	PREPARED BY:
Concentrations of components in liquid phase.	P.G.T. Fogg

EXPERIMENTAL VALUES:

T/K = 293.15

Concn. of H_2S in liquid* /mol dm^{-3}	Concn. of NH_3 in liquid* /mol dm^{-3}	Concn. of NH_4Cl /mol dm^{-3}	Partial pressure of H_2S /mmHg
0.202	0.753	0.374	6.2
0.201	1.102	0.748	9.0
0.200	1.495	1.122	12.9
0.200	1.852	1.496	14.4
0.198	2.215	1.870	17.3
0.195	2.600	2.244	21.4
0.198	2.946	2.618	24.6
0.201	3.315	2.992	27.6
0.194	3.670	3.366	33.7

* total H_2S and NH_3 including that equivalent to ionic species

AUXILIARY INFORMATION

METHOD/APPARATUS/PROCEDURE:	SOURCE AND PURITY OF MATERIALS:
Solutions of gases were prepared by passing H_2S, NH_3 and steam through water with exclusion of air. Samples of vapor in equilibrium with the solution under test, to which the salt had been added, were absorbed by solutions of cadmium acetate in acetic acid which were then analysed. These analyses together with P-V-T data were used to calculate partial pressures of NH_3 & H_2S. Another method consisted of passing a measured volume of N_2 at a measured total pressure through a series of vessels containing solution. H_2S in the effluent gas was determined iodimetrically. The two methods gave similar results.	No information
	ESTIMATED ERROR:
	REFERENCES:

COMPONENTS:	ORIGINAL MEASUREMENTS:
1. Hydrogen sulfide; H_2S; [7783-06-4] 2. Ammonia; NH_3; [7664-41-7] 3. Water; H_2O; [7732-18-5] 4. Ammonium sulfate $(NH_4)_2SO_4$; [7783-20-2]	van Krevelen, D.W.; Hoftijzer, P.J. Huntjens, F.J. *Recl. Trav. Chim. Pays-Bas* <u>1949</u>, *68*, 191-216.

VARIABLES:	PREPARED BY:
Concentrations of components in solution.	P.G.T. Fogg

EXPERIMENTAL VALUES:

T/K = 293.15

Concn. of H_2S in liquid[*] /mol dm^{-3}	Concn. of NH_3 in liquid[*] /mol dm^{-3}	Concn. of $(NH_4)_2SO_4$ /mol dm^{-3}	Partial pressure of H_2S /mmHg
0.202	0.705	0.1515	5.2
0.200	1.314	0.4545	7.3
0.201	1.651	0.606	9.2
0.201	1.968	0.7575	9.6
0.198	2.286	0.909	10.6
0.195	2.615	1.0605	11.8
0.194	2.851	1.212	12.9
0.194	3.213	1.3635	13.8

[*] total H_2S and NH_3 including that equivalent to ionic species

AUXILIARY INFORMATION

METHOD/APPARATUS/PROCEDURE:	SOURCE AND PURITY OF MATERIALS:
Solutions of gases were prepared by passing H_2S, NH_3 and steam through water with exclusion of air. Samples of vapor in equilibrium with the solution under test, to which the salt had been added, were absorbed by solutions of cadmium acetate in acetic acid which were then analysed. These analyses together with P-V-T data were used to calculate partial pressures of NH_3 & H_2S. Another method consisted of passing a measured volume of N_2 at a measured total pressure through a series of vessels containing solution. H_2S in the effluent gas was determined iodimetrically. The two methods gave similar results.	No information
	ESTIMATED ERROR:
	REFERENCES:

COMPONENTS:	ORIGINAL MEASUREMENTS:
1. Hydrogen sulfide; H_2S; [7783-06-4] 2. Ammonia; NH_3; [7664-41-7] 3. Carbon dioxide; CO_2; [124-38-9] 4. Water; H_2O; [7732-18-5]	Badger, E.H.M; Silver, L. *J. Soc. Chem. Ind.* <u>1938</u>, *57*, 110-112.

VARIABLES:	PREPARED BY:
Concentration in liquid.	P.G.T. Fogg

EXPERIMENTAL VALUES:

H_2S conc. in liquid* /mol dm^{-3}	NH_3 conc. in liquid* /mol dm^{-3}	CO_2 conc. in liquid* /mol dm^{-3}	P_{H_2S} /mmHg	P_{NH_3} /mmHg	P_{CO_2} /mmHg
0.189	1.189	0.41	3.21	4.11	1.45
0.1935	1.1935	0.49	5.12	2.92	3.50
0.390	1.390	0.495	12.56	2.42	3.70
0.045	1.045	0.50	0.00	2.72	3.65
0.380	1.380	0.6375	26.97	1.54	13.10
0.097	1.097	0.66	5.29	1.35	12.15
0.192	1.192	0.67	11.14	1.31	13.10
0.1955	1.1955	0.6875	15.25	1.30	19.00
0.192	1.192	0.70	15.95	0.89	20.45
0.1925	1.1925	0.745	27.38	0.90	29.20
0.0915	1.0915	0.77	12.20	0.71	35.15
0.088	1.088	0.79			42.45
0.0945	1.0945	0.80			45.10
0.088	1.088	0.815	13.06	0.55	
0.0945	1.0945	0.8175	16.91	0.57	

T/K = 293.2   760 mmHg = 1 atm = 1.01325 bar

*   total H_2S, NH_3 and CO_2 including that equivalent to ionic species.

AUXILIARY INFORMATION

METHOD/APPARATUS/PROCEDURE:	SOURCE AND PURITY OF MATERIALS:
A stream of nitrogen was passed through a series of thermostatted coils containing a solution of H_2S, CO_2 and NH_3 under test. The gases in the emergent stream were estimated by chemical methods.	No information
	ESTIMATED ERROR:
	REFERENCES:

COMPONENTS:	ORIGINAL MEASUREMENTS:
1. Hydrogen sulfide; H_2S; [7783-06-4] 2. Ammonia; NH_3; [7664-41-7] 3. Carbon dioxide; CO_2; [124-38-9] 3. Water; H_2O; [7732-18-5]	Dryden, I.G.C. *J.Soc. Chem. Ind.* <u>1947</u>, *66*. 59-64.

VARIABLES:	PREPARED BY:
Temperature	P.G.T. Fogg

EXPERIMENTAL VALUES:

Solutions were prepared by saturating aqueous ammonia with mixtures of H_2S and CO_2 under a total pressure equal to barometric.

T/K	$[NH_3]$ / mol dm^{-3}	K_3'	γ
293.2	0.51	1.40 ± 0.02	2.14
293.2	0.99	1.81 ± 0.05	2.04
293.2	1.95	2.50 ± 0.02	1.88
308.2	1.00	1.35 ± 0.015	1.95
323.2	1.00	1.07 ± 0.015	1.83

$$K_3' = \frac{[CO_2] \, P_{H_2S}}{[H_2S] \, P_{CO_2}}$$

The concentration term $[NH_3]$ includes the concentration of NH_4^+ and any other species derived from NH_3 which is present in solution. In the same way $[CO_2]$ and $[H_2S]$ include all species such as HCO_3^- and HS^- derived from CO_2 and H_2S respectively.

$$\gamma = \frac{2\,[CO_2] + 2\,[H_2S]}{[NH_3]}$$

AUXILIARY INFORMATION

METHOD/APPARATUS/PROCEDURE:	SOURCE AND PURITY OF MATERIALS:
Solutions under test were prepared by passing H_2S or H_2S/CO_2 into aqueous ammonium carbonate and/or ammonia. Nitrogen was passed through Drechsel bottles in series containing a solution under test. The H_2S in the emergent nitrogen stream was absorbed in cadmium acetate solution and estimated iodimetrically. The solution in the last Drechsel bottle was analysed for CO_2 by precipitation of $CaCO_3$ and for H_2S by precipitation of CdS and subsequent estimation of the CdS by iodimetry.	1. from a Kipp's apparatus.
	ESTIMATED ERROR:
	REFERENCES:

COMPONENTS:	ORIGINAL MEASUREMENTS:
1. Hydrogen sulfide; H_2S; [7783-06-4] 2. Ammonia; NH_3; [7664-41-7] 3. Carbon dioxide; CO_2; [124-38-9] 4. Water; H_2O; [7732-18-5]	Dryden, I.G.C. J. Soc. Chem. Ind. <u>1947</u>, 66, 59-64.

EXPERIMENTAL VALUES:

Solutions under test were prepared by passing an appropriate amount of H_2S into a mixture of ammonium carbonate (1.5 mol dm^{-3}) and ammonia (1.5 mol dm^{-3})

NH_3 concn. in liquid* /mol dm^{-3}	H_2S concn. in liquid* /mol dm^{-3}	R	k_{20}	CO_2 concn. in liquid*† /mol dm^{-3}	P_{H_2S}† /mmHg	P_{H_2S}† /bar
0.47	0.20	0.37	965	0.050	3.74	0.004989
0.47	0.20	1.33	210	0.180	17.20	0.022927
0.48	0.20	0.75	640	0.105	5.64	0.007523
0.49	0.205	1.10	363	0.157	10.20	0.013596
0.49	0.2075	1.47	148	0.208	25.32	0.033753
0.49	0.195	0.885	475	0.131	7.413	0.009883
0.49	0.21	0.19	985	0.027	3.850	0.005133
0.49	0.195	0.37	1005	0.055	3.504	0.004672
0.51	0.205	0.53	720	0.081	5.141	0.006854
0.955	0.20	1.165	342	0.439	10.560	0.014079
0.98	0.195	1.00	535	0.392	6.582	0.008775
0.98	0.21	1.50	124	0.577	30.581	0.040771
0.98	0.2175	0.765	1045	0.292	3.758	0.005010
0.98	0.20	0.295	2420	0.115	1.492	0.001989
1.00	0.2075	0.85	750	0.337	4.996	0.006661
1.00	0.2075	1.57	132	0.622	28.385	0.037843
1.01	0.05	1.28	310	0.614	2.912	0.003882
1.02	0.06	1.07	655	0.514	1.654	0.002205
1.02	0.185	0.70	1120	0.292	2.983	0.003977
1.02	0.05	0.83	1075	0.403	0.840	0.001120
1.02	0.20	0.55	1480	0.226	2.440	0.003253
1.03	0.05	1.07	650	0.524	1.389	0.001852
1.04	0.2075	1.35	202	0.562	18.549	0.024730
1.04	0.05	0.41	3120	0.203	0.289	0.000385
1.05	0.0525	0.61	1810	0.304	0.524	0.000699
1.05	0.055	0.21	6000	0.104	0.166	0.000221
1.10	0.105	0.58	1830	0.289	1.036	0.001381
1.10	0.105	0.795	1055	0.396	1.797	0.002396
1.10	0.10	1.24	335	0.620	5.390	0.007186
1.11	0.0525	1.50	150	0.793	6.320	0.008426

T/K = 293.2   Total pressure = 760 mmHg = 1.01325 bar

COMPONENTS:	ORIGINAL MEASUREMENTS:
1. Hydrogen sulfide; H_2S; [7783-06-4] 2. Ammonia; NH_3; [7664-41-7] 3. Carbon dioxide; CO_2; [124-38-9] 4. Water; H_2O; [7732-18-5]	Dryden, I.G.C. *J. Soc. Chem. Ind.* 1947, 66, 59-64.

EXPERIMENTAL VALUES:

Solutions under test were prepared by passing an appropriate amount of H_2S into a mixture of ammonium carbonate (1.5 mol dm^{-3}) and ammonia (1.5 mol dm^{-3})

NH_3 concn. in liquid* /mol dm^{-3}	H_2S concn. in liquid* /mol dm^{-3}	R	k_{20}	CO_2 concn. in liquid*† /mol dm^{-3}	P_{H_2S}† /mmHg	P_{H_2S}† /bar
1.12	0.105	0.205	5400	0.104	0.351	0.000468
1.13	0.115	1.59	124	0.807	16.747	0.022327
1.14	0.105	1.01	575	0.523	3.297	0.004396
1.16	0.205	0.64	1180	0.306	3.137	0.004182
1.18	0.205	0.85	680	0.414	5.444	0.007258
1.18	0.21	1.29	239	0.626	15.866	0.021153
1.18	0.19	1.48	152	0.733	22.571	0.030092
1.20	0.19	1.06	424	0.535	8.092	0.010788
1.20	0.21	0.20	2700	0.099	1.404	0.001872
1.21	0.195	0.40	2180	0.203	1.615	0.002153
1.46	0.2075	0.44	2440	0.276	1.536	0.002048
1.47	0.185	0.63	1560	0.405	2.141	0.002854
1.48	0.215	0.89	720	0.563	5.392	0.007189
1.48	0.2075	0.235	3750	0.150	0.999	0.001332
1.48	0.205	1.005	525	0.641	7.051	0.009400
1.48	0.19	1.54	117	0.993	29.324	0.039095
1.48	0.2075	1.32	208	0.840	18.014	0.024016
1.52	0.17	0.84	970	0.567	3.165	0.004220

T/K = 293.2   Total pressure = 760 mmHg = 1.01325 bar

$$R = \frac{\text{conc. of } CO_2}{\text{conc. of } NH_3 - \text{conc. of } H_2S}$$

All concentrations in this expression refer to the liquid phase and are measured in mol dm^{-3}. They include ionic and other species derived from the gases.

   k_{20} = total conc. of H_2S in the liquid phase / conc. in the gas phase

* total H_2S, NH_3 and CO_2 including that equivalent to ionic and other species derived from the gases.

† calculated by the compiler.

COMPONENTS:	ORIGINAL MEASUREMENTS:
1. Hydrogen sulfide; H_2S; [7783-06-4] 2. Ammonia; NH_3; [7664-41-7] 3. Carbon dioxide; CO_2; [124-38-9] 4. Water; H_2O; [7732-18-5]	van Krevelen, D.W.; Hoftijzer, P.J. Huntjens, F.J. *Recl. Trav. Chim. Pays-Bas* 1949, *68*, 191-216.
VARIABLES: Temperature, concentration of components in liquid phase.	PREPARED BY: 　　P.G.T. Fogg

EXPERIMENTAL VALUES:

T/K	H_2S concn. in liquid* /mol dm^{-3}	NH_3 concn. in liquid* /mol dm^{-3}	CO_2 concn. in liquid* /mol dm^{-3}	Partial pressure of H_2S/mmHg	Partial pressure of H_2S/bar**
293.2	0.180	0.79	0.25	4.5	0.0060
	0.180	1.14	0.41	4.2	0.0056
	0.210	2.25	1.40	18.0	0.0240
	0.291	1.24	0.75	49.0	0.0653
	0.360	2.16	0.95	14.8	0.0197
	0.60	2.15	0.40	8.4	0.0112
313.2	0.160	0.74	0.38	27.8	0.0371
	0.180	0.79	0.25	13.3	0.0177
	0.180	1.14	0.41	12.3	0.0164
	0.184	1.17	0.41	11.7	0.0156
	0.210	2.25	1.40	37.6	0.0501
	0.290	1.13	0.21	12.1	0.0161
	0.350	0.70	0.104	32.3	0.0431
	0.360	2.16	0.94	38.4	0.0512
	0.600	2.15	0.40	27.0	0.0360
333.2	0.124	1.298	0.656	25.8	0.0344
	0.140	1.02	0.62	63.7	0.0849
	0.150	0.74	0.36	53.5	0.0713
	0.180	0.79	0.25	32.0	0.0427
	0.180	1.14	0.41	30.0	0.0400
	0.184	1.17	0.41	29.5	0.0393
	0.200	2.25	1.34	59.5	0.0793
	0.234	1.289	0.645	77.0	0.1026
	0.290	1.130	0.21	32.8	0.0437
	0.350	0.700	0.104	80.4	0.1072
	0.360	2.160	0.95	84.4	0.1125
	0.60	2.150	0.40	77.5	0.1033

*　total H_2S, NH_3 and CO_2 including that equivalent to ionic species.

**　calculated by the compiler.

AUXILIARY INFORMATION

METHOD/APPARATUS/PROCEDURE	SOURCE AND PURITY OF MATERIALS
Samples of vapor in equilibrium with the solution under test, were analysed by chemical methods. These analyses together with P-V-T data were used to calculate the partial pressure of H_2S. Another method consisted of passing a measured volume of N_2 at a measured total pressure through a series of vessels containing solution. H_2S in the effluent gas was determined iodimetrically. The two methods gave similar results.	No information
	ESTIMATED ERROR:
	REFERENCES:

COMPONENTS:	ORIGINAL MEASUREMENTS:
1. Hydrogen sulfide; H_2S; [7783-06-4] 2. Ammonia; NH_3; [7664-41-7]; 3. Carbon dioxide; CO_2; [124-38-9] 4. Water; H_2O; [7732-18-5]	Wilson, G.M.; Gillespie, P.C.; Owens, J.L. *Proc. 64th Ann. Conv. Gas Processors Association*, <u>1985</u>, 282-288

VARIABLES:	PREPARED BY:
Temperature, composition	P.G.T. Fogg

EXPERIMENTAL VALUES:

T/°F	T/K*	Component	Composition /mole% liquid	vapor	Partial pressure /psia	Partial pressure /bar*
200	366.48	NH_3	2.00	6.09	1.50	0.103
		CO_2	0.57	33.99	8.38	0.578
		H_2S	0.51	17.45	4.30	0.296
		H_2O	96.92	42.47	10.49	0.723
		total	100.00	100.00	24.67	1.701
200	366.48	NH_3	2.06	13.25	2.20	0.152
		CO_2	0.26	8.80	1.45	0.100
		H_2S	0.51	11.85	1.96	0.135
		H_2O	97.17	66.10	11.09	0.765
		total	100.00	100.00	16.70	1.151
200	366.48	NH_3	17.76	35.52	23.44	1.616
		CO_2	3.84	17.36	11.46	0.790
		H_2S	4.08	32.65	21.55	1.486
		H_2O	74.32	14.47[+]	9.55	0.658
		total	100.00	100.00	66.00	4.551
300	422.04	NH_3	2.00	6.81	8.86	0.611
		CO_2	0.24	27.14	35.31	2.434
		H_2S	0.49	13.78	17.93	1.236
		H_2O	97.27	52.27	68.00	4.688
		total	100.00	100.00	130.10	8.970

*estimated by compiler.

[+]water content too low to measure but calculated by authors using Raoult's law with correction for pressure enhancement.

AUXILIARY INFORMATION

METHOD/APPARATUS/PROCEDURE:	SOURCE AND PURITY OF MATERIALS:
Rocked static equilibrium cells were used for measurements. These were electrically heated and insulated with fibreglass. Pressures were measured with a 3-D Instruments precision pressure gauge. Platinum resistance thermometers were used to measure and control temperature. The vapor phase was analysed either by gas chromatography or by absorbing and weighing the components with NH_3 absorbed by HCl, H_2S by $CdSO_4$, H_2O by *Drierite*® and CO_2 by *Ascarite*®. Samples of liquid phase were allowed to come to ambient temperature and pressure and the components estimated. Further details in refs. 1 & 2.	No information

ESTIMATED ERROR:

REFERENCES:

1. Owens, J.L.; Cunningham, J.R.; Wilson, G.M. *Research Report RR-65*, Gas Processors Association, Tulsa, Oklahoma <u>1983</u>.

2. Wilson, G.M.; Owens, J.L.; Cunningham, J.R. *Research Report RR-52*, Gas Processors Association, Tulsa, Oklahoma <u>1982</u>.

COMPONENTS:	ORIGINAL MEASUREMENTS:
1. Hydrogen sulfide; H_2S; [7783-06-4]	Wilson, G.M.; Gillespie, P.C.; Owens, J.L.
2. Ammonia; NH_3; [7664-41-7];	
3. Carbon dioxide; CO_2; [124-38-9]	*Proc. 64th Ann. Conv. Gas Processors*
4. Water; H_2O; [7732-18-5]	*Association*, <u>1985</u>, 282-288

EXPERIMENTAL VALUES:

T/°F	T/K*	Component	Composition /mole%		Partial pressure /psia	Partial pressure /bar*
			liquid	vapor		
300	422.04	NH_3	2.12	8.33	11.74	0.809
		CO_2	0.29	30.87	43.52	3.001
		H_2S	0.53	14.44	20.37	1.404
		H_2O	97.06	46.36	65.37	4.507
		total	100.00	100.00	141.00	9.722
300	422.04	NH_3	17.32	15.69	96.02	6.620
		CO_2	4.19	43.83	268.24	18.494
		H_2S	3.82	29.12	178.21	12.287
		H_2O	74.67	11.36[†]	69.53	4.794
		total	100.00	100.00	612.00	42.196
300	422.04	NH_3	2.16	0.336	4.54	0.313
		CO_2	1.44	67.80	915.21	63.102
		H_2S	1.73	23.84	321.84	22.190
		H_2O	94.72	8.03[†]	108.41	7.475
		total	100.00	100.00	1350.00	93.079
400	477.59	NH_3	2.03	6.56	32.47	2.239
		CO_2	0.26	28.55	141.3	9.742
		H_2S	0.57	15.50	76.71	5.289
		H_2O	97.14	49.39	244.5	16.858
		total	100.00	100.00	494.98	34.128
400	477.59	NH_3	2.14	5.06	31.37	2.163
		CO_2	0.44	35.86	222.3	15.327
		H_2S	0.83	20.16	125.0	8.618
		H_2O	96.59	38.92	241.3	16.637
		total	100.00	100.00	619.97	42.745
400	477.59	NH_3	17.30	14.70	426.30	29.392
		CO_2	3.43	49.77	1443.33	99.514
		H_2S	3.56	17.83[†]	517.07	35.651
		H_2O	75.70	17.70	513.30	35.391
		total	100.00	100.00	2900.00	199.948
400	477.59	NH_3	1.80	1.46	30.37	2.094
		CO_2	1.44	56.03	1165.42	80.353
		H_2S	1.84	21.70[†]	451.36	31.120
		H_2O	94.74	20.81	432.85	29.844
		total	100.00	100.00	2080.00	143.411
500	533.15	NH_3	2.12	5.89	70.39	4.853
		CO_2	0.313	15.04	179.73	12.392
		H_2S	0.523	8.63	103.13	7.111
		H_2O	97.05	70.44	841.75	58.037
		total	100.00	100.00	1195.00	82.392
500	533.15	NH_3	2.12	4.86	78.00	5.378
		CO_2	0.575	22.88	376.22	25.939
		H_2S	0.883	13.24	212.50	14.651
		H_2O	96.42	59.02	947.28	65.313
		total	100.00	100.00	1605.00	110.661
500	533.15	NH_3	1.86	2.31	66.30	4.571
		CO_2	2.13	38.04	1091.75	75.274
		H_2S	1.97	15.56[†]	446.57	30.790
		H_2O	94.04	44.09	1265.38	87.245
		total	100.00	100.00	2870.00	197.880

*estimated by compiler.
†water content too low to measure but calculated by authors using Raoult's law with correction for pressure enhancement.

COMPONENTS:	ORIGINAL MEASUREMENTS:
1. Hydrogen sulfide; H_2S; [7783-06-4] 2. Ammonia; NH_3; [7664-41-7]; 3. Carbon dioxide; CO_2; [124-38-9] 4. Water; H_2O; [7732-18-5] 5. Carbon monoxide; CO; [630-08-0] 6. Nitrogen; N_2; [7727-37-9] 7. Methane; CH_4; [74-82-8] 8. Hydrogen; H_2; [1333-74-0]	Wilson, G.M.; Gillespie, P.C.; Owens, J.L. *Proc. 64th Ann. Conv. Gas Processors Association*, 1985, 282-288.
VARIABLES:	PREPARED BY:
Temperature, composition of gas and liquid phase.	P.G.T. Fogg

EXPERIMENTAL VALUES:

T/°F	T/K*	Total pressure /psia	/kPa*	Component	Composition /mole% liquid	vapor	Partial pressure /psia	/kPa*
100	310.93	33.4	230	NH_3	2.11	0.85	0.28	1.93
				CO_2	0.27	0.065	0.022	0.152
				H_2S	0.53	0.47	0.16	1.10
				H_2O	97.09	2.33	0.78	5.38
				CO	a	25.99	8.68	59.85
				N_2	a	9.99	3.34	23.03
				CH_4	a	12.96	4.33	29.85
				H_2	a	47.35	15.81	109.00
100	310.93	500	3447	NH_3	2.11	0.067	0.34	2.34
				CO_2	0.27	0.0029	0.015	0.103
				H_2S	0.53	0.028	0.14	0.97
				H_2O	97.05	0.19	0.93	6.41
				CO	0.0101	26.89	134.5	927.3
				N_2	0.0025	10.32	51.60	355.77
				CH_4	0.0063	13.44	67.21	463.40
				H_2	0.0255	49.06	245.30	1691.28

[a] too small to measure.

* estimated by compiler.

AUXILIARY INFORMATION

METHOD/APPARATUS/PROCEDURE:	SOURCE AND PURITY OF MATERIALS:
Rocked static equilibrium cells were used for measurements. These were electrically heated and insulated with fibreglass. Pressures were measured with a 3-D Instruments precision pressure gauge. Platinum resistance thermometers were used to measure and control temperature. The vapor phase was analysed either by gas chromatography or by absorbing and weighing the components with NH_3 absorbed by HCl, H_2S by $CdSO_4$, H_2O by *Drierite®* and CO_2 by *Ascarite®*. Samples of liquid phase were allowed to come to ambient temperature and pressure and the components estimated. Further details in refs. 1 & 2.	No information
	ESTIMATED ERROR:
	REFERENCES: 1. Owens, J.L.; Cunningham, J.R.; Wilson, G.M. *Research Report RR-65, Gas Processors Association, Tulsa, Oklahoma,* 1983 2. Wilson, G.M.; Owens, J.L.; Cunningham, J.R. *Research Report RR-52, Gas Processors Association, Tulsa, Oklahoma,* 1982.

COMPONENTS:	ORIGINAL MEASUREMENTS:
1. Hydrogen sulfide; H_2S; [7783-06-4] 2. Ammonia; NH_3; [7664-41-7]; 3. Carbon dioxide; CO_2; [124-38-9] 4. Water; H_2O; [7732-18-5] 5. Carbon monoxide; CO; [630-08-0] 6. Nitrogen; N_2; [7727-37-9] 7. Methane; CH_4; [74-82-8] 8. Hydrogen; H_2; [1333-74-0]	Wilson, G.M.; Gillespie, P.C.; Owens, J.L. *Proc. 64th Ann. Conv. Gas Processors Association*, 1985, 282-288.

EXPERIMENTAL VALUES:

T/°F	T/K*	Total pressure /psia	/kPa*	Component	Composition /mole% liquid	vapor	Partial pressure /psia	/kPa*
100	310.93	1000	6895	NH_3	2.11	0.036	0.36	2.48
				CO_2	0.27	0.0012	0.012	0.083
				H_2S	0.53	0.015	0.15	1.03
				H_2O	97.01	0.093	0.93	6.41
				CO	0.021	26.93	269.3	1856.8
				N_2	0.0052	10.33	103.4	712.9
				CH_4	0.013	13.46	134.6	928.0
				H_2	0.047	49.13	491.3	3387.4
200	366.48	500	3447	NH_3	2.05	0.48	2.40	16.55
				CO_2	0.25	0.32	1.61	11.10
				H_2S	0.51	0.47	2.37	16.34
				H_2O	97.15	2.24	11.21	77.29
				CO	0.0080	24.60	123.0	848.1
				N_2	0.0025	9.53	47.6	328.2
				CH_4	0.0058	12.58	62.9	433.7
				H_2	0.0229	49.76	248.9	1716.1
200	366.48	500	3447	NH_3	2.07	0.083	0.41	2.83
				CO_2	1.62	28.46	142.3	981.1
				H_2S	0.54	4.74	23.72	163.54
				H_2O	95.74	2.16	10.79	74.39
				CO	–	13.17	65.85	454.02
				N_2	–	5.01	25.07	172.85
				CH_4	–	9.18	45.90	316.47
				H_2	–	37.20	185.9	1281.7
200	366.48	1000	6895	NH_3	2.02	0.28	2.75	18.96
				CO_2	0.26	0.18	1.75	12.07
				H_2S	0.52	0.26	2.64	18.20
				H_2O	97.13	1.12	11.20	77.22
				CO	0.0162	24.58	245.8	1694.7
				N_2	0.0048	9.84	98.4	678.4
				CH_4	0.0103	12.68	126.8	874.3
				H_2	0.0411	51.06	510.6	3520.5
300	422.04	500	3447	NH_3	2.11	1.86	9.32	64.26
				CO_2	0.25	9.90	49.52	341.43
				H_2S	0.50	4.19	20.96	144.51
				H_2O	97.10	13.07	65.35	450.57
				CO	0.0020	8.60	43.01	296.54
				N_2	0.0019	7.89	39.46	272.07
				CH_4	0.0053	11.69	58.45	403.00
				H_2	0.0284	42.80	213.9	1474.8

* estimated by compiler.

COMPONENTS:	ORIGINAL MEASUREMENTS:
1. Hydrogen sulfide; H_2S; [7783-06-4] 2. Ammonia; NH_3; [7664-41-7]; 3. Carbon dioxide; CO_2; [124-38-9] 4. Water; H_2O; [7732-18-5] 5. Carbon monoxide; CO; [630-08-0] 6. Nitrogen; N_2; [7727-37-9] 7. Methane; CH_4; [74-82-8] 8. Hydrogen; H_2; [1333-74-0]	Wilson, G.M.; Gillespie, P.C.; Owens, J.L. *Proc. 64th Ann. Conv. Gas Processors Association*, <u>1985</u>, 282-288.

EXPERIMENTAL VALUES:

T/°F	T/K*	Total pressure /psia	/kPa*	Component	Composition /mole% liquid	vapor	Partial pressure /psia	/kPa*
300	422.04	500	3447	NH_3	2.02	1.57	7.84	54.05
				CO_2	0.60	29.20	146.0	1006.6
				H_2S	0.52	4.68	23.40	161.34
				H_2O	96.83	13.93	69.64	480.15
				CO	0	0	0	0
				N_2	0.0024	6.40	32.02	220.77
				CH_4	0.0065	10.09	50.44	347.77
				H_2	0.0227	34.13	170.6	1176.2
300	422.04	1000	6895	NH_3	2.01	0.69	6.94	47.85
				CO_2	0.94	29.13	291.3	2008.4
				H_2S	0.48	2.72	27.24	187.81
				H_2O	96.49	6.69	66.94	461.54
				CO	0	0	0	0
				N_2	0.0056	7.57	75.72	522.07
				CH_4	0.0150	12.24	122.4	843.9
				H_2	0.0560	40.96	409.5	2823.4
300	422.04	1000	6895	NH_3	2.09	0.99	9.91	68.33
				CO_2	0.18	4.28	42.80	295.10
				H_2S	0.43	2.55	25.47	175.61
				H_2O	97.20	6.54	65.37	450.71
				CO	0.0119	17.52	175.2	1208.0
				N_2	0.0064	9.95	99.5	686.0
				CH_4	0.0163	14.19	141.9	978.4
				H_2	0.0637	43.98	439.9	3033.0
400	477.59	1000	6895	NH_3	2.13	3.27	32.70	225.46
				CO_2	0.45	22.28	222.8	1536.2
				H_2S	0.84	12.44	124.5	858.4
				H_2O	96.51	23.99	239.8	1653.4
				CO	0	0	0	0
				N_2	0.0057	8.13	81.3	560.5
				CH_4	0.0111	10.02	100.2	690.9
				H_2	0.0571	19.87	198.7	1370.0
400	477.59	1000	6895	NH_3	2.02	3.15	31.52	217.32
				CO_2	0.77	27.15	271.5	1871.9
				H_2S	0.61	8.18	81.77	563.78
				H_2O	96.52	23.87	238.7	1645.8
				CO	0	0	0	0
				N_2	0.0062	4.62	46.21	318.61
				CH_4	0.0155	7.27	72.72	501.39
				H_2	0.0556	25.76	257.6	1776.1

* estimated by compiler.

COMPONENTS:	ORIGINAL MEASUREMENTS:
1. Hydrogen sulfide; H_2S; [7783-06-4] 2. Carbon dioxide; CO_2; [124-38-9] 3. Ammonia; NH_3; [7664-41-7] 4. Phenol; C_6H_6O; [108-95-2] 5. Water; H_2O; [7732-18-5]	Golutvin, Yu. M.; Malysheva, T.V.; Skorobogatova, V.I. *Izvest. Sibir. Otdel. Akad. Nauk. S.S.S.R.* 1958, *No.8*, 83-7.
VARIABLES: Pressure of H_2S & CO_2	PREPARED BY: P.G.T. Fogg

EXPERIMENTAL VALUES:

Concentration of NH_3 before dissolution of H_2S & CO_2 = 0.0086 mol dm^{-3}
Concentration of phenol before dissolution of H_2S & CO_2 = 0.0194 mol dm^{-3}

Liquid phase		Gas phase
Conc. of H_2S	Conc. of CO_2	P_{H_2S} / P_{CO_2}
/mol dm^{-3}	/mol dm^{-3}	
0.000000	0.56960	0.0000
0.024410	0.51930	0.0934
0.058980	0.49870	0.3452
0.062910	0.53830	0.4248
0.063620	0.51370	0.2898
0.071900	0.50350	0.4859
0.078410	0.50330	0.4702
0.089720	0.48910	0.5620
0.109800	0.44220	0.5772
0.133600	0.43280	0.8758
0.136380	0.43070	0.7232
0.184500	0.40810	1.0330
0.199120	0.37210	1.3610
0.298700	0.25220	2.9180
0.317300	0.26090	2.8840
0.381900	0.19600	4.4830
0.404900	0.18320	4.6960
0.417800	0.15380	5.1484
0.477300	0.14178	7.1750
0.498700	0.08856	15.6900
0.599000	0.00000	

  T/K = 293.2;   Total pressure = 1.013 bar

Concentrations of H_2S, CO_2 & NH_3 include all ionic species derived from these compounds.

AUXILIARY INFORMATION

METHOD/APPARATUS/PROCEDURE:	SOURCE AND PURITY OF MATERIALS:
Premixed hydrogen sulfide and carbon dioxide were bubbled through an aqueous solution of ammonia and phenol in a thermostatted saturator for 4.5 to 5 hrs. The gas and liquid phases were analysed by chemical methods.	1. from a Kipp's apparatus. 2. by action of hydrochloric acid on marble chips.
	ESTIMATED ERROR: $\delta T/K$ = ± 0.1 (authors)
	REFERENCES:

COMPONENTS:	EVALUATOR:
1. Hydrogen sulfide; H_2S; [7783-06-4]	Peter G.T. Fogg,
	Department of Applied Chemistry
2. Carbon dioxide; CO_2; [124-38-9]	and Life Sciences,
	Polytechnic of North London,
3. Water; H_2O; [7732-18-5]	Holloway, London N7 8DB, U.K.
4. Alkanolamines	February 1987

CRITICAL EVALUATION:

Systems containing hydrogen sulfide, water and alkanolamines have been
extensively studied during the recent years because aqueous solutions of
alkanolamines are widely used to remove hydrogen sulfide from industrial gas
mixtures. Much of this work has been carried out by Mather and his co-workers.
Several models of the behaviour of alkanolamine/water systems in the presence of
hydrogen sulfide and/or carbon dioxide have been published (1-10). Some sets of
measurements (10-14) have been partially correlated with the model developed by
Kent and Eisenberg (8). Improved models have been published by Deshmukh and
Mather (9), and by Dingman et al. (10). Predictions based upon models are
still subject to some uncertainty.

Three component systems containing hydrogen sulfide, water and alkanolamines.

In all hydrogen sulfide/alkanolamine/water systems which have been investigated
the mole ratio of hydrogen sulfide to alkanolamine increases with partial
pressure of hydrogen sulfide and decreases with increase in temperature and
concentration of alkanolamine.

2-Aminoethanol (monoethanolamine); C_2H_7NO; [141-43-5]

     Data published by eight groups (1, 15-21) have been compiled. The
measurements were made over a range of concentrations of MEA (monoethanolamine)
from 0.6 to 5 mol dm^{-3}, pressures of hydrogen sulfide from 7 x 10^{-7}
to 42 bar and temperatures from 288 to 413 K. The different sets of
measurements tend to complement each other and, where there is overlap, there is
little evidence of inconsistency between one set and another. Lawson and Garst
(20) have shown that their data for absorption in 15 wt% solution of MEA (2.5
mol dm^{-3}) is in good agreement under most conditions with unpublished data
communicated to them by J.P. Bocard and with that published by Muhlbauer and
Monaghan (17), Leibush and Shneerson (16), and, Jones et al. (18). However
they have shown that there is disagreement with the measurements by Muhlbauer
and Monaghan at 373 K and pressures below 0.12 bar and with measurements by
Leibush and Shneerson at 323 K at partial pressures below 0.0014 bar.   The
evaluator has noted a discrepancy between the two measurements in each of the
following pairs of measurements:

Authors	Concentration of MEA / mol dm^{-3}	T/K	P_{H_2S} /bar	Mole ratio H_2S / MEA
Riegger et al. (15)	5	318.15	0.033	0.453
Lee et al. (19)	5	313.15	0.042	0.435
Jones et al. (18)	2.4 (15.3 wt%)	373.15	0.0089	0.066
Lawson & Garst (20)	2.4 (15.2 wt%)	373.15	0.0075	0.0793
Lawson & Garst (20)	2.4 (15.2%)	373.15	0.00041	0.0143
Isaacs et al. (21)	2.5	373.15	0.00031	0.0210

In the case of the last pair of measurements the difference is too great to be
due solely to the slight difference in concentrations of MEA.

2,2'-Iminobisethanol (diethanolamine); $C_4H_{11}NO_2$; [111-42-2]

     Five sets of measurements on this system have been compiled (1, 11, 16,
20, 22). Data reported by Bottoms (23) were in graphical form and have not been
compiled. Concentrations of DEA (diethanolamine) range from 0.5 to 5 mol dm^{-3},
temperatures from 288 to 422 K and partial pressures of hydrogen sulfide from
1.5 x 10^{-5} bar to 37 bar. The general pattern of data is very similar to
that for the monoethanolamine system. Different sets of measurements tend to
complement each other and there is usually good agreement when measurements
overlap. Lawson and Garst (20) have compared their measurements of solubilities

COMPONENTS:	EVALUATOR:
1. Hydrogen sulfide; H_2S; [7783-06-4] 2. Carbon dioxide; CO_2; [124-38-9] 3. Water; H_2O; [7732-18-5] 4. Alkanolamines	Peter G.T. Fogg, Department of Applied Chemistry and Life Sciences, Polytechnic of North London, Holloway, London N7 8DB, U.K. February 1987

CRITICAL EVALUATION:

in a 25 wt% solution of DEA and demonstrated that they are consistent with data for 298 K to 333 K given by Atwood et al. (1), Lee et al. (31) and also hitherto unpublished data in a private communication from J.P. Bocard. Lal et al. (11) have shown that their measurements of solubilities in DEA solution of concentration 2.0 mol dm^{-3} at 313 K are compatible with those of Lawson and Garst in a solution of concentration 2.45 mol dm^{-3} at 311 K.   The compiler has noted the following discrepancy between other published data:

Authors	Concentration of DEA / mol dm^{-3}	T/K	P_{H_2S} /bar	Mole ratio H_2S / MEA
Atwood et al. (1)	4.9 (50 wt%)	322	0.0476	0.153
Lee et al. (22)	5.0	323.15	0.0432	0.205

2-(2-Aminoethoxy)ethanol (diglycolamine); $C_4H_{11}NO_2$; [929-06-6]

Martin et al. (24) have reported measurements of the absorption of hydrogen sulfide by 60 wt% aqueous solutions of diglycolamine at temperatures from 323 to 373 K and partial pressures of 0.04 to 17.3 bar. Dingman et al. (10) have reported measurements by 65 wt% solutions at temperatures from 311 to 355 K and partial pressures from 3×10^{-5} to 1.8 bar. The two sets of measurements are consistent with each other in the pressure range over which measurements overlap i.e. 0.04 to 1.8 bar. There is some scatter of measurements at very low pressures close to the minimum pressure.

1,1'-Iminobis-2-propanol (diisopropanolamine); $C_6H_{15}NO_2$; [110-97-4]

Absorption in aqueous solutions of diisopropanolamine of concentration 2.5 mol dm^{-3} at 313 and 373 K and pressures from 0.02 to 32 bar was measured by Isaacs et al. (25). No other measurements on this system are available for comparison but there is no reason to doubt the reliability of this set of measurements. These authors have also measured absorption in solutions containing tetrahydrothiophene, 1,1-dioxide (sulfolane) (26) in addition to diisopropanolamine. These measurements also appear to be reliable but again no other sets of measurements are available for comparison.

2,2'-(Methylimino)bisethanol (methyldiethanolamine); $C_5H_{13}NO_2$; [105-59-9]

Jou et al. (12) measured absorption in solutions of methyldiethanolamine at concentrations from 1.0 to 4.3 mol dm^{-3} at 298 to 393 K and 2.1×10^{-5} to 19.6 bar. A consistent set of data was obtained and there is no reason to doubt the reliability of these measurements. No other measurements are available for comparison.

2,2',2"-Nitrilotrisethanol (triethanolamine); $C_6H_{15}NO_3$; [102-71-6]

Atwood et al. (1) measured absorption of hydrogen sulfide by 1 to 3.5 mol dm^{-3} (15 to 50 wt%) aqueous solutions of TEA (triethanolamine) at temperatures from 300 to 333 K and pressures from 4×10^{-6} to 0.92 bar. Jou et al. (13) measured absorption by solutions containing from 2 to 5 mol dm^{-3} of TEA at temperatures from 298 to 398 K and partial pressures from 1.2×10^{-4} to 62.7 bar. In the range of conditions in which the measurements overlap there is a good correlation between the two sets of measurements except for the following discrepancy:

COMPONENTS:	EVALUATOR:
1. Hydrogen sulfide; H_2S; [7783-06-4] 2. Carbon dioxide; CO_2; [124-38-9] 3. Water; H_2O; [7732-18-5] 4. Alkanolamines	Peter G.T. Fogg, Department of Applied Chemistry and Life Sciences, Polytechnic of North London, Holloway, London N7 8DB, U.K. February 1987

CRITICAL EVALUATION:

Authors	Concentration of TEA / mol dm^{-3}	T/K	P_{H_2S} /bar	Mole ratio H_2S / MEA
Atwood et al. (1)	2 (30 wt%)	310.9	0.000757	0.0124
Jou et al. (13)	2	323.2	0.000683	0.0133

Four component systems of hydrogen sulfide, carbon dioxide, water and alkanolamine.

These systems were briefly investigated by Leibush & Schneerson (16). The first detailed study was by Muhlbauer & Monaghan in 1957 (17). In each system there is the possibility of varying the partial pressure of each gas, the temperature and the concentration of alkanolamine. Measurements by different workers tend to complement each other with different groups having measured absorption under different sets of conditions. Direct comparison of one set of measurements with another is, in many cases, not possible. In all cases the mole ratio in solution of hydrogen sulfide to alkanolamine increases with partial pressure of hydrogen sulfide. It decreases with increase in temperature and concentration of alkanolamine and also with partial pressure of carbon dioxide. In the same way the mole ratio of carbon dioxide to alkanolamine increases with partial pressure of carbon dioxide and decreases with increase in temperature, concentration of alkanolamine and partial pressure of hydrogen sulfide.

2-Aminoethanol (monoethanolamine); C_2H_7NO; [141-43-5]

Authors	Range of measurements			
	Conc. of MEA /mol dm^{-3}	T/K	P_{H_2S}/bar	P_{CO_2}/bar
Muhlbauer & Monaghan (17)	2.4 - 2.6	298-373	0.0001 - 1.3	0 - 1.7
Jones et al. (18)	2.5 (15.3 wt%)	313-393	0.0015 - 2.0	0.0019 - 4.7
Lee et al. (27)	5.0	313-373	0 - 34.2	0 - 55.6
Lawson & Garst (20)	2.5 - 5 (15.2; 30 wt%)	298-393	0 - 2.9	0 - 2.3
Lee et al. (14)	2.5	313-373	0 - 60	0 - 70
Nasir & Mather (28)	5.0	373	0.00013 - 0.042	0.00002 - 0.030
Isaacs et al. (26)	2.5	373	0 - 0.034	0.00003 - 0.014

Lee et al. (14) have compared their own measurements with those of Muhlbauer and Monaghan and those of Jones et al. for a concentration of MEA of 2.5 mol dm^{-3}, a temperature of 373.15 K, partial pressures of hydrogen sulfide of 0.005 to 0.18 bar and partial pressures of carbon dioxide of 0.0009 to 0.17 bar. Agreement is within 10% under some conditions but there are wide differences under other conditions. This may be seen in the following examples given by Lee et al.:

COMPONENTS:	EVALUATOR:
1. Hydrogen sulfide; H_2S; [7783-06-4]	Peter G.T. Fogg, Department of Applied Chemistry and Life Sciences, Polytechnic of North London, Holloway, London N7 8DB, U.K.
2. Carbon dioxide; CO_2; [124-38-9]	
3. Water; H_2O; [7732-18-5]	
4. Alkanolamines	February 1987

CRITICAL EVALUATION:

Mole ratio in solution		P_{H_2S}/bar			P_{CO_2}/bar		
CO_2/MEA	H_2S/MEA	ref (14),	(17),	(18)	ref(14),	(17)	(18)
0.1	0.1	0.044	0.051	0.037	0.007		0.0075
0.2	0.3	0.62	0.57		0.19	0.22	
0.3	0.1	0.15	0.15	0.14	0.20	0.25	0.23
0.4	0.2	1.02		0.60	1.30		0.80

Lawson and Garst contrasted their measurements for 2.4 mol dm^{-3} (15 wt%) at 373 K with those of Muhlbauer and Monaghan. They noted that Muhlbauer and Monaghan's measurements of the partial pressures of hydrogen sulfide and carbon dioxide tend to be higher than their own values and than those of Jones *et al.* for similar concentrations of gases in solution. They pointed out that this discrepancy was greater at low gas concentrations than at higher concentrations.

2,2'-Iminobis-ethanol (*diethanolamine*); $C_4H_{11}NO_2$; [111-42-2]

Authors	Conc. of DEA /mol dm^{-3}	T/K	P_{H_2S}/bar	P_{CO_2}/bar
Lee *et al.* (29)	2.0 - 3.5	323-373	0.006 -16.5	0.002 - 57.6
Lawson & Garst (20)	2.4 (25 wt%) - 4.9 (50 wt%)	311-394	0.00013 -22.0	0 - 22.9
Lal *et al.* (11)	2.0	313-373	0.0006 - 0.048	0 - 0.065

Lal *et al.* showed that the solubilities at 313 K, in the absence of carbon dioxide, were consistent with data given by Lawson and Garst. In the presence of carbon dioxide there was poor correlation with predictions from the model published by Kent and Eisenberg (8).

1,1'-Iminobis-2-propanol (*diisopropanolamine*); $C_6H_{15}NO_2$; [110-97-4]

This system was investigated by Isaacs *et al.* (30) and tables of smoothed data have been prepared. No other measurements on this system are available for comparison but there is no reason to doubt the reliability of this smoothed data.

2-(2-Aminoethoxy)ethanol (*diglycolamine*); $C_4H_{11}NO_2$; [929-06-6]

This system was investigated by Dingman *et al.* (10). No other measurements of the four component system are available for comparison. However the data for the system when one or other of the gases is at zero concentration are consistent with data for the three component systems published by Martin *et al.* (24).

References

1. Atwood, K.; Arnold, M.R.; Kindrick, R.C.
    Ind. Eng. Chem. 1957, 49, 1439-44.

2. Astarita, G.; Gioia, F.; Balzano, C. *Chem. Eng. Sci.* 1965, 20, 1101-5.

3. Astarita, G.; Savage, D.W. *Chem. Eng. Sci.* 1982, 37, 677-86.

4. Danckwerts, P.V.; McNeil, K.M. *Trans. Inst. Chem. Eng.* 1967, 45, T32-T49.

COMPONENTS:	EVALUATOR:
1. Hydrogen sulfide; H_2S; [7783-06-4]	Peter G.T. Fogg, Department of Applied Chemistry and Life Sciences, Polytechnic of North London, Holloway, London N7 8DB, U.K.
2. Carbon dioxide; CO_2; [124-38-9]	
3. Water; H_2O; [7732-18-5]	
4. Alkanolamines	February 1987

CRITICAL EVALUATION:

5. Klyamer, S.D. *Gazov. Prom.* 1971, *16 (9)*, 38-42.

6. Klyamer, S.D.; Kolesnikova, T.L. *Zhur. Fiz. Khim.* 1972, *46*, 1056.

7. Klyamer, S.D.; Kolesnikova, T.L.; Rodin, Yu.A.
    Gazov. Prom., 1973, *18 (2)*, 44-48.

8. Kent, R.L.; Eisenberg, B. *Hydrocarbon Processing* 1976, *55 (2)*, 87-90.

9. Deshmukh, R.D.; Mather, A.E. *Chem. Eng. Sci.* 1981, *36*, 355-362.

10. Dingman, J.C.; Jackson, J.L.; Moore, T.F.
    Proc. 62nd Annual Convention of the Gas Processors Association 1983,
    256-68.

11. Lal, D.; Otto, F.D.; Mather, A.E. *Can. J. Chem. Eng.* 1985, *63*, 681-5.

12. Jou, F-Y.; Mather, A.E.; Otto, F.D.
    Ind. Eng. Chem. Process Des. Dev. 1982, *21*, 539-44.

13. Jou, F-Y.; Mather, A.E.; Otto, F.D. *Can. J. Chem. Eng.* 1985, *63*,
    122-5.

14. Lee, J.I.; Otto, F.D.; Mather, A.E. *Can. J. Chem. Eng.* 1976, *54*, 214-9.

15. Riegger, E.; Tartar, H.V.; Lingafelter, E.C.
    J. Amer. Chem. Soc. 1944, *66*, 2024-7.

16. Leibush, A.G.; Shneerson, A.L. *J. Applied Chem. USSR* 1950, *23*, 149-57.

17. Muhlbauer, H.G.; Monaghan, P.R. *Oil & Gas J.* 1957, *55 (17)*, 139-45.

18. Jones, J.H.; Froning, H.R.; Claytor, E.E. Jr.
    J. Chem. Eng. Data 1959, *4*, 85-92.

19. Lee, J.I.; Otto, F.D.; Mather, A.E. *Can. J. Chem. Eng.* 1974, *52*,
    803-805.

20. Lawson, J.D.; Garst, A.W. *J. Chem. Eng. Data* 1976, *21*, 20-30.

21. Isaacs, E.E.; Otto, F.D.; Mather, A.E.
    J. Chem. Eng. Data 1980, *25*, 118-120.

22. Lee, J.I.; Otto, F.D.; Mather, A.E.
    J. Chem. Eng. Data 1973, *18*, 71-3,420

23. Bottoms, R.R. *Ind. Eng. Chem.* 1931, *23*, 501-4.

24. Martin, J.L.; Otto, F.D.; Mather, A.E.
    J. Chem. Eng. Data 1978, *23*, 163-4.

25. Isaacs, E.E.; Otto, F.D.; Mather, A.E.
    J. Chem. Eng. Data 1977, *22*, 71-3.

26. Isaacs, E.E.; Otto, F.D.; Mather, A.E.
    J. Chem. Eng. Data 1977, *22*, 317-9.

27. Lee, J.I.; Otto, F.D.; Mather, A.E. *J. Chem. Eng. Data* 1975, *20*, 161-3.

28. Nasir, P.; Mather, A.E. *Can. J. Chem. Eng.* 1977, *55*, 715-7.

29. Lee, J.I.; Otto, F.D.; Mather, A.E. *Can. J. Chem. Eng.* 1974, *52*, 125-7.

30. Isaacs, E.E.; Otto, F.D.; Mather, A.E.
    Can. J. Chem. Eng. 1977, *55*, 210-2.

31. Lee, J.I.; Otto, F.D.; Mather, A.E. *Can. Gas J.* 1972, *(May-June)*, 34-39.

COMPONENTS:	ORIGINAL MEASUREMENTS:
1. Hydrogen sulfide; H_2S; [7783-06-4] 2. 2-Aminoethanol,(Monoethanol-amine); C_2H_7NO; [141-43-5] 3. Water; H_2O; [7732-18-5]	Riegger, E.; Tartar, H.V.; Lingafelter, E.C. *J. Amer. Chem. Soc.* <u>1944</u>, *66*, 2024-7.

VARIABLES:	PREPARED BY:
Temperature, pressure, composition	C.L. Young

EXPERIMENTAL VALUES:

p^+/mmHg	Molality of amine	$\alpha*$ Temperature		
		298.15K	318.15K	333.15K
700	0.6	1.148	1.124	1.083
	1.0	1.086	1.051	1.040
	1.5	1.050	1.011	0.998
	2.0	1.033	0.988	0.968
	3.0	1.011	0.958	0.934
	4.0	0.998	0.940	0.909
	5.0	0.991	0.927	0.891
600	0.6	1.126	1.097	1.056
	1.0	1.072	1.033	1.011
	1.5	1.041	0.996	0.970
	2.0	1.025	0.975	0.944
	3.0	1.004	0.948	0.910
	4.0	0.991	0.928	0.884
	5.0	0.984	0.914	0.865
500	0.6	1.101	1.070	1.027
	1.0	1.058	1.012	0.984
	1.5	1.032	0.980	0.945
	2.0	1.016	0.960	0.916
	3.0	0.996	0.934	0.880
	4.0	0.980	0.913	0.858
	5.0	0.974	0.899	0.837

AUXILIARY INFORMATION

METHOD/APPARATUS/PROCEDURE:	SOURCE AND PURITY OF MATERIALS:
Samples of liquid saturated with hydrogen sulfide. Liquid samples added to standard iodine soln., and excess back titrated with sodium thiosulfate.	1. Commercial sample. 2. Carbide and Carbon Chemicals Corp. sample distilled, b.p. 170.1°C. 3. No details given.
	ESTIMATED ERROR: $\delta T/K = \pm 0.1$; $\delta\alpha = 0.8\%$. (estimated by compiler).
	REFERENCES:

COMPONENTS	ORIGINAL MEASUREMENTS
1. Hydrogen sulfide; H_2S; [7783-06-4] 2. 2-Aminoethanol,(Monoethanolamine); C_2H_7NO: [141-43-5] 3. Water; H_2O; [7732-18-5]	Riegger, E.; Tartar, H.V. Lingafelter, E.C. *J. Amer. Chem. Soc.* <u>1944</u>, *66*, 2024-7.

EXPERIMENTAL VALUES:

p^+/mmHg	Molality of amine	$\alpha*$ Temperature 298.15K	318.15K	333.15K
400	0.6	1.080	1.045	0.995
	1.0	1.042	0.993	0.952
	1.5	1.020	0.961	0.912
	2.0	1.006	0.943	0.885
	3.0	0.985	0.918	0.848
	4.0	0.971	0.897	0.821
	5.0	0.963	0.880	0.801
300	0.6	1.053	1.011	0.960
	1.0	1.022	0.967	0.916
	1.5	1.002	0.939	0.876
	2.0	0.990	0.921	0.847
	3.0	0.970	0.891	0.810
	4.0	0.955	0.869	0.778
	5.0	0.945	0.850	0.753
200	0.6	1.027	0.971	0.908
	1.0	0.998	0.929	0.863
	1.5	0.979	0.900	0.822
	2.0	0.966	0.880	0.793
	3.0	0.946	0.846	0.751
	4.0	0.931	0.819	0.714
	5.0	0.918	0.800	0.683
100	0.6	0.986	0.908	0.811
	1.0	0.956	0.864	0.757
	1.5	0.934	0.826	0.708
	2.0	0.919	0.795	0.674
	3.0	0.893	0.748	0.624
	4.0	0.870	0.714	0.581
	5.0	0.852	0.684	0.547
50	0.6	0.934	0.826	0.694
	1.0	0.902	0.782	0.634
	1.5	0.876	0.742	0.576
	2.0	0.856	0.706	0.532
	3.0	0.819	0.648	0.474
	4.0	0.784	0.601	0.425
	5.0	0.758	0.564	0.386
25	0.6	0.866	0.731	0.551
	1.0	0.833	0.686	0.490
	1.5	0.802	0.631	0.433
	2.0	0.777	0.601	0.388
	3.0	0.730	0.533	0.331
	4.0	0.687	0.487	0.291
	5.0	0.643	0.453	0.285

+   partial pressure of hydrogen sulfide

*   Mole of hydrogen sulfide per mole of amine.

COMPONENTS:	ORIGINAL MEASUREMENTS:
1. Hydrogen sulfide; H_2S; [7783-06-4] 2. Water; H_2O; [7732-18-5] 3. 2-Aminoethanol, (*monoethanolamine*); C_2H_7NO; [141-43-5]	Leibush, A.G.; Shneerson, A.L. *Zhur. Prik. Khim.* 1950, *23*, 145-152. *J. Applied Chem. USSR* 1950, *23*, 149-157.

VARIABLES:	PREPARED BY:
Temperature, pressure, composition	P.G.T. Fogg

EXPERIMENTAL VALUES:

Conc.of MEA /mol dm^{-3}	T/K	P_{H_2S}/mmHg	P_{H_2S}/bar*	Mole ratio H_2S/MEA
0.93	288.2	0.050	0.000067	0.068
		1.53	0.00204	0.243
		1.85	0.00247	0.454
		8.6	0.0115	0.714
	298.2	0.100	0.000133	0.068
		0.87	0.00116	0.243
		3.14	0.00419	0.454
		17.2	0.0229	0.714
	323.2	0.378	0.000504	0.068
		3.38	0.00451	0.243
		13.4	0.0179	0.454
		65.6	0.0875	0.714
2.5	288.2	0.098	0.000131	0.061
		0.264	0.000352	0.119
		1.06	0.00141	0.250
		4.25	0.00567	0.450
		11.0	0.0147	0.614
		16.4	0.0219	0.688
		57.5	0.0767	0.846

*calculated by compiler.   MEA is 2-aminoethanol (monoethanolamine)

AUXILIARY INFORMATION

METHOD/APPARATUS/PROCEDURE:	SOURCE AND PURITY OF MATERIALS:
N_2 or a mixture of N_2 & H_2S was passed successively through two absorbers containing H_2S dissolved in an aqueous solution of MEA. The H_2S in the emerging gas was passed into cadmium or zinc acetate solution and the precipitated sulfides estimated by iodimetry. Hydrogen sulfide in the MEA solutions was also determined by iodimetry. CO_2 in the MEA solutions was estimated by reaction with 30% H_2SO_4 and absorption of CO_2 evolved in standard baryta.	1. From H_2SO_4 & Na_2S; no SO_2 detected. 3. Contained 3% of impurities including 0.6 to 1.5 volumes of CO_2 per unit volume of solution.

ESTIMATED ERROR:
P_{H_2S} likely to be 5 to 15% too high because of CO_2 content. (authors)

REFERENCES:

COMPONENTS:	ORIGINAL MEASUREMENTS:
1. Hydrogen sulfide; H_2S; [7783-06-4] 2. Water; H_2O; [7732-18-5] 3. 2-Aminoethanol, (*monoethanolamine*); C_2H_7NO; [141-43-5]	Leibush, A.G.; Shneerson, A.L. *Zhur. Prik. Khim.* 1950, *23*, 145-152. *J. Applied Chem. USSR* 1950, *23*, 149-157.

EXPERIMENTAL VALUES:

Conc.of MEA /mol dm^{-3}	T/K	P_{H_2S}/mmHg	P_{H_2S}/bar*	Mole ratio H_2S/MEA
2.5	298.2	0.160	0.000213	0.061
		0.452	0.000603	0.110
		0.45	0.00060	0.119
		1.14	0.00152	0.192
		1.71	0.00228	0.250
		2.53	0.00337	0.303
		4.0	0.0053	0.373
		7.15	0.00953	0.450
		20.4	0.0272	0.614
		31.4	0.0419	0.688
		104	0.139	0.846
		124	0.165	0.854
	323.	0.134	0.000179	0.022
		0.64	0.00085	0.061
		1.84	0.00245	0.119
		7.25	0.00967	0.250
		30.6	0.0408	0.450
		121	0.161	0.688
		348	0.464	0.846

*calculated by compiler.

MEA is 2-aminoethanol (monoethanolamine)

COMPONENTS:	ORIGINAL MEASUREMENTS:
1. Hydrogen sulfide; H_2S; [7783-06-4] 2. 2-Aminoethanol (Monoethanolamine); C_2H_7NO; [141-43-5] 3. Water; H_2O; [7732-18-5]	Atwood, K.; Arnold, M. R.; Kindrick, R. C. *Ind. Eng. Chem.* <u>1957</u>, *49*, 1439-44.
VARIABLES: Temperature, pressure, composition	PREPARED BY: C. L. Young

EXPERIMENTAL VALUES:

T/K	t/°F	Wt-% amine	P/mmHg	P/kPa	Conc. H_2S /mol ℓ^{-1}	Mole ratio[†]
299.8	80	5	0.000626	8.35×10^{-5}	0.00275	0.00336
			0.0518	0.00691	0.0483	0.0591
			73.0	9.7	0.758	0.945
310.9	100		0.0414	0.00552	0.0307	0.0376
			0.123	0.0164	0.0542	0.0664
			283	37.7	0.806	1.00
333.1	140		0.00804	0.00107	0.00476	0.00582
			106	14.1	0.578	0.714
310.9	100	15	0.0137	0.00183	0.0245	0.00995
			2.24	0.299	0.464	0.190
			275	36.7	2.116	0.895
322.0	120		0.0439	0.00585	0.0384	0.0156
			60.8	8.11	1.458	0.609
322.0	120	20	3370	449.3	–	0.999
			5850	779.9	–	1.102
344.3	160		5720	762.6	–	1.055
299.8	80	30	0.000525	0.0000700	0.00600	0.00121
			0.00465	0.000620	0.0276	0.00557
			1.50	0.200	0.757	0.155
			57.5	7.67	3.26	0.699
			289	38.5	4.16	0.908
322.0	120		0.0266	0.00355	0.0376	0.00758
			5.92	0.789	0.653	0.133
			264	35.2	3.41	0.733

† Moles of hydrogen sulfide per mole of amine.

AUXILIARY INFORMATION

METHOD/APPARATUS/PROCEDURE:	SOURCE AND PURITY OF MATERIALS:
Gas saturation method was used in which a known quantity of hydrogen was passed through a series of saturators containing a solution of known hydrogen sulfide concentration. At partial pressures of H_2S greater than 100 mmHg, solution saturation method was used. H_2S in soln. determined by iodimetry. Details in source.	1. Stated purity 99.7 mole per cent. 2. Analytical grade. 3. No details.
	ESTIMATED ERROR: $\delta T/K = \pm 0.12$; $\delta p/kPa = \pm 2\%$; δ (Mole ratio) = $\pm 3\%$ (estimated by compiler).
	REFERENCES:

COMPONENTS:	ORIGINAL MEASUREMENTS:
1. Hydrogen sulfide; H_2S; [7783-06-4] 2. 2-Aminoethanol (monoethanolamine); C_2H_7NO ; [141-43-5] 3. Water; H_2O; [7732-18-5]	Jones, J. H.; Froning, H. R.; Claytor, E. E. Jr. *J. Chem. Engng. Data* <u>1959</u>, *4*, 85-92.

VARIABLES:	PREPARED BY:
Temperature, pressure	C. L. Young

EXPERIMENTAL VALUES:

T/K	$P^§$/10^5Pa	Conc of MEA /Wt-%	$\alpha^†$	T/K	$P^§$/10^5Pa	Conc of MEA /Wt-%	$\alpha^†$
313.15	0.0013	15.3	0.125	353.15	0.0089	15.3	0.119
	0.0040		0.208		0.0428		0.251
	0.0121		0.362		0.118		0.403
	0.0575		0.643		0.304		0.569
	0.0796		0.729		0.687		0.711
	0.141		0.814	373.15	0.0028		0.041
	0.191		0.842		0.0089		0.066
	0.251		0.884		0.0640		0.199
	0.348		0.920		0.307		0.396
	0.545		0.948		0.959		0.612
	0.800		0.965	393.15	0.0068		0.036
333.15	0.0095		0.202		0.0295		0.088
	0.0149		0.251		0.173		0.206
	0.0484		0.419		0.684		0.394
	0.167		0.636	413.15	0.0065		0.025
	0.577		0.849		0.0460		0.077
	0.581		0.842		0.395		0.197
	1.13		0.952				

$P^§$ partial pressure of hydrogen sulfide

$\alpha^†$ mole ratio in liquid, moles of hydrogen sulfide/ moles of monoethanolamine

AUXILIARY INFORMATION

METHOD/APPARATUS/PROCEDURE:	SOURCE AND PURITY OF MATERIALS:
Rocking static equilibrium cell fitted with liquid and gas sampling ports. Pressure measured with Bourdon gauge. Concentration of hydrogen sulfide in gas phase determined by mass spectrometry. Concentration of hydrogen sulfide in liquid determined by adding to excess of acidified iodine solution and back titrating with standard thiosulfate solution. Details in source.	1. Purity 99.9 mole per cent. Mass spectrometry showed trace amounts of methyl mercaptan, carbon disulfide and carbon dioxide. 2. Purity 99.2 mole per cent. 3. Distilled.

	ESTIMATED ERROR:
	$\delta T/K = \pm 0.1$ at 313.15 K, ± 0.5 at 413.15 K; $\delta P/kPa = \pm 1\%$; $\delta\alpha = \pm 3\%$ (estimated by compiler).

	REFERENCES:

COMPONENTS:	ORIGINAL MEASUREMENTS:
1. Hydrogen sulfide; H_2S; [7783-06-4] 2. 2-Aminoethanol, (Monoethanolamine); C_2H_7NO: [141-43-5] 3. Water; H_2O; [7732-18-5]	Lee, J.I.; Otto, F.D.; Mather, A.E. *Can. J. Chem. Engng.* 1974, *52*, 803-5.

VARIABLES:	PREPARED BY:
Temperature, pressure, composition	C.L. Young

EXPERIMENTAL VALUES:

T/K	Conc of MEA /mol dm^{-3}	p_{H_2S}/kPa	α
313.15	2.5	2252.5	1.505
		2049.1	1.49
		1323.1	1.23
		1304.5	1.28
		1257.6	1.27
		1213.5	1.26
		1046.6	1.19
		1020.4	1.22
		968.7	1.21
		959.7	1.19
		439.9	1.04
		428.2	1.01
		154.4	0.93
		18.4	0.78
		16.0	0.81
		15.7	0.781
373.15	2.5	4247.1	1.44
		4205.8	1.55
		4192.0	1.47
		3412.9	1.41
		2778.5	1.31
		2140.8	1.18
		1896.0	1.145
		1379.6	1.04
		730.8	0.965

AUXILIARY INFORMATION

METHOD/APPARATUS/PROCEDURE:	SOURCE AND PURITY OF MATERIALS:
Recirculating vapor flow apparatus with Jerguson liquid level gauge cell with magnetic pump. Nitrogen added to vapor to increase pressure to at least 200 kPa. Vapor analysed by gas chromatography. Partial pressure of hydrogen sulfide estimated from knowledge of pressure and vapor pressure of monoethanolamine. Liquid samples passed into sulfuric acid and displaced hydrogen sulfide collected in buret.	1. Matheson C.P. grade purity 99.91 mole per cent. 2. Fisher Scientific sample, purity 99.95 mole per cent. 3. Distilled Nitrogen used as carrier, purity 99.997 mole per cent.

	ESTIMATED ERROR:
	$\delta T/K = \pm 0.5$; $\delta p/kPa = \pm 1\%$ $\delta\alpha = \pm 3-5\%$

	REFERENCES:

COMPONENTS:	ORIGINAL MEASUREMENTS:
1. Hydrogen sulfide; H_2S; [7783-06-4]	Lee, J.I.; Otto, F.D.; Mather, A.E.
2. 2-Aminoethanol, (Monoethanolamine); C_2H_7NO: [141-43-5]	*Can. J. Chem. Engng.* <u>1974</u>, *52*, 803-5.
3. Water; H_2O; [7732-18-5]	

EXPERIMENTAL VALUES:

T/K	Conc of MEA / mol dm^{-3}	p_{H_2S}/kPa	α
373.15	2.5	726.0	0.957
		546.1	0.936
		410.2	0.884
		283.4	0.805
		194.3	0.751
		148.2	0.694
		127.5	0.602
		122.7	0.62
		103.4	0.56
		84.8	0.53
313.15	5.0	2220.1	1.322
		915.6	1.025
		723.9	0.974
		497.8	0.900
		402.6	0.840
		146.2	0.865
		144.8	0.895
		128.2	0.855
		115.1	0.767
		30.7	0.677
		4.21	0.435
		2.55	0.376
		1.63	0.278
373.15	5.0	3176.4	1.032
		1983.6	0.895
		679.1	0.785
		430.9	0.700
		282.7	0.615
		221.3	0.580
		194.4	0.515
		51.9	0.344
		42.1	0.288
		7.24	0.123

α = Mole ratio in liquid phase, H_2S/Monoethanolamine.

COMPONENTS:	ORIGINAL MEASUREMENTS:
1. Hydrogen sulfide; H_2S; [7783-06-4] 2. Water; H_2O; [7732-18-5] 3. 2-Aminoethanol, (Monoethanol-amine); C_2H_7NO; [141-43-5]	Lawson, J.D.; Garst, A.W. J. Chem. Engng. Data, 1976, 21, 20-30.

VARIABLES:	PREPARED BY:
Temperature, pressure	C.L. Young

EXPERIMENTAL VALUES:

T/K	P/bar	Conc of MEA /wt%	Liquid comp. mol H_2S/mol amine	Mole fraction of hydrogen sulfide in liquid, x_{H_2S}
313.15	0.000016	15.2	0.0140	0.000702
	0.000019		0.0147	0.000737
	0.00011		0.0329	0.00165
	0.00025		0.0396	0.00198
	0.00041		0.0590	0.00295
	0.00036		0.0591	0.00296
	0.00073		0.0789	0.00394
	0.00065		0.0795	0.00397
	0.015		0.373	0.0184
	0.012		0.380	0.0187
	2.558		1.026	0.0490
	2.398		1.049	0.0500
	5.675		1.099	0.0523
	5.782		1.116	0.0530
	24.78		1.633	0.0757
333.15	0.000012	15.2	0.0052	0.000261
	0.000099		0.0141	0.000707
	0.00037		0.0339	0.00170
	0.00059		0.0406	0.00203
	0.00099		0.0593	0.00297
	0.0015		0.0805	0.00402
	0.045		0.384	0.0189
	0.047		0.392	0.0193

AUXILIARY INFORMATION

METHOD/APPARATUS/PROCEDURE:	SOURCE AND PURITY OF MATERIALS:
Rocking equilibrium cell fitted with liquid and vapor sampling valves. Pressure measured with Bourdon gauge. Cell charged with amine then hydrogen sulfide and methane added as an inert gas to "achieve the desired total pressure". Vapor phases analysed by mass spectrometry. Liquid samples analysed by electrometric titration, details in source. Additional analytical methods were used for some samples.	1. Purity 99.5 mole per cent. 2. Distilled. 3. Commercial sample purity better than 99 mole per cent as determined by acid titration.

ESTIMATED ERROR:
$\delta T/K$ = ±0.15 at 300K increasing to ±0.6 at 413K; $\delta P/bar$ =±0.5%; δx_{H_2S} = ±3%.

REFERENCES:

COMPONENTS:	ORIGINAL MEASUREMENTS:
1. Hydrogen sulfide; H_2S; [7783-06-4] 2. Water; H_2O; [7732-18-5] 3. 2-Aminoethanol, (Monoethanolamine); C_2H_7NO; [141-43-5]	Lawson, J.D.; Garst, A.W. *J. Chem. Engng. Data,* 1976, *21,* 20-30.

EXPERIMENTAL VALUES:

T/K	P/bar	Conc of MEA /wt%	Liquid comp mol H_2S/mol amine	Mole fraction of hydrogen sulfide in liquid, x_{H_2S}
333.15	3.450	15.2	0.997	0.0476
	5.995		1.114	0.0529
	32.64		1.526	0.0711
353.15	0.000037		0.0055	0.000276
	0.00013		0.0143	0.000717
	0.00099		0.0349	0.00175
	0.00155		0.0418	0.00209
	0.113		0.382	0.0188
	0.120		0.385	0.0190
	4.663		0.993	0.0475
	8.660		1.028	0.0490
	32.37		1.468	0.0686
373.13	0.000063		0.0058	0.00029
	0.00041		0.0143	0.00072
	0.0063		0.0633	0.00317
	0.0075		0.0793	0.00396
	0.0184		0.121	0.00603
	0.306		0.376	0.0185
	0.306		0.384	0.0189
	9.299		0.952	0.0456
	10.10		1.015	0.0485
	38.50		1.358	0.0638
393.15	0.023		0.0653	0.00327
	0.024		0.0812	0.00406
	0.043		0.122	0.00608
	0.639		0.383	0.0189
	0.673		0.386	0.0190
	0.986		0.923	0.0443
	1.239		0.939	0.0450
	38.24		1.36	0.0639
413.15	0.045		0.0610	0.00305
	0.053		0.0749	0.00374
299.81	0.00076	30	0.081	0.00900
	0.0052		0.231	0.0252
	0.0092		0.311	0.0337
	0.017		0.406	0.0435
310.93	0.00001		0.0083	0.000930
	0.00083		0.048	0.00535
	0.0075		0.199	0.0218
	0.107		0.616	0.0646
366.48	0.00036		0.0089	0.000997
	0.0013		0.0509	0.00568
	0.142		0.207	0.0227
	1.667		0.570	0.0601

COMPONENTS:	ORIGINAL MEASUREMENTS:
1. Hydrogen sulfide; H_2S; [7783-06-4] 2. Water; H_2O; [7732-18-5] 3. 2-Aminoethanol (Monoethanolamine); C_2H_7NO; [141-43-5]	Isaacs, E. E.; Otto, F. D.; Mather, A. E. *J. Chem. Eng. Data* <u>1980</u>, *25*, 118-120.

VARIABLES:	PREPARED BY:
Pressure	C. L. Young

EXPERIMENTAL VALUES:

T/K	Partial pressure of hydrogen sulfide /kPa	Mole ratio H_2S/C_2H_7NO in liquid
373.2	0.0090	0.016
	0.0310	0.0210
	0.140	0.0300
	0.354	0.0430
	1.67	0.090
	1.50	0.092

Concentration of 2-Aminoethanol was 2.5 kmol m^{-3} (soln.).

AUXILIARY INFORMATION

METHOD/APPARATUS/PROCEDURE:	SOURCE AND PURITY OF MATERIALS:
Nitrogen passed through three 250 ml vessels in series which contained amine solution with a certain amount of dissolved carbon dioxide and hydrogen sulfide. Emerging gas from last vessel analysed by GC. Liquid sample taken from last vessel. Hydrogen sulfide content determined by iodine-thiosulfate titration.	No details given.

	ESTIMATED ERROR:
	$\delta T/K = \pm 0.5$; $\delta x = \pm 6\%$ (estimated by compiler)

	REFERENCES:

COMPONENTS:	ORIGINAL MEASUREMENTS:
1. Hydrogen sulfide; H_2S; [7783-06-4] 2. Water; H_2O; [7732-18-5] 3. Ethanol, 2,2'-iminobis-, (*diethanolamine*); $C_4H_{11}NO_2$; [111-42-2]	Leibush, A.G.; Shneerson, A.L. *Zhur. Prik. Khim.* <u>1950</u>, *23*, 145-152. *J. Applied Chem. USSR* <u>1950</u>, *23*, 149-157.
VARIABLES: Temperature, pressure, composition of liquid phase	PREPARED BY: P.G.T. Fogg

EXPERIMENTAL VALUES:

Conc.of DEA /mol dm^{-3}	T/K	P_{H_2S}/mmHg	P_{H_2S}/bar*	Mole ratio H_2S/DEA
0.97	288.2	0.076	0.000101	0.040
		0.29	0.00039	0.091
		1.27	0.00169	0.194
		5.73	0.00764	0.420
		24.0	0.0320	0.670
		43.0	0.0573	0.734
		94.5	0.1260	0.838
	298.2	0.147	0.000196	0.040
		0.50	0.00067	0.091
		1.97	0.00263	0.194
		7.92	0.01056	0.420
		38.0	0.0507	0.670
		70.8	0.0944	0.734
		145	0.193	0.838

*calculated by compiler.   DEA = diethanolamine

AUXILIARY INFORMATION

METHOD/APPARATUS/PROCEDURE:	SOURCE AND PURITY OF MATERIALS:
N_2 or a mixture of N_2 & H_2S was passed successively through two absorbers containing H_2S dissolved in an aqueous solution of DEA. The H_2S in the emerging gas was passed into cadmium or zinc acetate solution and the precipitated sulfides estimated by iodimetry. Hydrogen sulfide in the DEA solutions was also determined by iodimetry.	1. From H_2SO_4 & Na_2S: no SO_2 detected 3. Contained 0.2 to 0.6 volumes of CO_2 per unit volume of solution; less than 1% by weight of total impurities.
	ESTIMATED ERROR: $\delta T/K$ = ± 0.1 (authors) P_{H_2S} likely to be 5 to 15% too high because of CO_2 content. (authors)
	REFERENCES:

COMPONENTS:	ORIGINAL MEASUREMENTS:
1. Hydrogen sulfide; H_2S; [7783-06-4]	Leibush, A.G.; Shneerson, A.L.
2. Water; H_2O; [7732-18-5]	*Zhur. Prik. Khim.* 1950, *23*, 145-152.
3. Ethanol, 2,2'-iminobis-, (*diethanolamine*); $C_4H_{11}NO_2$; [111-42-2]	*J. Applied Chem. USSR* 1950, *23*, 149-157.

EXPERIMENTAL VALUES:

Conc.of DEA /mol dm^{-3}	T/K	P_{H_2S}/mmHg	P_{H_2S}/bar[*]	Mole ratio H_2S/DEA
0.97	323.2	0.34	0.00045	0.040
		1.59	0.00212	0.091
		6.17	0.00823	0.194
		32.7	0.0436	0.420
		143	0.191	0.670
		197	0.263	0.734
		312	0.416	0.838
2.0	288.2	0.319	0.000425	0.073
		3.0	0.0040	0.221
		6.9	0.0092	0.342
		14.6	0.0195	0.465
		40.5	0.0540	0.624
		99	0.132	0.780
	298.2	0.167	0.000223	0.035
		0.71	0.00095	0.073
		3.3	0.0044	0.175
		5.1	0.0068	0.221
		12.1	0.0161	0.342
		12.5	0.0167	0.353
		13.5	0.0180	0.368
		23.7	0.0316	0.465
		68.5	0.0913	0.624
		219	0.292	0.780
	323.2	0.048	0.000064	0.006
		0.276	0.000368	0.022
		0.565	0.000753	0.035
		11.3	0.0151	0.175
		15.1	0.0201	0.221
		40.8	0.0544	0.342
		86.5	0.1153	0.465

[*]calculated by compiler.   DEA = diethanolamine

COMPONENTS:	ORIGINAL MEASUREMENTS:
1. Hydrogen sulfide; H_2S; [7783-06-4] 2. 2,2^L-Iminobisethanol (Diethanol-amine); $C_4H_{11}NO_2$; [111-42-2] 3. Water; H_2O; [7732-18-5]	Atwood, K.; Arnold, M. R.; Kindrick, R. C. *Ind. Eng. Chem.* <u>1957</u>, *49*, 1439-44.

VARIABLES:	PREPARED BY:
Temperature, pressure, composition	C. L. Young

EXPERIMENTAL VALUES:

T/K	t/°F	Wt-% amine	P/mmHg	P/kPa	Conc. H_2S /mol ℓ^{-1}	Mole ratio[†]
310.9	100	10	0.346	0.0462	0.0520	0.0543
			712	94.9	0.939	0.997
333.1	140		0.587	0.0783	0.0337	0.352
			178	23.7	0.675	0.713
299.8	80	25	0.00870	0.001160	0.0123	0.00505
			2.89	0.385	0.316	0.130
			320	42.7	2.024	0.862
322.0	120		0.131	0.0175	0.0327	0.0134
			682	90.9	2.002	0.852
310.9	100	50	0.0566	0.00755	0.0281	0.00560
			63.1	8.42	1.915	0.394
			422	56.3	3.54	0.751
322.0	140		0.135	0.0180	0.0281	0.00559
			35.7	4.76	0.757	0.153
			646	86.1	3.11	0.653

[†] Moles of hydrogen sulfide per mole of amine.

AUXILIARY INFORMATION

METHOD/APPARATUS/PROCEDURE:	SOURCE AND PURITY OF MATERIALS:
Gas saturation method was used in which a known quantity of hydrogen was passed through a series of saturators containing a solution of known hydrogen sulfide concentration. At partial pressures of H_2S greater than 100 mmHg, solution saturation method was used. H_2S in soln. determined by iodimetry. Details in source.	1. Stated purity 99.7 mole per cent. 2. Analytical grade. 3. No details.

ESTIMATED ERROR:

$\delta T/K = \pm 0.12$; $\delta p/kPa = \pm 2\%$;

δ(Mole ratio) $= \pm 3\%$ (estimated by compiler).

REFERENCES:

COMPONENTS:	ORIGINAL MEASUREMENTS:

1. Hydrogen sulfide; H_2S;
   [7783-06-4]
 	
2. 2,2´-Iminobisethanol (Diethanol-
   amine); $C_4H_{11}NO_2$; [111-42-2]
 	
3. Water; H_2O; [7732-18-5]
	Lee, J.I.; Otto, F.D.; Mather, A.E.
	
	J. Chem. Eng. Data, 1973, 18, 71-3,
	420.

VARIABLES:
 	
   Temperature, pressure, composition
	PREPARED BY:
	
	         C. L. Young

EXPERIMENTAL VALUES:

T/K	Conc of 2 /mol dm^{-3}	p/kPa	α
298.15	2.0	2.45	0.440
		22.1	0.731
		27.5	0.761
		99.8	0.888
		222.0	0.944
		282.0	0.994
		362.9	1.005
		459.5	1.087
		516.4	1.111
		651.5	1.140
		703.3	1.188
		899.8	1.256
		937.7	1.326
		1267.2	1.398
		1421.7	1.467
		1525.1	1.547
323.15	2.0	5.41	0.331
		21.9	0.542
		32.4	0.635
		34.0	0.641
		42.6	0.646
		65.9	0.724
		186.1	0.842
		300.6	0.910

AUXILIARY INFORMATION

METHOD/APPARATUS/PROCEDURE:	SOURCE AND PURITY OF MATERIALS:

Recirculating vapor flow apparatus
with Jerguson liquid level gauge
cell and magnetic pump. Nitrogen
added to vapor to increase pressure
to at least 200 kPa. Vapor analysed
by gas chromatography. Partial
pressure of hydrogen sulfide
estimated from knowledge of pressure
and vapor pressure of diethanolamine.
Liquid samples passed into sulfuric
acid and displaced hydrogen sulfide
collected in buret.
	1. Matheson, C.P. grade purity 99.91
	   mole per cent.
	
	2. Purity 99.8 mole per cent.
	
	3. (Nitrogen used as carrier, purity
	   99.997 mole per cent.)

	ESTIMATED ERROR:
	   $\delta T/K = \pm 0.5$; $\delta p/kPa = \pm 1\%$;
	   $\delta\alpha = \pm 3\text{-}5\%$.

	REFERENCES:

T/K	Conc. of 2/mol dm^{-3}	p/kPa	α
323.15	2.0	614.0	0.978
		903.2	1.028
		910.1	1.030
		1001.1	1.023
		1138.3	1.180
		1332.7	1.255
		1373.4	1.227
		1655.4	1.310
		1728.5	1.275
		1892.6	1.347
348.15	2.0	3.25	0.159
		7.72	0.226
		13.4	0.316
		15.1	0.354
		37.2	0.445
		63.0	0.530
		128.1	0.639
		228.9	0.783
		476.4	0.905
		710.1	0.938
		974.2	1.012
		1139.7	1.079
		1152.1	1.080
		1365.1	1.062
		1475.5	1.117
		1778.8	1.200
		2000.1	1.274
373.15	2.0	11.0	0.150
		57.6	0.314
		176.5	0.506
		429.2	0.694
		889.4	0.873
		1434.8	1.053
		1896.0	1.127
		1968.4	1.131
393.15	2.0	25.2	0.162
		74.5	0.256
		342.0	0.521
		688.8	0.680
		1042.5	0.812
		1516.8	0.975
		1916.7	1.062
		1925.7	1.070
298.15	3.5	0.834	0.201
		3.89	0.352
		6.76	0.449
		16.9	0.613
		35.8	0.681
		48.5	0.711
		100.0	0.785
		111.0	0.790
		266.1	0.875
		389.7	0.948
		732.9	1.020
		1037.0	1.103
		1357.0	1.120
		1756.1	1.206
323.15	3.5	0.738	0.064
		5.07	0.200
		21.8	0.387
		66.8	0.616
		123.4	0.713
		219.9	0.776
		392.3	0.858
		672.2	0.918
		935.6	0.975
		1385.8	1.039
		1985.7	1.097

T/K	Conc of 2./mol dm^{-3}	p/kPa	α
348.15	3.5	2.36	0.072
		12.8	0.198
		59.4	0.396
		166.8	0.614
		307.5	0.723
		589.5	0.831
		871.5	0.855
		1441.0	0.945
		2104.9	1.042
373.15	3.5	5.47	0.069
		33.9	0.193
		139.3	0.391
		226.1	0.478
		357.1	0.582
		512.3	0.675
		601.2	0.688
		921.1	0.762
		1362.4	0.830
		2076.0	0.927
393.15	3.5	13.3	0.069
		78.2	0.187
		253.0	0.365
		361.3	0.440
		677.7	0.618
		750.1	0.625
		1238.3	0.725
		2003.6	0.836

α = mole ratio in liquid phase H_2S/Diethanolamine

COMPONENTS:	ORIGINAL MEASUREMENTS:
1. Hydrogen sulfide; H_2S; [7783-06-4] 2. 2,2'-iminobisethanol, (Diethanolamine); $C_4H_{11}NO_2$; [111-42-2] 3. Water; H_2O; [7732-18-5]	Lee, J.I.; Otto, F.D.; Mather, A.E. *J. Chem. Eng. Data.* <u>1973</u>, *18*, 71-3, 420. Some data taken from Amer. Chem. Soc. deposited document.
VARIABLES:	PREPARED BY:
Temperature, pressure, composition	C.L. Young.

EXPERIMENTAL VALUES:

T/K	Conc.of DEA/mol dm^{-3}	P/kPa	α
298.15	0.5	0.119	0.2080
		0.390	0.2965
		1.76	0.4028
		15.2	0.6506
		71.7	0.8642
		161.7	1.015
		201.0	1.034
		275.4	1.194
		464.0	1.493
		647.8	1.896
		658.4	1.813
		1034.2	2.567
		1272.1	2.893
		1561.6	3.293
323.15	0.5	0.0779	0.0946
		0.883	0.208
		19.2	0.549
		70.2	0.7242
		70.3	0.7471
		102.0	0.8296
		244.1	1.015
		315.8	1.048
		414.4	1.196
		589.5	1.380
		775.6	1.581

AUXILIARY INFORMATION

METHOD/APPARATUS/PROCEDURE:	SOURCE AND PURITY OF MATERIALS:
Recirculating vapor flow apparatus with Jerguson liquid level gauge cell and magnetic pump. Nitrogen added to vapor to increase pressure to at least 200 kPa. Vapor analysed by gas chromatography. Partial pressure of hydrogen sulfide estimated from knowledge of pressure and vapor pressure of diethanolamine. Liquid samples passed into sulfuric acid and displaced hydrogen sulfide collected in buret. System contained a trace of carbon dioxide.	1. Matheson C.P. grade purity 99.91 mole per cent. 2. Purity 99.8 mole per cent. 3. Distilled nitrogen used as carrier, purity 99.997 mole per cent.
	ESTIMATED ERROR: $\delta T/K = \pm 0.5$; $\delta P/kPa = \pm 1\%$; $\delta\alpha = \pm 3-5\%$.
	REFERENCES:

COMPONENTS:	ORIGINAL MEASUREMENTS

1. Hydrogen sulfide; H_2S: [7783-06-4]
 	
2. 2,2'-iminobisethanol,
   (Diethanolamine); $C_4H_{11}NO_2$;
   [111-42-2]
 	
3. Water; H_2O; [7732-18-5]
	Lee, J.I.; Otto, F.D.;
	Mather, A.E.
	
	J. Chem. Eng. Data. <u>1973</u>, 18,
	71-3,420.

EXPERIMENTAL VALUES:

T/K	Conc of DEA/mol dm^{-3}	P/kPa	α
323.15	0.5	1132.8	2.0492
		1521.0	2.431
348.15	0.5	0.800	0.0960
		2.46	0.1691
		4.73	0.2491
		33.3	0.5094
		88.2	0.642
		311.6	0.980
		353.7	0.9753
		854.2	1.3702
		1081.1	1.588
		1622.3	1.932
373.15	0.5	2.66	0.105
		9.60	0.202
		35.6	0.368
		74.5	0.486
		191.5	0.6858
		552.3	1.068
		1125.2	1.455
		1603.0	1.682
393.15	0.5	4.96	0.105
		17.1	0.202
		36.9	0.2876
		206.3	0.6252
		518.5	0.8714
		978.4	1.256
		1425.1	1.5412
		1585.8	1.662
298.15	5.0	0.476	0.1156
		0.820	0.1605
		1.10	0.2028
		2.72	0.3172
		14.6	0.5418
		18.0	0.5894
		19.0	0.5821
		42.7	0.7166
		76.2	0.789
		262.0	0.879
		559.2	0.942
		755.7	0.998
		868.7	1.085
		1104.5	1.102
		1671.9	1.160
323.15	5.0	0.896	0.0852
		4.32	0.205
		14.9	0.329
		46.2	0.506
		131.3	0.6845
		273.9	0.820
		343.4	0.8322
		930.8	0.901
		1246.8	0.992
		1635.4	1.043

COMPONENTS	ORIGINAL MEASUREMENTS:

1. Hydrogen sulfide; H_2S; [7783-06-4]
 	
2. 2,2'-iminobisethanol,
   (Diethanolamine); $C_4H_{11}NO_2$;
   [111-42-2]
 	
3. Water; H_2O: [7732-18-5]
	Lee, J.I.; Otto, F.D.;
	Mather, A.E.
	
	J. Chem. Eng. Data. <u>1973</u>, *18*,
	71-3,420

EXPERIMENTAL VALUES:

T/K	Conc. of DEA /mol dm^{-3}	P/kPa	α
348.15	5.0	0.618	0.0223
		1.43	0.0511
		8.96	0.1226
		17.8	0.187
		53.6	0.3385
		77.2	0.386
		144.8	0.4974
		274.4	0.661
		354.0	0.6963
		620.5	0.7922
		1173.5	0.9181
		1359.6	0.942
		1776.1	0.9729
373.15	5.0	5.82	0.0484
		48.9	0.1869
		124.1	0.3167
		219.2	0.415
		384.5	0.5827
		819.1	0.7412
		1542.3	0.8241
393.15	5.0	14.9	0.0489
		71.8	0.1408
		248.2	0.3284
		400.2	0.438
		495.0	0.5111
		761.2	0.5559
		971.5	0.6126
		1338.9	0.6623
		1661.6	0.7274

α = Mole ratio in liquid, H_2S/Diethanolamine

COMPONENTS:	ORIGINAL MEASUREMENTS:
1. Hydrogen sulfide; H_2S; [7783-06-4] 2. Water; H_2O; [7732-18-5] 3. 2,2'-iminobisethanol, Diethanolamine; $C_4H_{11}NO_2$; [111-42-2]	Lawson, J.D. and Garst, A.W. J. Chem. Engng. Data. 1976, 21, 20-30.

VARIABLES:	PREPARED BY:
Temperature, pressure	C.L. Young

EXPERIMENTAL VALUES:

T/K	P_{H_2S}/bar	Conc. of DEA Wt. %	Liquid comp. mol H_2S /mol amine.	Mole fraction of hydrogen sulfide in liquid, x_{H_2S}
310.93	0.000015	25	0.0038	0.00021
	0.000009		0.0043	0.00023
	0.000012		0.0044	0.00024
	0.000097		0.0112	0.00060
	0.000071		0.0157	0.00085
	0.0011		0.0526	0.00283
	0.0017		0.0755	0.00406
	0.0044		0.133	0.00712
	0.0087		0.177	0.00946
	0.0095		0.196	0.0105
	0.0160		0.197	0.0106
	0.0180		0.217	0.0116
	0.0205		0.230	0.0123
	0.0184		0.230	0.0123
	0.0153		0.234	0.0125
	0.0145		0.258	0.0137
	0.0289		0.319	0.0169
	0.0786		0.457	0.0241
	0.0919		0.481	0.0253
	0.213		0.652	0.0340
	1.013		0.855	0.0441
	1.070		0.830	0.0429
	2.944		0.975	0.0500

AUXILIARY INFORMATION

METHOD/APPARATUS/PROCEDURE:	SOURCE AND PURITY OF MATERIALS:
Rocking equilibrium cell fitted with liquid and vapor sampling valves. Pressure measured with Bourdon gauge. Cell charged with amine then hydrogen sulfide and methane added as an inert gas to "achieve the desired total pressure" Vapor phases analysed by mass spectrometry. Liquid samples analysed by electrometric titration, details in source. Additional analytical methods were used for some samples.	1. Purity 99.5 mole per cent. 2. Distilled. 3. Commercial sample purity better than 99 mole per cent as determined by acid titration.

ESTIMATED ERROR:
    $\delta T/K$ = ±0.15 at 310 K increasing to ±0.6 at 422 K;
    δx_{H_2S} = ±10% increasing to ±16% at low pressures

REFERENCES:

COMPONENTS:	ORIGINAL MEASUREMENTS:
1. Hydrogen sulfide; H_2S; [7783-06-4]	Lawson, J.D. and Garst, A.W.
2. Water; H_2O; [7732-18-5]	*J. Chem. Engng. Data.* <u>1976</u>,
3. 2,2'-iminobisethanol, Diethanolamine; $C_4H_{11}NO_2$; [111-42-2]	*21*, 20-30.

EXPERIMENTAL VALUES:

T/K	P_{H_2S}/bar	Conc of DEA Wt. %	Liquid comp. mol H_2S/mol amine	Mole fraction of hydrogen sulfide in liquid, x_{H_2S}
310.93	2.758	25	1.082	0.0552
	2.678		1.043	0.0533
	13.988		1.107	0.0564
	20.783		1.395	0.0700
	22.781		1.412	0.0708
	26.645		1.582	0.0787
324.82	0.0165		0.180	0.00962
	0.0560		0.317	0.0168
	0.264		0.588	0.0308
	2.558		0.958	0.0492
	17.852		1.241	0.0628
338.71	0.00011		0.0045	0.00024
	0.00008		0.0069	0.00037
	0.00013		0.0092	0.00050
	0.00035		0.0100	0.00054
	0.00028		0.0151	0.00081
	0.00160		0.0301	0.00162
	0.00214		0.0488	0.00263
	0.0262		0.124	0.00665
	0.0274		0.143	0.00766
	0.0362		0.179	0.00957
	0.0617		0.205	0.0109
	0.0610		0.207	0.0111
	0.0857		0.252	0.0134
	0.0763		0.255	0.0136
	0.0508		0.256	0.0137
	0.0719		0.257	0.0137
	0.0466		0.264	0.0141
	0.113		0.318	0.0169
	0.280		0.446	0.0235
	0.261		0.462	0.0243
	0.402		0.515	0.0271
	0.477		0.591	0.0309
	0.666		0.644	0.0336
	2.545		0.894	0.0460
	5.276		0.942	0.0484
	3.997		0.991	0.0508
	18.651		1.230	0.0623
	23.847		1.25	0.0632
	26.831		1.315	0.0663
	31.441		1.432	0.0718
352.59	0.075	25	0.180	0.00962
	0.204		0.314	0.0167
	1.047		0.582	0.0304
	4.570		0.918	0.0472
	19.580		1.212	0.0614
366.48	0.00013		0.0039	0.00021
	0.0015		0.0153	0.00083
	0.0023		0.0208	0.00112
	0.0043		0.0319	0.00172
	0.087		0.132	0.00708
	0.124		0.178	0.00952
	0.181		0.201	0.0107

COMPONENTS:	ORIGINAL MEASUREMENTS
1. Hydrogen sulfide; H_2S; [7783-06-4]	Lawson, J.D. and Garst, A.W.
2. Water; H_2O; [7732-18-5]	J. Chem. Engng. Data. 1976,
3. 2,2'-iminobisethanol, Diethanolamine; $C_4H_{11}NO_2$; [111-42-2]	21, 20-30.

T/K	P_{H_2S}/bar	Conc.of DEA Wt. %	Liquid comp. mol H_2S/mol amine.	Mole fraction of hydrogen sulfide in liquid, x_{H_2S}
366.48	0.217	25	0.211	0.0123
	0.175		0.212	0.0113
	0.366		0.315	0.0167
	1.412		0.565	0.0296
	5.928		0.894	0.0460
	6.128		0.896	0.0462
	23.847		1.116	0.0568
	35.038		1.295	0.0653
380.37	0.222		0.169	0.00904
	0.580		0.306	0.0163
	1.932		0.563	0.0295
	7.354		0.856	0.0442
	25.712		1.134	0.0577
394.25	0.0045		0.0143	0.000771
	0.0041		0.0151	0.000815
	0.396		0.169	0.00904
	0.465		0.192	0.0103
	0.493		0.202	0.0108
	0.986		0.295	0.0157
	3.304		0.531	0.0279
	5.023		0.600	0.0314
	8.420		0.768	0.0398
	8.260		0.841	0.0434
	28.110		1.095	0.0558
	37.036		1.241	0.0628
408.15	0.613		0.162	0.00867
	4.476		0.496	0.0261
	10.112		0.766	0.0397
422.04	1.092		0.151	0.00809
	4.983		0.455	0.0240
	10.698		0.587	0.0307
310.93	0.00012	50	0.0111	0.00162
	0.0039		0.0522	0.00757
	0.032		0.212	0.0301
	0.253		0.553	0.0748
333.15	0.00009	54.5	0.0053	0.000774
	0.044		0.124	0.0178
	0.866		0.587	0.0790
366.48	0.0045	50	0.011	0.00161
	0.041		0.055	0.00798
	0.386		0.211	0.0299
	2.758		0.566	0.0764

COMPONENTS:	ORIGINAL MEASUREMENTS:
1. Hydrogen sulfide; H_2S; [7783-06-4] 2. Water; H_2O; [7732-18-5] 3. Ethanol, 2,2'-iminobis-(*diethanolamine*); $C_4H_{11}NO_2$; [111-42-2]	Lal, D.; Otto, F.D.; Mather, A.E. *Can.J.Chem.Eng.* 1985, *63*, 681-685.

VARIABLES:	PREPARED BY:
Temperature, pressure	P.G.T. Fogg

EXPERIMENTAL VALUES:

T/K	P_{H_2S}/kPa	Mole ratio in liquid, H_2S/DEA	T/K	P_{H_2S}/kPa	Mole ratio in liquid, H_2S/DEA
313.15	0.042	0.043	373.15	0.0075	0.0071
	0.051	0.051		0.011	0.0085
	0.053	0.057		0.012	0.0081
	0.053	0.051		0.016	0.0098
	0.064	0.054		0.025	0.0094
	0.106	0.063		0.044	0.0109
	0.149	0.083		0.061	0.0138
	0.162	0.081		0.092	0.0134
	0.190	0.087		0.137	0.0179
	0.216	0.097		0.195	0.0208
	0.245	0.102		0.238	0.0194
	0.276	0.117		0.405	0.0327
	0.384	0.141		0.653	0.041
	0.551	0.189		0.752	0.042
	1.355	0.219		0.895	0.048
				1.293	0.057
				1.407	0.064
				3.181	0.102

Concentration of diethanolamine = 2.0 kmol m^{-3} (2.0 mol dm^{-3})

AUXILIARY INFORMATION

METHOD/APPARATUS/PROCEDURE:	SOURCE AND PURITY OF MATERIALS:
Apparatus described in ref. (1) was used. Nitrogen was passed through three 500 cm^3 stainless steel vessels in series. These contained amine solution and dissolved H_2S. Temperatures were controlled to ± 0.5 K by an oil-bath. The gas emerging from the last vessel was analysed by gas chromatography. Samples of liquid from this vessel were analysed for H_2S by iodine-thiosulfate titration.	3. supplied by Dow chemicals.

ESTIMATED ERROR:
$\delta T/K$ = ± 0.5 (authors)
$\delta P/P$ = ± 0.15 at P_{H_2S} > 0.05 kPa; larger at lower partial pressures (authors).

REFERENCES:

1. Isaacs, E.E.; Otto, F.D.; Mather, A.E. *J.Chem.Eng.Data* 1980, *25*, 118.

COMPONENTS:	ORIGINAL MEASUREMENTS:
1. Hydrogen sulfide; H_2S; [7783-06-4]	Martin, J.L.; Otto, F.D.; Mather, A.E.
2. Ethanol, 2-(2-aminoethoxy)-(Diglycolamine); $C_4H_{11}NO_2$; [929-06-6]	*J. Chem. Engng. Data.* <u>1978</u>, *23*, 163-4.
3. Water; H_2O; [7732-18-5]	

VARIABLES:	PREPARED BY:
Temperature, pressure	C.L. Young

EXPERIMENTAL VALUES:

T/K	$P^§$/kPa	Conc of DGA /Wt%	$\alpha^†$	T/K	$P^§$/kPa	Conc of DGA /Wt%	$\alpha^†$
323.15	1730	60.0	1.091	323.15	2.52	60.0	0.221
	1490		1.068				
	1730		1.052	373.15	1890	60.0	0.909
	1480		1.043		1660		0.849
	1140		1.001		1460		0.806
	679		0.929		987		0.712
	414		0.877		662		0.640
	217		0.778		620		0.639
	121		0.711		509		0.571
	121		0.699		396		0.522
	86.1		0.657		273		0.456
	72.9		0.654		126		0.316
	63.1		0.608		65.3		0.217
	46.2		0.577		55.1		0.210
	53.1		0.543		48.2		0.182
	28.3		0.508		17.5		0.116
	18.9		0.452		16.5		0.100
	11.5		0.381		13.4		0.095
	8.87		0.344		9.29		0.071
	5.36		0.291		4.87		0.057
	3.91		0.262				

$P^§$ partial pressure of hydrogen sulfide

$\alpha^†$ mole ratio in liquid, hydrogen sulfide/diglycolamine

AUXILIARY INFORMATION

METHOD/APPARATUS/PROCEDURE:	SOURCE AND PURITY OF MATERIALS:
Recirculating vapor flow apparatus with Jerguson liquid level gauge cell and magnetic pump. Nitrogen added to vapor to increase pressure to at least 350 kPa. Vapor analyzed by gas chromatography. Partial pressure of hydrogen sulfide estimated from knowledge of pressure and vapor pressure of diethanolamine. Liquid samples passed into sulfuric acid and displaced carbon dioxide collected in buret.	1. No details given. 2. Jefferson Chemical Co. sample 3. No details given.
	ESTIMATED ERROR: $\delta T/K = \pm 0.5$; $\delta P/bar = \pm 1.0\%$; $\delta\alpha_{H_2S} = \pm 0.02$ or $\pm 4\%$ (whichever is greater).
	REFERENCES:

COMPONENTS:	ORIGINAL MEASUREMENTS:
1. Hydrogen sulfide; H_2S; [7783-06-4] 2. Water; H_2O; [7732-18-5] 3. 2,2'-(Methylimino)bisethanol (*methyldiethanolamine*); $C_5H_{13}NO_2$; [105-59-9]	Jou, F-Y.; Mather, A.E.; Otto, F.D. *Ind.Eng.Chem.Process Des.Dev.* 1982, *21*, 539 - 544.

VARIABLES:	PREPARED BY:
Temperature, pressure, composition.	P.G.T. Fogg

EXPERIMENTAL VALUES:

Concentration of methyldiethanolamine (MDEA) = 1.0 kmol m^{-3} (1.0 mol dm^{-3})

T/K	P_{H_2S}/kPa	Mole ratio in liquid H_2S/MDEA	T/K	P_{H_2S}/kPa	Mole ratio in liquid H_2S/MDEA
298.2	0.00674	0.0293	313.2	1.99	0.297
	0.0273	0.0539		10.9	0.604
	0.170	0.137		43.3	0.866
	1.10	0.315		102.0	0.994
	8.37	0.658		417.0	1.308
	38.9	0.911		1290.0	1.917
	182.0	1.183		1800.0	2.250
	413.0	1.424		2730.0	2.902
	830.0	1.850	343.2	0.00210	0.00540
	1380.0	2.364		0.00336	0.00661
	1960.0	2.935		0.00810	0.0109
313.2	0.00230	0.0111		0.0110	0.0130
	0.00409	0.0144		0.0232	0.0183
	0.0109	0.0225		0.0391	0.0225
	0.0910	0.0613		0.0836	0.0300
	0.513	0.149		0.269	0.0537

AUXILIARY INFORMATION

METHOD/APPARATUS/PROCEDURE:	SOURCE AND PURITY OF MATERIALS:
An equilibrium cell consisted of a Jerguson gage with a 250 cm^3 tubular gas reservoir mounted at the top. Gas was circulated with a magnetic pump. Temperatures were measured by thermopiles and controlled to ±0.5 °C by an air-bath. Pressures were measured by a Heise bourdon tube gage. The cell was charged with MDEA and H_2S to the appropriate pressure. Nitrogen was added, where necessary, to ensure that the total pressure was above atmospheric. Gases were circulated for at least 8 h. The gas phase was analysed by gas chromatography. H_2S in the liquid phase was determined by iodimetry and amine by titration with H_2SO_4. Solubilities at partial pressures less than 100 kPa were measured by a flow method.	1. of high purity from Linde & Matheson. 3. from Aldrich Chemical Co.; purity > 99%
	ESTIMATED ERROR:
	$\delta T/K$ = ±0.5 (authors)
	REFERENCES:

COMPONENTS:	ORIGINAL MEASUREMENTS:
1. Hydrogen sulfide; H_2S; [7783-06-4] 2. Water; H_2O; [7732-18-5] 3. 2,2'-(Methylimino)bisethanol (*methyldiethanolamine*); $C_5H_{13}NO_2$; [105-59-9]	Jou, F-Y.; Mather, A.E.; Otto, F.D. *Ind.Eng.Chem.Process Des.Dev.* <u>1982</u>, *21*, 539 - 544.

EXPERIMENTAL VALUES:

Concentration of methyldiethanolamine (MDEA) = 1.0 kmol m^{-3} (1.0 mol dm^{-3})

T/K	P_{H_2S}/kPa	Mole ratio in liquid H_2S/MDEA	T/K	P_{H_2S}/kPa	Mole ratio in liquid H_2S/MDEA
343.2	11.0	0.354	373.2	994.0	1.251
	14.3	0.384		2090.0	1.641
	146.0	0.886		2930.0	2.298
	511.0	1.185		4920.0	2.665
	925.0	1.405		5890.0	3.000
	1650.0	1.751	393.2	0.050	0.0098
	2420.0	2.085		0.52	0.0324
	3690.0	2.591		1.70	0.0566
	5030.0	3.229		11.5	0.145
373.2	0.090	0.0179		129.0	0.495
	0.236	0.0283		496.0	0.871
	1.03	0.0561		1290.0	1.233
	4.22	0.118		2710.0	1.750
	35.8	0.358		3730.0	2.078
	113.0	0.593		5230.0	2.627
	414.0	0.936			

Concentration of methyldiethanolamine = 2.0 kmol m^{-3} (2.0 mol dm^{-3})

T/K	P_{H_2S}/kPa	Mole ratio in liquid H_2S/MDEA	T/K	P_{H_2S}/kPa	Mole ratio in liquid H_2S/MDEA
313.2	0.00260	0.00725	313.2	1010.0	1.489
	0.00871	0.0127		2260.0	1.906
	0.0308	0.0238	373.2	0.745	0.029
	0.171	0.0576		16.43	0.156
	0.370	0.0871		29.07	0.203
	1.20	0.162		72.53	0.357
	5.76	0.368		146.9	0.474
	8.98	0.443		266.3	0.660
	27.3	0.674		483.3	0.846
	107.0	0.965		1020	1.076
	258.0	1.063		1550	1.256

Concentration of methyldiethanolamine = 4.28 kmol m^{-3} (4.28 mol dm^{-3})

T/K	P_{H_2S}/kPa	Mole ratio in liquid H_2S/MDEA	T/K	P_{H_2S}/kPa	Mole ratio in liquid H_2S/MDEA
298.2	0.00593	0.000960	298.2	296.0	1.088
	0.0180	0.0171		476.0	1.173
	0.114	0.0446		765.0	1.273
	0.195	0.0611		1060.0	1.373
	0.464	0.0972		1370.0	1.506
	0.603	0.106		1670.0	1.588
	36.9	0.780		1830.0	1.686
	88.8	0.921		1960.0	1.699

COMPONENTS:	ORIGINAL MEASUREMENTS:

1. Hydrogen sulfide; H_2S;
   [7783-06-4]
2. Water; H_2O; [7732-18-5]
3. 2,2'-(Methylimino)bisethanol
   (*methyldiethanolamine*);
   $C_5H_{13}NO_2$; [105-59-9]
	Jou, F-Y.; Mather, A.E.; Otto, F.D.
	
	Ind.Eng.Chem.Process Des.Dev.
	<u>1982</u>, *21*, 539 - 544.

EXPERIMENTAL VALUES:

Concentration of methyldiethanolamine (MDEA) = 4.28 kmol m^{-3} (4.28 mol dm^{-3})

T/K	P_{H_2S}/kPa	Mole ratio in liquid H_2S/MDEA	T/K	P_{H_2S}/kPa	Mole ratio in liquid H_2S/MDEA
313.2	0.00314	0.00508	343.2	1420.0	1.163
	0.00502	0.00580		2530.0	1.355
	0.00714	0.00734		3460.0	1.521
	0.0102	0.00805		4120.0	1.616
	0.0387	0.0165		4990.0*	1.727
	0.271	0.0446	373.2	0.0417	0.00434
	1.42	0.103		0.133	0.00763
	8.71	0.268		0.383	0.0142
	28.5	0.499		1.66	0.0305
	107.0	0.849		26.1	0.130
	500.0	1.083		240.0	0.435
	949.0	1.210		765.0	0.763
	1540.0	1.369		1690.0	1.004
	2140.0	1.520		2300.0	1.104
	2360.0	1.576		3630.0	1.272
	2800.0*	1.723		4720.0	1.409
343.3	0.00130	0.00129		5680.0	1.518
	0.00274	0.00177	393.2	0.342	0.00950
	0.00451	0.00258		25.1	0.0895
	0.00714	0.00311		252.0	0.303
	0.00985	0.00367		1130.0	0.677
	0.353	0.0253		2510.0	0.969
	16.8	0.188		3400.0	1.084
	23.7	0.233		4690.0	1.221
	132.0	0.549		5390.0	1.285
	528.0	0.953		5840.0	1.328

* Liquid H_2S phase present.

COMPONENTS:	ORIGINAL MEASUREMENTS:
1. Hydrogen sulfide; H_2S; [7783-06-4] 2. 2,2',2''-Nitrilotrisethanol, (Triethanolamine); $C_6H_{15}NO_3$; [102-71-6] 3. Water; H_2O; [7732-18-5]	Atwood, K.; Arnold, M. R.; Kindrick, R. C. *Ind. Eng. Chem.* 1957, *49*, 1439-44.

VARIABLES:	PREPARED BY:
Temperature, pressure, composition	C. L. Young

EXPERIMENTAL VALUES:

T/K	t/°F	Wt-% amine	P/mmHg	P/kPa	Conc. H_2S /mol ℓ^{-1}	Mole ratio[†]
299.8	80	15	1.31	0.175	0.0404	0.0395
			241	32.1	0.539	0.531
322.0	120		0.00424	0.000565	0.000795	0.000776
			4.19	0.559	0.0396	0.0387
			115	15.3	0.264	0.259
333.1	140	20	5.66	0.755	0.0446	0.0324
			624	83.2	0.562	0.412
310.9	100	30	0.568	0.0757	0.0261	0.0124
			55.3	7.373	0.315	0.151
333.1	140		0.0222	0.00296	0.00224	0.00107
			624	83.2	0.714	0.345
299.8	80	50	0.655	0.0873	0.0381	0.0106
			22.9	3.05	0.265	0.0739
			242	32.3	1.105	0.313
322.0	120		0.00295	0.000393	0.000976	0.000271
			57.2	7.63	0.297	0.0829
			693	92.4	1.179	0.334

[†] Mole of hydrogen sulfide per mole of amine.

AUXILIARY INFORMATION

METHOD/APPARATUS/PROCEDURE:	SOURCE AND PURITY OF MATERIALS:
Gas saturation method was used in which a known quantity of hydrogen was passed through a series of saturators containing a solution of known hydrogen sulfide concentration. At partial pressures of H_2S greater than 100 mmHg, solution saturation was used. H_2S in soln. determined by iodimetry. Details in source.	1. Stated purity 99.7 mole per cent. 2. Analytical grade. 3. No details.
	ESTIMATED ERROR: $\delta T/K = \pm 0.12$; $\delta p/kPa = \pm 2\%$; δ(Mole ratio) = $\pm 3\%$ (estimated by compiler).
	REFERENCES:

COMPONENTS:	ORIGINAL MEASUREMENTS:
1. Hydrogen sulfide; H_2S; [7783-06-4] 2. Water; H_2O; [7732-18-5] 3. 2,2',2"-Nitrilotrisethanol, (*triethanolamine*); $C_6H_{15}NO_3$; [102-71-6]	Jou, F-Y.; Mather, A.E.; Otto, F.D. *Can.J.Chem.Eng.* 1985, *63*, 122-125. (Numerical data deposited in the Depository for Unpublished Data, Ottawa, Canada)[*]
VARIABLES: Temperature, pressure, composition of liquid phase.	PREPARED BY: P.G.T. Fogg

EXPERIMENTAL VALUES:

Conc. of TEA /mol dm^{-3}	T/K	P_{H_2S} /kPa	Total pressure /kPa	Mole ratio in liquid H_2S/TEA
2.0	298.2	0.0267	403	0.0164
		0.184	303	0.0412
		1.10	348	0.102
		4.81	438	0.208
		42.2	752	0.526
		263	266	0.995
		720	723	1.350
		1480	1485	1.751
		1980	2030	1.989
2.0	323.2	0.0683	483	0.0133
		0.275	417	0.0284
		22.6	542	0.258
		340	352	0.836
		885	901	1.164
		1810	1838	1.527
		2700	2720	1.828
		3420	3431	1.993

TEA = triethanolamine

[*]Depository for Unpublished Data, CISTI, National Research Council of Canada, Ottawa, Ontario, K1A 0S2, Canada.

AUXILIARY INFORMATION

METHOD/APPARATUS/PROCEDURE:	SOURCE AND PURITY OF MATERIALS:
The equilibrium cell consisted of a Jerguson gauge (Model 19-T-20) with a 250 cm^3 tubular gas reservoir mounted at the top. Gas was circulated with a magnetic pump (1) Temperatures were measured by thermopiles and controlled to ± 0.5 K by a air-bath. Pressures were measured by a Heise bourdon tube gauge. The cell was charged with TEA solution and H_2S to the desired pressure. If necessary nitrogen was added to ensure that the total pressure was above atmospheric. Gases were circulated for at least 8 h. The gas phase was analysed by gas chromatography. H_2S in the liquid phase was determined by iodimetry and amine by titration with H_2SO_4.	1. from Linde & Matheson; purity > 99% 2. distilled. 3. from Fisher Scientific; purity 99.4%
	ESTIMATED ERROR: $\delta T/K$ = ± 0.5 (authors)
	REFERENCES: 1. Ruska, W.E.A.; Hurt, L.J.; Kobayashi, R. *Rev.Sci.Instrum.* 1970, *41*, 1444.

COMPONENTS:	ORIGINAL MEASUREMENTS:
1. Hydrogen sulfide; H_2S; [7783-06-4] 2. Water; H_2O; [7732-18-5] 3. 2,2',2"-Nitrilotrisethanol (*triethanolamine*); $C_6H_{15}NO_3$; [102-71-6]	Jou, F-Y.; Mather, A.E.; Otto, F.D. *Can.J.Chem.Eng.* <u>1985</u>, *63*, 122-125.

EXPERIMENTAL VALUES:

Conc. of TEA /mol dm^{-3}	T/K	P_{H_2S} /kPa	Total pressure /kPa	Mole ratio in liquid H_2S/TEA
2.0	348.2	0.0238	480	0.00402
		0.821	442	0.0237
		8.54	487	0.0813
		197	233	0.431
		1010	1045	0.960
		2250	2286	1.361
		4110	4162	1.776
		5500	5562	2.076
2.0	373.2	0.0172	387	0.00205
		0.0585	343	0.00328
		0.590	528	0.0125
		1.35	425	0.0193
		4.88	490	0.0401
		72.7	410	0.176
		708	800	0.620
		3120	3231	1.334
		4940	5052	1.651
		6160	6272	1.912
2.0	398.2	0.0466	492	0.00174
		0.095	490	0.00220
		0.383	501	0.00464
		1.45	452	0.0114
		11.4	480	0.036
		165	391	0.203
		1730	1955	0.760
		3550	3775	1.153
		5190	5417	1.465
		6270	6505	1.627
3.5	298.2	0.0124	334	0.0070
		0.219	478	0.0256
		16.0	287	0.242
		299	320	0.881
		729	732	1.161
		1500	1504	1.493
		1984	1988	1.670
3.5	323.2	0.00832	349	0.00236
		0.33	358	0.0163
		3.40	314	0.0638
		35.2	499	0.224
		466	478	0.780
		1740	1759	1.229
		2500	2508	1.404
		3430	3436	1.641
3.5	348.2	0.0281	344	0.00241
		0.372	473	0.0107
		12.5	409	0.0746
		57.8	485	0.172
		494.0	528	0.509
		2000	2042	1.022
		3270	3306	1.281
		4290	4329	1.506
		5530	5568	1.714

COMPONENTS:	ORIGINAL MEASUREMENTS:
1. Hydrogen sulfide; H_2S; [7783-06-4] 2. Water; H_2O; [7732-18-5] 3. 2,2',2"-Nitrilotrisethanol, (*triethanolamine*); $C_6H_{15}NO_3$; [102-71-6]	Jou, F-Y.; Mather, A.E.; Otto, F.D. *Can.J.Chem.Eng.* 1985, *63*, 122-125.

EXPERIMENTAL VALUES:

Conc. of TEA /mol dm^{-3}	T/K	P_{H_2S} /kPa	Total pressure /kPa	Mole ratio in liquid H_2S/TEA
3.5	373.2	0.0418	442	0.00154
		1.09	452	0.00975
		19.1	489	0.0485
		329	412	0.274
		1730	1811	0.710
		3020	3122	0.992
		4690	4797	1.224
		5910	6024	1.382
3.5	398.2	0.0183	453	0.00545
		0.886	511	0.00498
		10.3	652	0.0224
		137	344	0.101
		1190	1402	0.416
		3570	3786	0.844
		5440	5648	1.098
5.0	298.2	0.0171	349	0.00514
		0.201	452	0.0202
		2.30	509	0.0739
		66.8	473	0.406
		312.0	314	0.798
		1090.0	1100	1.285
		1580.0	1579	1.440
		1990.0	1997	1.600
5.0	323.2	0.0315	344	0.00376
		0.995	479	0.0243
		9.97	427	0.0842
		361.0	369	0.567
		1660.0	1686	1.135
		2760.0	2769	1.432
		3430.0	3444	1.587
5.0	348.2	0.0152	290	0.00099
		0.158	492	0.00386
		2.13	321	0.0167
		73.9	325	0.133
		279.0	693	0.296
		1340.0	1370	0.701
		4060.0	4092	1.290
		5490.0	5526	1.566
5.0	373.2	0.0269	403	0.00069
		0.381	487	0.00298
		6.07	476	0.0160
		324.0	393	0.178
		1100.0	1170	0.421
		4050.0	4126	0.922
		5450.0	5519	1.097
5.0	398.2	0.0872	438	0.00074
		1.84	496	0.00456
		45.7	513	0.0354
		296.0	469	0.128
		2480.0	2657	0.483
		5450.0	5622	0.782

COMPONENTS:	ORIGINAL MEASUREMENTS:

1. Hydrogen sulfide; H_2S;
     [7783-06-4]
 	
2. 1,1'-Iminobis-2-Propanol,
     (Diisopropanolamine); $C_6H_{15}NO_2$;
     [110-97-4]
3. Water; H_2O; [7732-18-5]
	Isaacs, E.E.; Otto, F.D. Mather, A.E.
	
	J. Chem. Engng. Data. <u>1977</u>, *22*, 71-3.

VARIABLES:	PREPARED BY:
Temperature, pressure, composition	C.L. Young.

EXPERIMENTAL VALUES: T/K   Conc. of 2 /mol dm^{-3} P/kPa α

T/K	Conc. of 2 /mol dm^{-3}	P/kPa	α
313.15	2.5	2152.5	1.414
		1199.6	1.215
		903.2	1.092
		450.9	0.962
		179.9	0.824
		97.2	0.750
		73.7	0.728
		72.3	0.706
		47.5	0.671
		27.5	0.562
		22.7	0.473
		15.8	0.409
		9.6	0.349
		6.8	0.289
		2.0	0.132
373.15	2.5	3207.4	1.249
		2209.7	1.100
		880.4	0.849
		462.6	0.651
		211.6	0.478
		155.8	0.388
		82.7	0.300
		50.4	0.221
		31.6	0.176
		38.1	0.167
		12.1	0.098

α = mole ratio in liquid phase H_2S/
            Diisopropanolamine.

AUXILIARY INFORMATION

METHOD/APPARATUS/PROCEDURE:	SOURCE AND PURITY OF MATERIALS:
Recirculating vapor flow apparatus with Jerguson liquid level gauge cell and magnetic pump. Nitrogen added to vapor to increase pressure to at least 600 kPa. Vapor analysed by gas chromatography. Partial pressure of hydrogen sulfide calculated from pressure and composition. Liquid samples passed into sulfuric acid. Hydrogen sulfide collected in buret.	No details given.
	ESTIMATED ERROR: $\delta T/K = \pm0.5$; $\delta P/kPa = \pm1\%$; $\delta\alpha = \pm4\%$ or ±0.02 whichever is greater.
	REFERENCES:

COMPONENTS:	ORIGINAL MEASUREMENTS:
1. Hydrogen sulfide; H_2S; [7783-86-4] 2. Water; H_2O; [7732-18-5] 3. 1,1'-iminobis-2-propanol, (*diisopropanolamine*); $C_6H_{15}NO_2$; [110-97-4] 4. Tetrahydrothiophene, 1,1-dioxide, (*sulfolane*); $C_4H_8O_2S$; [126-33-0]	Isaacs, E.E; Otto, F.D.; Mather, A.E. *J. Chem. Eng. Data* 1977, 22, 317-319.
VARIABLES: Temperature, concentration	PREPARED BY: P.G.T. Fogg

EXPERIMENTAL VALUES:

The solvent consisted of 40 wt% of diisopropanolamine (DIPA), 40 wt% of sulfolane and 20 wt% of water. The authors indicated that this is a typical *Sulfinol* solution as used in the *Sulfinol* process patented by Shell.

T/K	P_{H_2S}/kPa	Mole ratio in liquid H_2S/DIPA	T/K	P_{H_2S}/kPa	Mole ratio in liquid H_2S/DIPA
313.2	4.6	0.152	373.2	63.7	0.074
	5.2	0.175		76.4	0.083
	13.8	0.297		71.7	0.119
	20.3	0.308		165.0	0.150
	25.3	0.424		262.1	0.243
	55.9	0.582		419.5	0.352
	277.6	0.901		658.7	0.510
	502.3	1.091		1122.6	0.733
	585.3	1.173		1748.8	0.929
	865.6	1.492		2405.9	1.283
	1081.9	1.598		3862.3	1.988
	1410.3	2.022			
	2051.2	3.339			
	2291.3	4.429			

AUXILIARY INFORMATION

METHOD/APPARATUS/PROCEDURE:	SOURCE AND PURITY OF MATERIALS:
The equilibrium cell consisted of a Jerguson gage with a 250 cm³ tubular gas reservoir mounted at the top. Gas was circulated with a magnetic pump (1). Temperatures were measured by thermopiles and controlled to ± 0.5 °C by an air-bath. Pressures were measured by a Heise bourdon tube gage. The cell was purged with nitrogen and charged with 50-150 cm³ of *Sulfinol* solution. An appropriate quantity of hydrogen sulfide was then added. Nitrogen was added, when necessary, to ensure that the total pressure was above 350 kPa. Gases were circulated for at least 8 h. The gas phase was analysed by gas chromatography and the liquid phase by treating samples with H_2SO_4 (2.5 mol dm-3), recording P-V-T data for the gases evolved and analysing them by gas chromatography.	3. minimum purity 97% 4. minimum purity 99% ESTIMATED ERROR: $\delta T/K$ = ± 0.5 δmole ratio = ± 0.02 or ± 4% whichever is the larger. (authors) REFERENCES: 1. Ruska, W.E.A.; Hurt, L.J.; Kobayashi, R. *Rev. Sci. Instrum.* 1979, 41, 1444.

COMPONENTS:	ORIGINAL MEASUREMENTS:
1. Hydrogen sulfide; H_2S; [7783-06-4] 2. Carbon dioxide; CO_2; [124-38-9] 3. Water; H_2O; [7732-18-5] 4. 2-Aminoethanol; (*monoethanolamine*); C_2H_7NO; [141-43-5]	Leibush, A.G.; Shneerson, A.L. *Zhur. Prik. Khim.* 1950, *23*, 145-152. *J. Applied Chem. USSR* 1950, *23*, 149-157.
VARIABLES: Temperature, pressure, composition of liquid phase	PREPARED BY: P.G.T. Fogg

EXPERIMENTAL VALUES:

Concentration of monoethanolamine (MEA) = 2.5 mol dm^{-3}; T/K = 298.2

P_{H_2S}/mmHg	P_{CO_2}/mmHg	Mole ratios in liquid phase	
		H_2S/MEA	CO_2/MEA
1.94	0.00	0.265	0.025
2.78	0.00	0.265	0.092
4.1	0.00	0.270	0.155
7.15	0.26	0.268	0.255
20.2	1.77	0.262	0.311
70.4	75	0.262	0.385
0.69	0.00	0.145	0.029
2.08	0.00	0.145	0.216
3.62	0.26	0.145	0.300
19.8	4.57	0.145	0.416

760 mmHg = 1 atm = 1.01325 bar

AUXILIARY INFORMATION

METHOD/APPARATUS/PROCEDURE:	SOURCE AND PURITY OF MATERIALS:
N_2 or a mixture of N_2 & H_2S was passed successively through two absorbers containing H_2S & CO_2 dissolved in an aqueous solution of MEA. The H_2S in the emerging gas was passed into cadmium or zinc acetate solution and the precipitated sulfides estimated by iodimetry. Hydrogen sulfide in the MEA solutions was also determined by iodimetry. CO_2 in the gas phase was absorbed in standard baryta solution after removal of H_2S by acidified potassium permanganate solution. CO_2 in the liquid phase was estimated by reaction with 30% H_2SO_4, removal of evolved H_2S by permanganate and absorption of evolved CO_2 in standard baryta solution.	1. From H_2SO_4 & Na_2S: no SO_2 detected 4. Contained 0.6 to 1.5 volumes of CO_2 per unit volume of solution; less than 1% by weight of total impurities.
	ESTIMATED ERROR: $\delta T/K$ = ± 0.1 (authors)
	REFERENCES:

COMPONENTS:	ORIGINAL MEASUREMENTS:
1. Hydrogen sulfide; H_2S; [7783-06-4] 2. Carbon dioxide; CO_2; [124-38-9] 3. Water; H_2O; [7732-18-5] 4. 2-Aminoethanol, (*monoethanol-amine*); C_2H_7NO; [141-43-5]	Muhlbauer, H.G.; Monaghan, P.R. *Oil & Gas J.* <u>1957</u>, *55(17)*, 139-145.
VARIABLES: Temperature, pressure, composition of liquid phase.	PREPARED BY: P.G.T. Fogg

EXPERIMENTAL VALUES:

T/K	Conc. of MEA /mol dm^{-3}	P_{H_2S} /mmHg	P_{CO_2} /mmHg	Mole ratio in liquid H$_2$S/MEA	CO$_2$/MEA
298.15	2.52	0.09	0.16	0.112	0.100
	2.60	0.48	0.52	0.022	0.333
	2.56	0.82	0.24	0.038	0.309
	2.50	0.99	0	0.196	0
	2.53	1.18	0.10	0.114	0.174
	2.49	1.26	0	0.272	0
	2.62	1.27	0.40	0.043	0.347
	2.55	1.81	0.36	0.074	0.304
	2.63	1.88	0.70	0.066	0.348
	2.53	1.94	0.21	0.146	0.174
	2.51	2.23	0	0.308	0
	2.53	2.49	0.16	0.226	0.100
	2.50	3.37	0	0.323	0
	2.59	3.38	13.0	0.019	0.487
	2.54	4.15	0.47	0.120	0.302
	2.53	4.23	0.23	0.220	0.175

MEA = monoethanolamine   760 mmHg = 1 atm = 1.01325 bar

AUXILIARY INFORMATION

METHOD/APPARATUS/PROCEDURE:	SOURCE AND PURITY OF MATERIALS:
Data for 25 °C were obtained by passing a mixture of N_2, CO_2 & H_2S through two thermostatted wash-bottles in series containing monoethanolamine plus dissolved CO_2 & H_2S. After 4 h liquid samples were analysed by chemical methods and gas samples by chromatography. Data for 100 °C were obtained by agitating gas and liquid samples in a thermostatted steel bomb for about 1 h before analysis of the two phases.	
	ESTIMATED ERROR: $\delta T/K = \pm 0.02$ (authors)
	REFERENCES:

COMPONENTS:	ORIGINAL MEASUREMENTS:
1. Hydrogen sulfide; H_2S; [7783-06-4] 2. Carbon dioxide; CO_2; [124-38-9] 3. Water; H_2O; [7732-18-5] 4. 2-Aminoethanol, (*monoethanol-amine*); C_2H_7NO; [141-43-5]	Muhlbauer, H.G.; Monaghan, P.R. *Oil & Gas J.* 1957, *55(17)*, 139-145.

EXPERIMENTAL VALUES:

T/K	Conc. of MEA /mol dm^{-3}	P_{H_2S} /mmHg	P_{CO_2} /mmHg	Mole ratio in liquid	
				H_2S/MEA	CO_2/MEA
298.15	2.52	4.40	0.15	0.280	0.104
	2.62	4.80	1.06	0.090	0.356
	2.60	6.0	0.60	0.155	0.286
	2.53	7.41	0.31	0.372	0.108
	2.47	7.67	0	0.450	0
	2.53	7.8	0.34	0.284	0.172
	2.63	8.3	1.70	0.112	0.352
	2.59	9.28	19.5	0.036	0.472
	2.55	10.9	0.78	0.185	0.296
	2.51	11.4	0.27	0.449	0.105
	2.62	13.9	2.6	0.134	0.346
	2.44	16.5	0	0.597	0
	2.59	17.3	24.4	0.057	0.445
	2.52	17.4	0.29	0.384	0.165
	2.51	27.0	0.46	0.543	0.098
	2.58	28.0	33.5	0.075	0.447
	2.53	29.2	2.27	0.256	0.290
	2.54	30.4	7.50	0.171	0.357
	2.46	31.0	0	0.706	0
	2.52	36.0	1.06	0.451	0.180
	2.63	38.7	49.2	0.092	0.464
	2.58	47.6	47.5	0.098	0.430
	2.51	52.2	1.13	0.488	0.172
	2.50	52.8	0.82	0.614	0.112
	2.45	53.2	0	0.786	0
	2.53	55.6	13.8	0.214	0.358
	2.56	59.5	34.1	0.142	0.410
	2.62	65.2	66.8	0.117	0.471
	2.52	68.6	8.61	0.289	0.307
	2.53	92.7	32.4	0.234	0.364
	2.52	93.8	11.4	0.336	0.292
	2.61	109.4	104	0.138	0.460
	2.42	111	0	0.866	0
	2.49	140	2.76	0.684	0.112
	2.53	142	19.9	0.364	0.286
	2.49	163	9.31	0.556	0.181
	2.42	165.2	0	0.913	0
	2.52	166.2	55.4	0.266	0.357
	2.61	172	157	0.170	0.456
	2.56	230	83.8	0.300	0.350
	2.42	306	0	0.934	0
	2.52	330	62.4	0.403	0.280
	2.53	340	94.2	0.383	0.317
	2.49	342	19.6	0.614	0.167
	2.49	443	16.0	0.742	0.104
	2.47	711	20.4	0.820	0.108
	2.41	720	0	0.939	0

COMPONENTS:	ORIGINAL MEASUREMENTS:
1. Hydrogen sulfide; H_2S; [7783-06-4] 2. Carbon dioxide; CO_2; [124-38-9] 3. Water; H_2O; [7732-18-5] 4. 2-Aminoethanol, (*monoethanol-amine*); C_2H_7NO; [141-43-5]	Muhlbauer, H.G.; Monaghan, P.R. *Oil & Gas J.* <u>1957</u>, *55(17)*, 139-145.

EXPERIMENTAL VALUES:

T/K	Conc. of MEA /mol dm^{-3}	P_{H_2S} /mmHg	P_{CO_2} /mmHg	Mole ratio in liquid H_2S/MEA	CO_2/MEA
373.15	2.45	9.2	0	0.060	0
	2.44	22.6	9.1	0.075	0.076
	2.44	24.8	0	0.102	0
	2.47	34.4	462	0.023	0.424
	2.43	50.7	14.2	0.127	0.072
	2.45	52.5	22.8	0.108	0.155
	2.45	60.9	26.1	0.113	0.156
	2.44	75.3	90.7	0.093	0.251
	2.45	76.3	500	0.043	0.413
	2.44	85.3	30.3	0.146	0.158
	2.44	87	10.3	0.188	0.076
	2.45	90	0	0.248	0
	2.44	111	289	0.079	0.344
	2.39	118	18.1	0.224	0.084
	2.43	123.8	0	0.294	0
	2.40	129	49.2	0.177	0.167
	2.44	136	104.5	0.141	0.244
	2.43	139	37.7	0.196	0.153
	2.42	151	14.5	0.262	0.074
	2.44	170	566	0.081	0.385
	2.44	195	0	0.372	0
	2.43	204	690	0.090	0.398
	2.44	219	125	0.187	0.230
	2.39	224	24.5	0.299	0.075
	2.40	228	74.9	0.236	0.168
	2.44	239	388	0.139	0.325
	2.43	241	451	0.126	0.351
	2.42	260	924	0.099	0.408
	2.44	263	0	0.420	0
	2.43	269	428	0.144	0.331
	2.39	316	28.6	0.381	0.073
	2.39	319	84.5	0.286	0.171
	2.42	324	1290	0.106	0.422
	2.43	360	179	0.252	0.229
	2.43	373	565	0.169	0.328
	2.43	397	568	0.177	0.321
	2.40	403	42.4	0.412	0.081
	2.39	410	109.0	0.326	0.172
	2.42	420	0	0.491	0
	2.42	447	0	0.511	0
	2.42	466	237	0.275	0.234
	2.42	470	643	0.191	0.324
	2.42	518	0	0.546	0
	2.42	525	740	0.198	0.333
	2.40	540	52.6	0.453	0.082
	2.42	545	829	0.193	0.339
	2.39	564	155	0.375	0.169
	2.40	632	57.1	0.489	0.083
	2.42	638	312	0.306	0.239
	2.41	644	0	0.578	0
	2.41	700	385	0.308	0.246
	2.37	742	63	0.532	0.073
	2.39	760	169	0.422	0.156
	2.40	804	0	0.639	0
	2.40	833	168	0.444	0.146
	2.40	899	0	0.649	0
	2.38	947	77.0	0.549	0.076
	2.39	952	0	0.660	0
	2.40	964	551	0.342	0.241

COMPONENTS:	ORIGINAL MEASUREMENTS:
1. Carbon dioxide; CO_2; [124-38-9] 2. Hydrogen sulfide; H_2S; [7783-06-4] 3. 2-Aminoethanol (Monoethanolamine); $C_2H_7NO_2$; [141-43-5] 4. Water; H_2O; [7732-18-5]	Jones, J. H.; Froning, H. R.; Claytor, E. E. Jr. *J. Chem. Engng. Data* <u>1959</u>, *4*, 85-92.

VARIABLES:	PREPARED BY:
Temperature, pressure	C. L. Young

EXPERIMENTAL VALUES:

T/K	$P^{§}_{H_2S}/10^5$ Pa	$P^{§}_{CO_2}/10^5$ Pa	Conc. of MEA /wt%	$\alpha^{†}_{H_2S}$	$\alpha^{†}_{CO_2}$
313.15	0.0015	0.0019	15.3	0.0164	0.387
	0.0028	0.0041		0.0214	0.424
	0.0029	0.0025		0.0282	0.400
	0.0069	0.0179		0.0248	0.480
	0.0072	0.0177		0.0272	0.472
	0.0076	0.0033		0.0612	0.392
	0.0077	0.0072		0.0436	0.436
	0.0101	0.0019		0.101	0.349
	0.0116	0.579		0.0056	0.652
	0.0117	0.0576		0.0199	0.527
	0.0203	0.0349		0.0488	0.488
	0.0224	0.0100		0.106	0.412
	0.0261	0.607		0.0123	0.658
	0.0573	0.0237		0.149	0.424
	0.0744	0.0040		0.361	0.200
	0.1019	0.0104		0.351	0.293
	0.144	0.0247		0.314	0.335
	0.156	0.0608	(cont.)	0.235	0.415

$^{§}P_{H_2S}$ partial pressure of hydrogen sulfide

$^{§}P_{CO_2}$ partial pressure of carbon dioxide

AUXILIARY INFORMATION

METHOD/APPARATUS/PROCEDURE:	SOURCE AND PURITY OF MATERIALS:
Rocking static equilibrium cell fitted with liquid and gas sampling ports. Pressure measured with Bourdon gauge. Concentration of carbon dioxide and hydrogen sulfide in gas phase determined by mass spectrometry. Concentration of hydrogen sulfide in liquid determined by iodimetry and concentration of carbon dioxide in liquid determined by precipitation as barium carbonate or by stripping out carbon dioxide and reabsorbing on a solid absorbent. Details in source.	1. Bone dry sample. 2. Purity 99.9 mole per cent. Mass spectrometry showed trace amounts of methyl mercaptan, carbon disulfide and carbon dioxide. 3. Purity 99.3 mole per cent. 4. Distilled.
	ESTIMATED ERROR: $\delta T/K = \pm 0.1$ at 313.15 K, ± 0.5 at 413.15 K; $\delta P/kPa = \pm 1\%$; $\delta\alpha = \pm 3\%$ (estimated by compiler).
	REFERENCES:

COMPONENTS:	ORIGINAL MEASUREMENTS:

1. Carbon dioxide; CO_2; [124-38-9]
 	
2. Hydrogen sulfide; H_2S;
   [7783-06-4]
 	
3. 2-Aminoethanol (Monoethanolamine);
   $C_2H_7NO_2$; [141-43-5]
 	
4. Water; H_2O; [7732-18-5]
	Jones, J. H.; Froning, H. R.;
	Claytor, E. E. Jr.
	J. Chem. Engng. Data
	<ins>1959</ins>, *4*, 85-92.

EXPERIMENTAL VALUES:

T/K	$P^§_{H_2S}/10^5 Pa$	$P^§_{CO_2}/10^5 Pa$	Conc. of MEA /wt%	$\alpha^†_{H_2S}$	$\alpha^†_{CO_2}$
313.15	0.171	0.696	15.3	0.0712	0.620
	0.175	0.709		0.0716	0.620
	0.449	0.197		0.327	0.414
	1.123	0.470		0.425	0.406
	1.879	0.771		0.495	0.398
333.15	0.0013	0.0143		0.0074	0.406
	0.0069	0.0037		0.0597	0.290
	0.0079	0.0011		0.107	0.166
	0.0136	0.580		0.0070	0.594
	0.0192	0.173		0.0200	0.529
	0.0292	0.0559		0.0569	0.452
	0.0309	0.0189		0.0878	0.381
	0.0953	0.0128		0.255	0.274
	0.1069	1.004		0.0396	0.606
	0.528	0.499		0.238	0.460
	1.995	4.672		0.235	0.570
373.15	0.0037	0.0765		0.0053	0.293
	0.0069	0.429		0.0050	0.420
	0.0093	0.0173		0.0272	0.171
	0.0205	0.375		0.0140	0.412
	0.0341	0.0791		0.0460	0.272
	0.0500	0.760		0.0245	0.456
	0.0765	0.0403		0.112	0.180
	0.199	0.1033		0.178	0.238
	0.373	0.853		0.140	0.405
	0.623	0.131		0.366	0.175
393.15	0.0088	0.437		0.0059	0.322
	0.0135	0.010		0.0405	0.0393
	0.0197	0.0385		0.0344	0.106
	0.0204	0.0884		0.0264	0.178
	0.0524	0.381		0.0361	0.298
	0.0972	0.188		0.104	0.202
	0.1033	0.205		0.104	0.200
	0.144	0.421		0.0940	0.280

$\alpha^†_{H_2S}$ = mole H_2S/mole 2-Aminoethanol

$\alpha^†_{CO_2}$ = mole CO_2/mole 2-Aminoethanol

MEA = 2-Aminoethanol (monoethanolamine)

COMPONENTS:	ORIGINAL MEASUREMENTS:
1. Carbon dioxide; CO_2; [124-38-9]	Jones, J. H.; Froning, H. R.;
2. Hydrogen sulfide; H_2S; [7783-06-4]	Claytor, E. E. Jr.
3. 2-Aminoethanol (Monoethanolamine); $C_2H_7NO_2$; [141-43-5]	*J. Chem. Engng. Data* 1959, *4*, 85-92.
4. Water; H_2O; [7732-18-5]	

EXPERIMENTAL VALUES: Smoothed data for 15.3 wt% of MEA

$P^§_{H_2S}$ /mmHg	\multicolumn{6}{c}{Moles H_2S per Mole MEA}
	$R_L^* = 0.01$	$R_L = 0.05$	$R_L = 0.10$	$R_L = 0.50$	$R_L = 1.0$	$R_L = \infty$

$P^§_{H_2S}$ /mmHg	$R_L^* = 0.01$	$R_L = 0.05$	$R_L = 0.10$	$R_L = 0.50$	$R_L = 1.0$	$R_L = \infty$
\multicolumn{7}{c}{T/K = 313.15}						
1	0.0047	0.0190	0.0327	0.0863	0.1140	0.128
3	0.0055	0.0225	0.0395	0.1160	0.1630	0.212
10	0.0066	0.0263	0.0468	0.1510	0.2220	0.374
30	0.0077	0.0301	0.0540	0.1820	0.2720	0.579
100	0.0092	0.0351	0.0619	0.2120	0.3260	0.802
300	–	0.0399	0.0710	0.2350	0.3720	0.931
1000	–	0.0464	0.0830	0.2700	0.4250	1.00
\multicolumn{7}{c}{T/K = 333.15}						
1	0.0037	0.0145	0.0237	0.0650	0.0775	0.085
3	0.0046	0.0184	0.0304	0.0845	0.1130	0.137
10	0.0059	0.0234	0.0396	0.1125	0.1600	0.240
30	0.0074	0.0288	0.0492	0.1450	0.2120	0.386
100	0.0092	0.0355	0.0605	0.1840	0.2750	0.600
300	–	0.0431	0.0730	0.2190	0.3230	0.790
1000	–	–	0.0910	0.2620	0.3840	0.970
\multicolumn{7}{c}{T/K = 373.15}						
1	0.0024	0.0067	0.0103	0.0220	0.0247	0.029
3	0.0036	0.0101	0.0155	0.0340	0.0407	0.050
10	0.0056	0.0155	0.0239	0.0540	0.0675	0.091
30	0.0082	0.0228	0.0349	0.0810	0.1040	0.160
100	–	0.0343	0.0524	0.1250	0.1650	0.279
300	–	0.0503	0.0762	0.1800	0.2430	0.439
1000	–	–	–	0.2480	0.3340	0.680
\multicolumn{7}{c}{T/K = 413.15}						
1	0.0016	0.0031	0.0040	0.0072	0.0088	0.012
3	0.0030	0.0059	0.0078	0.0146	0.0184	0.025
10	0.0059	0.0120	0.0163	0.0312	0.0393	0.056
30	0.0110	0.0228	0.0308	0.0590	0.0750	0.101
100	–	0.0424	0.0558	0.1075	0.1400	0.182
300	–	–	0.0935	0.1800	0.2325	0.312
1000	–	–	–	0.3120	0.4050	0.520

R_L^* moles H_2S/mole CO_2

$^§P_{H_2S}$ partial pressure of hydrogen sulfide

COMPONENTS:	ORIGINAL MEASUREMENTS:
1. Carbon dioxide; CO_2; [124-38-9]	Jones, J. H.; Froning, H. R.; Claytor, E. E. Jr.
2. Hydrogen sulfide; H_2S; [7783-06-4]	
3. 2-Aminoethanol (Monoethanolamine); $C_2H_7NO_2$; [141-43-5]	*J. Chem. Engng. Data* 1959, *4*, 85-92.
4. Water; H_2O; [7732-18-5]	

EXPERIMENTAL VALUES: Smoothed data for 15.3 wt% of MEA

$P^{\S}_{H_2S}$ Moles H_2S per Mole MEA

/mmHg	$R_V^* = 0.01$	$R_V = 0.05$	$R_V = 0.10$	$R_V = 0.50$	$R_V = 1.0$	$R_V = 10$	$R_V = \infty$
			$T/K = 313.15$				
1	0.0013	0.0035	0.0050	0.0120	0.0178	0.0500	0.128
3	0.0022	0.0057	0.0084	0.0208	0.0300	0.0825	0.212
10	0.0039	0.0100	0.0149	0.0380	0.0540	0.1450	0.374
30	0.0064	0.0166	0.0250	0.0630	0.0910	0.2400	0.579
100	0.0107	0.0279	0.0415	0.1050	0.1510	0.3900	0.802
300	0.0167	0.0430	0.0638	0.1550	0.2200	0.5500	0.931
1000	-	0.0625	0.0920	0.2170	0.3050	0.7300	1.00
			$T/K = 333.15$				
1	0.0019	0.0049	0.0070	0.0172	0.0239	0.0643	0.085
3	0.0029	0.0074	0.0108	0.0260	0.0363	0.0940	0.137
10	0.0044	0.0115	0.0172	0.0414	0.0565	0.1420	0.240
30	0.0066	0.0175	0.0260	0.0621	0.0850	0.2080	0.386
100	0.0102	0.0272	0.0405	0.0980	0.1360	0.3140	0.600
300	-	0.0410	0.0610	0.1480	0.2040	0.4320	0.790
1000	-	-	0.0940	0.2170	0.2900	0.5500	0.970
			$T/K = 373.15$				
1	0.0017	0.0034	0.0046	0.0095	0.0118	0.0224	0.029
3	0.0030	0.0061	0.0082	0.0163	0.0207	0.0390	0.050
10	0.0056	0.0114	0.0155	0.0301	0.0381	0.0720	0.091
30	0.0098	0.0200	0.0270	0.0525	0.0665	0.1260	0.160
100	0.0176	0.0360	0.0483	0.0945	0.1200	0.2250	0.270
300	-	0.0585	0.0780	0.1510	0.1910	0.3700	0.439
1000	-	-	-	0.2250	0.2880	0.5820	0.680
			$T/K = 413.15$				
1	0.0013	0.0024	0.0031	0.0058	0.0078	0.0115	0.012
3	0.0026	0.0050	0.0065	0.0122	0.0160	0.0245	0.025
10	0.0056	0.0107	0.0140	0.0265	0.0352	0.0520	0.056
30	0.0110	0.0210	0.0278	0.0535	0.0705	0.0980	0.101
100	-	0.0429	0.0573	0.1110	0.1380	0.1800	0.182
300	-	-	0.1010	0.1850	0.2250	0.3020	0.312
1000	-	-	-	0.3000	0.3630	0.5000	0.520

$^{\S}P_{H_2S}$ partial pressure of hydrogen sulfide

$^{\S}P_{CO_2}$ partial pressure of carbon dioxide

R_V^* P_{H_2S}/P_{CO_2}

COMPONENTS:	ORIGINAL MEASUREMENTS:

1. Carbon dioxide; CO_2; [124-38-9]
2. Hydrogen sulfide; H_2S; [7783-06-4]
3. 2-Aminoethanol,
    (Monoethanolamine); C_2H_7NO;
    [141-43-5]
4. Water; H_2O; [7732-18-5]
	Lee, J.I.; Otto, F.D.; Mather, A.E.
	
	J. Chem. Eng. Data, <u>1975</u>, *20*, 161-163.

VARIABLES:	PREPARED BY:
Temperature, pressure, composition	C.L. Young

EXPERIMENTAL VALUES: Solubility of H_2S-CO_2 Mixtures in 5.0 mol dm^{-3} Monoethanolamine

T/K	P_{H_2S}/kPa	P_{CO_2}/kPa	H_2S/MEA a	CO_2/MEA
313.15	622.8	3.1	0.962	0.014
	588.4	3.0	0.923	0.014
	757.0	20.5	0.887	0.0459
	883.9	46.2	0.801	0.0868
	961.8	92.4	0.758	0.134
	1064.5	137.9	0.729	0.164
	1150.7	179.3	0.716	0.194
	1254.8	246.1	0.680	0.236
	1423.7	470.2	0.631	0.302
	1611.3	710.1	0.584	0.366
	1634.0	1213.5	0.505	0.430
	1730.6	1729.9	0.438	0.522
	1489.3	3564.6	0.276	0.749
	1372.0	4226.4	0.215	0.799
	384.0	952.2	0.150	0.640
	418.5	1200.4	0.139	0.670
	457.1	1402.4	0.128	0.688
	589.5	1890.5	0.110	0.715
	456.4	5332.4	0.052	0.894
	180.6	189.6	0.259	0.431
	379.9	193.7	0.243	0.492
	451.6	337.2	0.220	0.535
	324.0	297.9	0.157	0.602
	157.2	568.1	0.077	0.680
	168.2	857.7	0.068	0.698

AUXILIARY INFORMATION

METHOD/APPARATUS/PROCEDURE:	SOURCE AND PURITY OF MATERIALS:
Recirculating vapor flow apparatus with Jerguson liquid level gauge and magnetic pump, nitrogen added to vapor to increase pressure to at least 200 kPa. Vapor analysed by gas chromatography. Partial pressure of carbon dioxide and hydrogen sulfide calculated from pressure and vapor pressure of monoethanolamine. Liquid samples passed into sulfuric acid. Carbon dioxide and hydrogen sulfide collected in buret and then analysed by gas chromatography.	1. Purity 99.7 mole per cent. 2. Purity 99.9 mole per cent. 3. Purity 99.5 mole per cent. 4. Distilled. Nitrogen used as carrier, purity 99.99 mole per cent.

	ESTIMATED ERROR:
	  $\delta T/K = \pm 0.5$; $\delta P/kPa = \pm 1\%$;
	  $\delta a = \pm 3-5\%$.

	REFERENCES:

COMPONENTS:	ORIGINAL MEASUREMENTS:
1. Carbon Dioxide; CO_2; [124-38-9] 2. Hydrogen sulfide; H_2S; [7783-06-4] 3. 2-Aminoethanol, (Monoethanolamine);C_2H_7NO; [141-43-5] 4. Water; H_2O; [7732-18-5]	Lee, J.I.; Otto, F.D.; Mather, A.E. *J. Chem. Eng. Data.* <u>1975</u>,*20*,161-163

EXPERIMENTAL VALUES: Solubility of H_2S-CO_2 Mixtures in 5.0 mol dm^{-3} Monoethanolamine

T/K	P_{H_2S}/kPa	P_{CO_2}/kPa	H_2S/MEA α	CO_2/MEA
313.15	184.1	1206.6	0.057	0.740
	243.4	2582.8	0.043	0.817
	290.3	3473.6	0.037	0.857
	261.3	4239.6	0.029	0.902
	216.5	5359.3	0.024	0.941
	9.9	437.1	0.0044	0.674
	21.9	638.4	0.0036	0.746
	25.2	890.1	0.0027	0.746
	153.1	27.2	0.348	0.294
	233.7	65.8	0.356	0.317
	278.5	97.9	0.341	0.332
	146.2	39.3	0.304	0.326
	157.9	43.8	0.290	0.334
	104.1	26.2	0.263	0.327
	69.6	19.1	0.272	0.339
	75.8	17.6	0.249	0.336
	65.1	14.2	0.325	0.245
	69.6	12.3	0.259	0.319
	167.0	15.2	0.354	0.283
	26.0	3.1	0.366	0.196
	24.8	0.35	0.556	0.053
	46.4	0.62	0.631	0.043
	69.3	0.74	0.693	0.037
	144.8	1.2	0.764	0.034
	152.4	1.3	0.710	0.040
	280.6	7.0	0.716	0.088
	516.4	56.2	0.653	0.167
	229.6	6.3	0.664	0.101
	433.7	27.9	0.687	0.138
	307.5	9.2	0.733	0.067
	313.7	29.3	0.599	0.141
	418.5	68.3	0.522	0.245
	617.1	215.8	0.473	0.322
	917.0	937.0	0.348	0.517
	2104.3	994.2	0.595	0.398
	2525.5	1399.6	0.439	0.528
	2054.6	3397.0	0.340	0.680
	2633.8	3176.4	0.416	0.601
	1390.7	64.1	0.900	0.090
	2755.8	89.6	1.015	0.080
	2922.7	1219.0	0.900	0.250
	70.3	1.9	0.595	0.111
	255.8	27.4	0.537	0.181
	314.4	56.9	0.507	0.283
	383.3	126.2	0.398	0.338
	5.8	0.9	0.312	0.199
	4.9	0.65	0.216	0.256
	8.4	0.76	0.203	0.302
	15.9	2.0	0.215	0.325
	26.8	5.0	0.213	0.343
	48.3	15.5	0.209	0.376
	58.7	21.9	0.214	0.410
	45.1	21.2	0.220	0.365
	3.3	2.6	0.073	0.419
	10.0	11.2	0.049	0.478
	1.1	0.13	0.121	0.239
	1.4	0.33	0.116	0.281
	2.9	0.6	0.122	0.322
	0.0	0.06	0.000	0.218
	6.1	1.5	0.097	0.388

COMPONENTS:	ORIGINAL MEASUREMENTS:
1. Carbon dioxide; CO_2; [124-38-9] 2. Hydrogen sulfide; H_2S; [7783-06-4] 3. 2-Aminoethanol, (Monoethanolamine); C_2H_7NO; [141-43-5] 4. Water; H_2O; [7732-18-5]	Lee, J.I.; Otto, F.D.; Mather, A.E. *J. Chem. Eng. Data.* <u>1975</u>,*20*, 161-163.

EXPERIMENTAL VALUES: Solubility of H_2S-CO_2 Mixtures in 5.0 mol dm^{-3} Monoethanolamine.

T/K	P_{H_2S}/kPa	P_{CO_2}/kPa	H_2S/MEA α	CO_2/MEA
373.15	31.1	89.3	0.048	0.408
	31.6	123.3	0.039	0.427
	22.1	103.2	0.030	0.422
	192.4	345.4	0.147	0.417
	251.7	421.3	0.152	0.433
	238.6	812.9	0.125	0.474
	211.0	1392.0	0.086	0.532
	241.3	2444.2	0.067	0.588
	241.3	5564.0	0.040	0.686
	271.0	32.5	0.375	0.194
	479.9	416.4	0.294	0.371
	618.4	1605.1	0.183	0.494
	613.6	3230.2	0.127	0.599
	553.6	5137.9	0.124	0.611
	527.4	68.9	0.472	0.173
	1114.9	1246.6	0.358	0.412
	1030.8	4433.3	0.158	0.587
	1438.2	75.5	0.734	0.079
	1477.5	332.3	0.623	0.187
	1565.1	1320.3	0.402	0.333
	1406.5	5360.6	0.227	0.517
	2277.3	187.5	0.753	0.091
	2118.0	1058.3	0.571	0.252
	2096.0	1954.6	0.483	0.352
	2020.1	3802.4	0.350	0.460
	3301.9	62.1	0.884	0.029
	3501.1	639.1	0.808	0.133
	3416.3	1978.8	0.544	0.330
	55.4	0.6	0.297	0.028
	48.7	1.1	0.289	0.035
	49.0	1.3	0.276	0.044
	66.3	4.4	0.274	0.118
	107.6	19.3	0.276	0.218
	8.0	7.7	0.064	0.242
	12.1	10.6	0.059	0.287
	40.3	70.9	0.063	0.398
	141.7	262.7	0.115	0.439
	trace	0.04	0.006	0.027
	20.3	5.0	0.127	0.180

α = Mole ratio in liquid phase.

COMPONENTS:	ORIGINAL MEASUREMENTS:

1. Carbon dioxide; CO_2; [124-38-9]
2. Hydrogen sulfide; H_2S;
   [7783-06-4]
3. Water; H_2O; [7732-18-5]
4. 2-Aminoethanol, (Monoethanolamine)
   C_2H_7NO; [141-43-5]
	Lawson, J.D.; Garst, A.W.
	
	J. Chem. Engng. Data, 1976, *21*, 20-30.

VARIABLES:
 	
   Temperature, pressure
	PREPARED BY:
	
	   C.L. Young

EXPERIMENTAL VALUES:

T/K	$P^{\dagger}_{H_2S}$/bar	$P^{\dagger}_{CO_2}$/bar	Conc. of MEA /wt%	Liquid comp. mol/mol amine H_2S	CO_2
298.15	0.0068	–	15.2	0.155	0.413
	0.193	0.031		0.303	0.398
	0.125	–		0.431	0.321
	1.592	0.469		0.533	0.505
313.15	0.000055	–		0.0059	0.174
	0.00015	–		0.0135	0.191
	0.053	–		0.367	0.233
	1.985	–		0.493	0.394
333.15	0.00023	–		0.0057	0.176
	0.0032	–		0.0064	0.386
	0.00071	–		0.0130	0.191
	1.212	1.985		0.244	0.595
373.15	0.011	0.289		0.0054	0.389
	0.00081	–		0.0057	0.167
	0.00028	0.020		0.0073	0.186
	0.00017	–		0.0131	0.194
	0.023	0.040		0.0411	0.233
	0.033	–		0.0757	0.082
393.15	0.037	0.016		0.0545	0.0675
	0.334	0.340		0.133	0.166

AUXILIARY INFORMATION

METHOD/APPARATUS/PROCEDURE:	SOURCE AND PURITY OF MATERIALS:
Rocking equilibrium cell fitted with liquid and vapor sampling valves. Pressure measured with Bourdon gauge. Cell charged with amine then gases and methane added as an inert gas to "achieve the desired total pressure". Vapor phases analysed by mass spectrometry. Liquid samples analysed by electrometric titration, details in source. Additional analytical methods were used for some samples.	1. Purity 99.99 mole per cent. 2. Purity 99.5 mole per cent. 3. Distilled. 4. Commercial sample purity better than 99 mole per cent as determined by acid titration.
	ESTIMATED ERROR: $\delta T/K = \pm0.15$ at 293K increasing to ±0.5 at 393K; $\delta P/bar = \pm0.5\%$ (liquid comp) = $\pm3\%$.
	REFERENCES:

COMPONENTS:	ORIGINAL MEASUREMENTS:
1. Carbon dioxide; CO_2; [124-38-9] 2. Hydrogen sulfide; H_2S; [7783-06-4] 3. Water; H_2O; [7732-18-5] 4. 2-Aminoethanol, (Monoethanolamine); C_2H_7NO; [141-43-5]	Lawson, J.D. and Garst, A.W. *J. Chem. Engng. Data*, <u>1976</u>, *21*, 20-30.

EXPERIMENTAL VALUES:

T/K	$P^+_{H_2S}$/bar	$P^+_{CO_2}$/bar	Conc. of MEA /wt%	Liquid comp. mol/mol amine	
				H_2S	CO_2
310.93	0.00005	–	30	0.0128	0.119
	0.086	–		0.197	0.331
	0.826	0.233		0.277	0.405
338.71	0.00017	–		0.0128	0.113
	0.213	–		0.205	0.324
	1.732	0.826		0.291	0.403
366.48	0.0043	–		0.0130	0.116
	0.626	–		0.196	0.304
	2.891	2.291		0.280	0.392

† partial pressure

COMPONENTS:	ORIGINAL MEASUREMENTS:
1. Hydrogen sulfide; H_2S; [7783-06-4] 2. Carbon dioxide; CO_2; [124-38-9] 3. Water; H_2O; [7732-18-5] 4. 2-Aminoethanol (*monoethanolamine*) C_2H_7NO ; [141-43-5]	Lee, J.I.; Otto, F.D.; Mather,A.E. *Can. J. Chem. Eng.* 1976, *54*, 214-219. (Original data deposited in the National Depository of Unpublished Data, Ottawa, Canada)*
VARIABLES: Temperature, pressure, composition	PREPARED BY: P.G.T. Fogg

EXPERIMENTAL VALUES:

Concentration of monoethanolamine (MEA) = 2.5 kmol m^{-3} (2.5 mol dm^{-3})

T/K	P_{H_2S}/kPa	P_{CO_2}/kPa	H$_2$S/MEA	CO$_2$/MEA
313.2	234	117	0.347	0.453
	740	234	0.689	0.339
	1372	4385	0.253	1.030
	276	126.9	0.392	0.428
	1241	2013	0.429	0.731
	760	4985	0.128	1.080
	14.1	176.5	0.018	0.731
	1193	373.3	0.795	0.315
	1207	1882	0.502	0.673
	712	5102	0.150	1.090
	636	5205	0.123	1.117
	256	2096	0.062	0.978
	122	241.3	0.140	0.655
	327	135.8	0.498	0.374
	97.9	92.87	0.233	0.507
	37.7	29.85	0.170	0.494
	4.18	4.25	0.061	0.473
	46.5	27.30	0.216	0.440
	131.7	18.27	0.539	0.216
	215.1	30.27	0.615	0.197
	4.30	0.067	0.436	0.053
	12.76	0.096	0.570	0.046
	59.85	0.539	0.687	0.036
	110.3	0.082	0.815	0.004
	97.0	141.0	0.191	0.552
	80.7	137.9	0.181	0.562
	191.7	115.8	0.349	0.420
	242.7	98.6	0.468	0.355
	1.81	0.070	0.380	0.045
	0.335	0.000	0.181	0.021
	8.20	0.000	0.550	0.043
	7.39	0.000	0.552	0.049
	45.9	0.296	0.715	0.039

(column header: Mole ratio in the liquid)

* National Depository of Unpublished Data, National Science Library, National Research Council, Ottawa, K1A OS2, Canada.

AUXILIARY INFORMATION

METHOD/APPARATUS/PROCEDURE	SOURCE AND PURITY OF MATERIALS
A windowed equilibrium cell was charged with the solution under test. CO_2 and H_2S were added to give appropriate pressures as measured by a bourdon gage. The vapor phase was circulated through the liquid by a magnetic pump. When equilibrium was reached samples of the liquid and gas phases were withdrawn and analysed as indicated in refs. (1) - (3).	3. distilled. 4. commercial sample; purity 99.95%
	ESTIMATED ERROR: $\delta T/K$ = ± 0.5 $\delta(CO_2/MEA)$; $\delta(H_2S/MEA)$ = ± 0.02 or ± 4%, whichever is the larger (authors)
	REFERENCES: 1. Lee, J.I.; Otto, F.D.; Mather, A.E. *J. Chem. Eng. Data* 1972, *17*, 465. 2. *id. ib.* 1973, *18*, 71. 3. *id. Can. J. Chem. Eng.* 1974, *52*, 125

COMPONENTS:	ORIGINAL MEASUREMENTS:
1. Hydrogen sulfide; H_2S; [7783-06-4]	Lee, J.I.; Otto, F.D.; Mather,A.E.
2. Carbon dioxide; CO_2; [124-38-9] 3. Water; H_2O; [7732-18-5] 4. 2-Aminoethanol (*monoethanolamine*) C_2H_7NO; [141-43-5]	*Can. J. Chem. Eng.* <u>1976</u>, *54*, 214-219. (Original data deposited in the National Depository of Unpublished Data, Ottawa, Canada)*

EXPERIMENTAL VALUES:

Concentration of monoethanolamine (MEA) = 2.5 kmol m^{-3} (2.5 mol dm^{-3})

T/K	P_{H_2S}/kPa	P_{CO_2}/kPa	Mole ratio in the liquid H_2S/MEA	CO_2/MEA
313.2	66.9	0.215	0.780	0.015
	56.6	0.096	0.780	0.011
	25.5	0.000	0.711	0.011
	276.1	145.1	0.486	0.349
	57.5	21.9	0.313	0.372
	54.6	7.45	0.443	0.255
	14.1	1.13	0.358	0.249
	7.65	0.548	0.317	0.248
	5.25	0.679	0.273	0.255
	4.11	1.12	0.161	0.361
	57.2	12.7	0.366	0.306
	0.333	0.000	0.147	0.112
	1.06	0.180	0.144	0.258
	3.47	1.81	0.135	0.384
	6.45	6.62	0.091	0.464
	3.79	17.0	0.038	0.549
	1.90	8.81	0.025	0.537
	0.000	16.1	0.000	0.569
	0.000	9.38	0.000	0.537
	0.000	5.14	0.000	0.522
	0.000	0.413	0.000	0.381
	0.000	0.379	0.000	0.388
	904.6	7.653	1.084	0.026
	213.0	0.916	0.863	0.012
	29.03	0.000	0.769	0.011
	4.16	0.000	0.538	0.011
	2.84	0.000	0.432	0.011
	0.261	0.000	0.162	0.005
	0.060	0.000	0.077	0.003
	16.3	0.413	0.545	0.121
	10.0	0.274	0.494	0.132
	12.3	0.792	0.429	0.193
	15.24	2.58	0.306	0.312
	4.41	0.528	0.212	0.323
	3.25	0.979	0.128	0.408
	2.07	1.83	0.090	0.437
	5.21	0.623	0.237	0.294
	3.45	0.548	0.226	0.284
	0.537	0.210	0.151	0.237
	0.723	0.137	0.146	0.238
	2024	1.24	1.366	0.001
	2035	1.24	1.351	0.001
	2565	1.24	1.493	0.001
	2551	0.785	1.441	0.001
	2413	0.000	1.482	0.000
	1620	0.000	1.320	0.000
	779.1	0.000	1.096	0.000
	758.4	0.000	1.119	0.000
	451.6	0.000	0.994	0.000
	444.7	0.000	0.994	0.000
	173.7	0.000	0.912	0.000
	172.4	0.000	0.930	0.000
	589.5	621.9	0.399	0.569
	568.8	1193	0.227	0.779
	452.3	5219	0.015	1.162
	404.0	191.0	0.440	0.409
	377.1	189.6	0.455	0.415
	643.3	912.2	0.314	0.658
	717.0	4830	0.095	1.095
	761.9	4702	0.115	1.096

COMPONENTS:	ORIGINAL MEASUREMENTS:
1. Hydrogen sulfide; H_2S; [7783-06-4]	Lee, J.I.; Otto, F.D.; Mather,A.E.
2. Carbon dioxide; CO_2; [124-38-9] 3. Water; H_2O; [7732-18-5] 4. 2-Aminoethanol (*monoethanolamine*) C_2H_7NO; [141-43-5]	*Can. J. Chem. Eng.* 1976, *54*, 214-219. (Original data deposited in the National Depository of Unpublished Data, Ottawa, Canada)*

EXPERIMENTAL VALUES:

Concentration of monoethanolamine (MEA) = 2.5 kmol m^{-3} (2.5 mol dm^{-3})

T/K	P_{H_2S}/kPa	P_{CO_2}/kPa	Mole ratio in the liquid H_2S/MEA	CO_2/MEA
313.2	1265	268.9	0.853	0.238
	1583	1520	0.621	0.595
	1578	1489	0.597	0.612
	1559	1453	0.550	0.626
	207.5	10.48	0.716	0.110
	208.9	34.47	0.594	0.214
	1460	164.8	1.003	0.131
	1433	1846	0.453	0.725
	113.8	221	0.110	0.660
	489.5	312	0.450	0.460
	66.2	228	0.061	0.694
	78.6	779	0.029	0.838
	972.1	804	0.564	0.504
	388.2	1868	0.119	0.905
	832.9	1887	0.285	0.833
	1470	4169	0.325	0.950
	1394	120.0	1.037	0.102
	1442	2306	0.494	0.721
	1222	4539	0.214	1.031
	2523	124	1.227	0.075
	2621	3285	0.604	0.761
	0.461	0.071	0.122	0.217
	0.551	0.048	0.120	0.220
	3.12	1.15	0.120	0.369
	31.7	64.8	0.086	0.569
	42.4	6.01	0.390	0.262
	65.4	13.6	0.404	0.280
	461.3	53.1	0.748	0.144
	814.3	77.91	0.876	0.117
	146.5	7.31	0.662	0.112
	132.6	2.22	0.741	0.046
	21.7	0.079	0.652	0.032
	3.72	0.000	0.477	0.003
	3.10	0.000	0.464	0.003
	0.620	0.000	0.269	0.002
	0.641	0.000	0.283	0.002
	1.44	0.000	0.382	0.002
	22.1	0.000	0.785	0.002
	12.0	0.000	0.700	0.002
	8.27	0.000	0.610	0.002
	4.96	0.000	0.573	0.002
	0.827	0.000	0.276	0.001
	0.758	0.000	0.299	0.001
	10.5	5.24	0.558	0.038
	66.6	9.17	0.496	0.241
	134.4	50.3	0.457	0.366
	102.7	7.03	0.579	0.178
	182.7	24.0	0.619	0.208
	81.4	7.93	0.482	0.208
	108.2	16.4	0.535	0.251
	38.3	5.58	0.350	0.328
	4.62	0.000	0.420	0.039
373.2	215.5	56.9	0.387	0.243
	277.2	319.2	0.309	0.401
	304.1	696.4	0.251	0.493
	304.1	1020	0.226	0.568
	437.8	2017	0.177	0.675
	472.3	5092	0.079	0.902

COMPONENTS:	ORIGINAL MEASUREMENTS:
1. Hydrogen sulfide; H_2S; [7783-06-4] 2. Carbon dioxide; CO_2; [124-38-9] 3. Water; H_2O; [7732-18-5] 4. 2-Aminoethanol (*monoethanolamine*) C_2H_7NO; [141-43-5]	Lee, J.I.; Otto, F.D.; Mather,A.E. *Can. J. Chem. Eng.* 1976, *54*, 214-219. (Original data deposited in the National Depository of Unpublished Data, Ottawa, Canada)*

EXPERIMENTAL VALUES:

Concentration of monoethanolamine (MEA) = 2.5 kmol m^{-3} (2.5 mol dm^{-3})

T/K	P_{H_2S}/kPa	P_{CO_2}/kPa	Mole ratio in the liquid H_2S/MEA	CO_2/MEA
373.2	185.5	126.9	0.330	0.332
	233.0	256.5	0.305	0.405
	226.1	197.2	0.331	0.389
	366.8	224.8	0.434	0.334
	402.6	422.6	0.364	0.430
	425.4	1136	0.323	0.537
	52.05	15.7	0.236	0.235
	108.9	101.4	0.224	0.348
	211.7	912.4	0.162	0.567
	244.8	1893	0.119	0.700
	599.8	1076	0.365	0.481
	671.5	2679	0.252	0.653
	556.4	5109	0.156	0.815
	1129	339.9	0.685	0.216
	1270	1150	0.589	0.380
	3779	54.3	1.432	0.009
	682.9	41.2	0.783	0.036
	1751	2166	0.345	0.591
	1160	4385	0.228	0.774
	12.9	243.4	0.013	0.550
	102.7	82.7	0.176	0.411
	175.1	186.8	0.294	0.365
	248.9	224.8	0.296	0.378
	283.4	795.6	0.220	0.510
	3603	18.55	1.425	0.004
	3157	157.2	1.160	0.032
	3098	134.4	1.117	0.027
	3414	660.5	1.083	0.097
	3027	1339	0.936	0.222
	130.3	7.10	0.560	0.078
	113.8	20.4	0.396	0.159
	244.8	33.2	0.539	0.129
	299.2	28.8	0.636	0.096
	418.5	167.5	0.505	0.227
	3.03	1877	0.000	0.732
	136.5	2849	0.043	0.777
	385.4	13.9	0.764	0.031
	1084	38.6	0.935	0.029
	1004	38.4	0.864	0.028
	376.4	0.496	0.688	0.002
	244.8	0.296	0.689	0.002
	580.5	0.861	0.846	0.002
	544.7	0.785	0.836	0.002
	1842	0.958	1.088	0.000
	2744	1.406	1.144	0.000
	410.2	0.172	0.734	0.000
	403.3	0.137	0.743	0.000
	355.8	0.206	0.738	0.002
	5.63	2963	0.030	0.759
	0.000	467.5	0.000	0.611
	0.000	1993	0.000	0.773
	0.000	1965	0.000	0.779
	0.000	5616	0.000	0.943
	0.000	5523	0.000	0.944
	33.4	5634	0.003	0.930
	295.1	13.8	0.678	0.023
	180.0	0.372	0.630	0.004
	616.4	1.30	0.848	0.004
	354.4	0.220	0.740	0.001

COMPONENTS:	ORIGINAL MEASUREMENTS:
1. Hydrogen sulfide; H_2S; [7783-06-4] 2. Carbon dioxide; CO_2; [124-38-9] 3. Water; H_2O; [7732-18-5] 4. 2-Aminoethanol (*monoethanolamine*) C_2H_7NO; [141-43-5]	Lee, J.I.; Otto, F.D.; Mather,A.E. *Can. J. Chem. Eng.* <u>1976</u>, *54*, 214-219. (Original data deposited in the National Depository of Unpublished Data, Ottawa, Canada)*

EXPERIMENTAL VALUES:

Concentration of monoethanolamine (MEA) = 2.5 kmol m^{-3} (2.5 mol dm^{-3})

The tables below, given in the paper, contain smoothed values of the mole ratio H_2S/MEA in the liquid phase for various partial pressures of H_2S in the gas phase and mole ratios CO_2/MEA in the liquid phase.

Mole ratio CO_2/MEA		0.000	0.100	0.200	0.300	0.400	0.500	0.600
T/K	P_{H_2S}/kPa							
313.2	0.1	0.102	0.072	0.050	0.026	0.010		
	0.316	0.202	0.150	0.111	0.073	0.043	0.013	
	1.00	0.333	0.247	0.187	0.133	0.082	0.038	0.010
	3.16	0.527	0.373	0.281	0.202	0.128	0.053	0.022
	10.0	0.730	0.510	0.388	0.283	0.182	0.097	0.047
	31.6	0.866	0.633	0.483	0.360	0.250	0.155	0.086
	100	0.966	0.752	0.578	0.453	0.342	0.243	0.161
	316	1.056	0.892	0.713	0.590	0.470	0.371	0.290
	1000	1.227	1.078	0.891	0.767	0.662	0.570	0.500
	3000	1.620	1.325	1.106	0.968	0.883	0.817	0.750

Mole ratio CO_2/MEA		0.700	0.800	0.900	1.000	1.100	1.200
T/K	P_{H_2S}/kPa						
313.2	10.0	0.021					
	31.6	0.041	0.019				
	100	0.090	0.054	0.032	0.018	0.008	
	316	0.200	0.137	0.103	0.072	0.050	0.031
	1000	0.410	0.315	0.243	0.187	0.150	0.110
	3000	0.683	0.600	0.510	0.420	0.322	0.250*

Mole ratio CO_2/MEA		0.000	0.100	0.200	0.300	0.400
T/K	P_{H_2S}/kPa					
373.2	0.1	0.028	0.017	0.010		
	0.316	0.041	0.023	0.011	0.004	
	1.0	0.065	0.042	0.023	0.015	0.003
	3.16	0.118	0.080	0.053	0.037	0.022
	10.0	0.227	0.162	0.117	0.078	0.051
	31.6	0.400	0.293	0.225	0.158	0.100
	100	0.615	0.455	0.360	0.274	0.197
	316	0.853	0.637	0.520	0.420	0.333
	1000	0.990	0.835	0.710	0.600	0.502
	3000	1.336	1.110	0.942	0.822	0.707
	6000	1.770	1.410	1.132	0.993*	0.870*

Mole ratio CO_2/MEA		0.500	0.600	0.700	0.800	0.900
T/K	P_{H_2S}/kPa					
373.2	10.0	0.027	0.010			
	31.6	0.066	0.037	0.020		
	100.0	0.142	0.097	0.060	0.037	0.017
	316	0.252	0.185	0.133	0.095	0.058
	1000	0.405	0.307	0.241	0.187	0.133
	3000	0.585	0.470	0.379	0.303	0.236
	6000	0.740*	0.600*	0.487*	0.392*	0.316*

* extrapolated values given by the authors.

COMPONENTS:	ORIGINAL MEASUREMENTS:
1. Hydrogen sulfide; H_2S; [7783-06-4] 2. Carbon dioxide; CO_2; [124-38-9] 3. Water; H_2O; [7732-18-5] 4. 2-Aminoethanol, (Monoethanolamine) C_2H_7NO; [141-43-5]	Nasir, P.; Mather, A.E. *Can. J. Chem. Eng.* 1977,*55,* 715-717.

VARIABLES:	PREPARED BY:
Pressure	C.L. Young

EXPERIMENTAL VALUES: T/K = 373.15

Conc. of monoethanolamine = 5.0 mol dm^{-3}

P_{H_2S}/kPa	P_{CO_2}/kPa	$\alpha^\dagger_{H_2S}$	$\alpha^\dagger_{CO_2}$
0.0131	0.0103	0.013	0.0405
0.0152	0.0200	0.0083	0.0475
0.0455	0.0607	0.010	0.072
0.179	0.00214	0.0264	0.009
0.324	0.0165	0.025	0.026
0.372	0.0262	0.0463	0.023
0.531	0.490	0.0133	0.096
0.586	0.669	0.013	0.119
0.669	0.0145	0.036	0.012
0.703	0.331	0.018	0.078
0.814	1.131	0.011	0.116
0.896	0.434	0.021	0.082
1.03	0.0648	0.032	0.028
1.10	0.00896	0.042	0.006
1.22	0.607	0.027	0.083
1.36	0.386	0.033	0.06
1.76	0.0244	0.053	0.012
2.10	1.21	0.022	0.14
3.03	0.0138	0.06	0.004
3.17	0.896	0.035	0.199
3.24	0.193	0.053	0.038
3.38	0.848	0.046	0.092
3.84	0.793	0.037	0.103
4.14	0.0276	0.084	0.004
4.20	0.227	0.066	0.037

AUXILIARY INFORMATION

METHOD/APPARATUS/PROCEDURE:	SOURCE AND PURITY OF MATERIALS:
Nitrogen passes through three 250 ml vessels in series which contained amine solution with a certain amount of dissolved hydrogen sulfide and carbon dioxide. Emerging gas from last vessel analysed by GC. Hydrogen sulfide content determined by iodine-thiosulfate titration and carbon dioxide by barium carbonate precipitation.	No details given.
	ESTIMATED ERROR: $\delta T/K = \pm 0.5$; $\delta\alpha = \pm 6\%$ (estimated by compiler).
	REFERENCES:

COMPONENTS:	ORIGINAL MEASUREMENTS:
1. Hydrogen sulfide; H_2S; [7783-06-4] 2. Carbon dioxide; CO_2; [124-38-9] 3. Water; H_2O; [7732-18-5] 4. 2-Aminoethanol, (Monoethanolamine); C_2H_7NO; [141-43-5]	Nasir, P.; Mather, A.E. *Can. J. Chem. Eng.* <u>1977</u>,*55*, 715-717.

EXPERIMENTAL VALUES:

P_{H_2S}/kPa	P_{CO_2}/kPa	$\alpha^{\dagger}_{H_2S}$	$\alpha^{\dagger}_{CO_2}$
4.41	0.262	0.063	0.042
4.41	0.0690	0.08	0.0124
4.50	0.103	0.0796	0.02
4.94	0.0910	0.081	0.0164
5.24	0.276	0.069	0.043
6.41	1.10	0.071	0.115
6.81	0.483	0.071	0.051
7.17	1.17	0.069	0.116
7.28	0.476	0.083	0.057
7.65	0.0413	0.119	0.006
8.89	0.0565	0.128	0.006
9.31	2.96	0.064	0.175

† α_i = Mole of species i/mole of monoethanolamine.

COMPONENTS:	ORIGINAL MEASUREMENTS:
1. Hydrogen sulfide; H_2S; [7783-06-4] 2. Carbon dioxide; CO_2; [124-38-9] 3. Water; H_2O; [7732-18-5] 4. 2-Aminoethanol (MEA), (Mono-ethanolamine); C_2H_7NO; [141-43-5]	Isaacs, E. E.; Otto, F. D.; Mather, A. E. *J. Chem. Eng. Data* <u>1980</u>, *25*, 118-120.

VARIABLES:	PREPARED BY:
Pressure	C. L. Young

EXPERIMENTAL VALUES:

T/K	Conc. MEA /kmol m^{-3}	P/kPa		$\alpha^{\dagger}$	
		H_2S	CO_2	H_2S/MEA	CO_2/MEA
373.15	2.5	trace	0.886	trace	0.149
		trace	0.0337	0.0067	0.0600
		0.0353	0.00714	0.0181	0.0182
		0.0590	0.0198	0.0187	0.0357
		0.0842	0.0297	0.0205	0.0310
		0.0601	0.0140	0.0213	0.0217
		0.0242	0.0408	0.0219	0.0347
		0.138	0.0490	0.0231	0.0312
		0.149	0.0453	0.0232	0.0295
		0.120	0.0331	0.0238	0.0261
		0.121	0.0242	0.0239	0.0197
		0.102	0.0214	0.0247	0.0227
		0.140	0.0163	0.0248	0.0210
		0.211	0.0511	0.0263	0.0297
		0.499	0.565	0.0265	0.108
		0.237	0.0834	0.0269	0.0467
		0.138	0.0270	0.0272	0.0213
		0.133	0.0222	0.0277	0.0211
		0.131	0.0067	0.0325	0.0100
		0.314	0.0568	0.0326	0.0310
		0.583	0.244	0.0338	0.0791
		0.432	0.155	0.0340	0.0492
		0.544	0.246	0.0365	0.0718

(cont.)

AUXILIARY INFORMATION

METHOD/APPARATUS/PROCEDURE:	SOURCE AND PURITY OF MATERIALS:
Nitrogen passed through three 250 ml vessels in series which contained amine solution with a certain amount of dissolved carbon dioxide and hydrogen sulfide. Emerging gas from last vessel analysed by GC. Liquid sample taken from last vessel. Hydrogen sulfide content determined by iodine-thiosulfate titration and carbon dioxide by barium carbonate precipitation.	No details given.

	ESTIMATED ERROR:
	$\delta T/K = \pm 0.5$; $\delta\alpha = \pm 6\%$ (estimated by compiler).

	REFERENCES:

COMPONENTS:	ORIGINAL MEASUREMENTS:
1. Hydrogen sulfide; H_2S; [7783-06-4]	Isaacs, E. E.; Otto, F. D.; Mather, A. E.
2. Carbon dioxide; CO_2; [124-38-9]	*J. Chem. Eng. Data*
3. Water; H_2O; [7732-18-5]	1980, *25*, 118-120.
4. 2-Aminoethanol (MEA), (Mono-ethanolamine); C_2H_7NO; [141-43-5]	

EXPERIMENTAL VALUES:

T/K	Conc. MEA /kmol m^{-3}	P/kPa		$\alpha^{\dagger}$	
		H_2S	CO_2	H_2S/MEA	CO_2/MEA
373.15	2.5	0.329	0.0714	0.0369	0.0375
		0.424	0.0906	0.0382	0.0357
		0.260	0.0780	0.0388	0.0406
		0.476	0.0962	0.0400	0.0400
		0.548	0.140	0.0400	0.0492
		0.620	0.137	0.0409	0.0485
		0.458	0.0545	0.0418	0.0254
		0.561	0.126	0.0420	0.0462
		0.778	0.135	0.0425	0.0457
		0.391	0.0048	0.0459	0.0085
		0.823	0.220	0.0504	0.0593
		0.752	0.211	0.0507	0.0610
		0.949	0.242	0.0518	0.0711
		0.928	0.206	0.0536	0.0606
		0.851	0.204	0.0537	0.0516
		1.28	0.438	0.0555	0.0851
		0.932	0.248	0.0567	0.0545
		1.18	0.239	0.0570	0.0686
		0.966	0.116	0.0612	0.0332
		1.84	0.523	0.0654	0.0941
		1.24	0.132	0.0735	0.0333
		2.56	0.838	0.0748	0.115
		1.21	0.0027	0.0755	0.0062
		1.63	0.164	0.0846	0.0327
		2.10	0.195	0.0872	0.0401
		2.03	0.169	0.0924	0.0363
		3.92	1.36	0.0954	0.141
		2.81	0.355	0.122	0.0407
		3.36	0.0800	0.136	0.0062

† Mole ratio

COMPONENTS:	ORIGINAL MEASUREMENTS:
1. Hydrogen sulfide; H_2S; [7783-06-4] 2. Carbon dioxide; CO_2; [124-38-9] 3. Water; H_2O; [7732-18-5] 4. Ethanol, 2,2'-iminobis-, (*diethanolamine*); $C_4H_{11}NO_2$; [111-42-2]	Leibush, A.G.; Shneerson, A.L. *Zhur. Prik. Khim.* 1950, *23*, 145-152. *J. Applied Chem. USSR* 1950, *23*, 149-157.
VARIABLES:	PREPARED BY:
Temperature, pressure, composition of liquid phase	P.G.T. Fogg

EXPERIMENTAL VALUES:

Concentration of diethanolamine (DEA) = 2.0 mol dm^{-3}; T/K = 298.2

P_{H_2S}/mmHg	P_{CO_2}/mmHg	Mole ratios in liquid phase	
		H_2S/DEA	CO_2/DEA
1.72	0.00	0.126	0.016
5.13	1.17	0.126	0.198
18.7	12.4	0.128	0.342
23.6	25.0	0.123	0.406
0.77	–	0.027	0.186
6.0	0.8	0.149	0.186
23.0	2.6	0.302	0.186
67.5	5.4	0.425	0.189
92	12.0	0.517	0.195
141	18.5	0.556	0.203
1.3	4.5	0.017	0.363
6.1	6.8	0.067	0.363
37	14.4	0.206	0.355
61	24.6	0.277	0.363
145	56	0.404	0.324
240	80	0.491	0.270

760 mmHg = 1 atm = 1.01325 bar

AUXILIARY INFORMATION

METHOD/APPARATUS/PROCEDURE:	SOURCE AND PURITY OF MATERIALS:
N_2 or a mixture of N_2 & H_2S was passed successively through two absorbers containing H_2S & CO_2 dissolved in an aqueous solution of DEA. The H_2S in the emerging gas was passed into cadmium or zinc acetate solution and the precipitated sulfides estimated by iodimetry. Hydrogen sulfide in the DEA solutions was also determined by iodimetry. CO_2 in the gas phase was absorbed in standard baryta solution after removal of H_2S by acidified potassium permanganate solution. CO_2 in the liquid phase was estimated by reaction with 30% H_2SO_4, removal of evolved H_2S by permanganate and absorption of evolved CO_2 in standard baryta solution.	1. From H_2SO_4 & Na_2S: no SO_2 detected 4. Contained 0.2 to 0.6 volumes of CO_2 per unit volume of solution; less than 1% by weight of total impurities.
	ESTIMATED ERROR: $\delta T/K = \pm 0.1$ (authors)
	REFERENCES:

COMPONENTS:	ORIGINAL MEASUREMENTS:

1. Carbon dioxide; CO_2; [124-38-9]
2. Hydrogen sulfide; H_2S; [7783-06-4]
3. 2,2'-Iminobisethanol,
   (Diethanolamine); $C_4H_{11}NO_2$;
   [111-42-2]
4. Water; H_2O; [7732-18-5]
	Lee, J.I.; Otto F.D.; Mather, A.E.
	
	Can. J. Chem. Engng. <u>1974</u>, *52*, 125-7
	
	(Complete data in the Centre for
	Unpublished Data, National Science
	NRC, Ottawa, Ontario, K1A 0S2,
	Canada)

VARIABLES: Temperature, pressure, concentration	PREPARED BY: C.L. Young

EXPERIMENTAL VALUES: Solubility of H_2S-CO_2 Mixtures in 2.0 mol dm^{-3} DEA

T/K	P_{H_2S}/kPa	P_{CO_2}/kPa	α H_2S/DEA	α CO_2/DEA
323.15	1505.80	144.51	1.006	0.106
	1081.08	2800.62	0.447	0.693
	732.90	843.22	0.570	0.433
	703.94	2149.76	0.381	0.719
	682.57	404.02	0.703	0.308
	675.68	5135.17	0.188	0.994
	632.24	274.40	0.784	0.239
	629.48	33.57	0.989	0.048
	600.52	167.54	0.826	0.170
	508.13	67.56	0.881	0.094
	488.83	3433.56	0.173	0.977
	415.75	5343.39	0.125	1.140
	399.89	1292.75	0.294	0.704
	359.90	29.16	0.851	0.063
	340.59	71.01	0.785	0.125
	299.91	1974.64	0.176	0.902
	298.54	152.37	0.604	0.285
	289.23	18.13	0.831	0.053
	274.40	384.72	0.515	0.447
	242.69	92.38	0.634	0.250
	194.43	277.16	0.361	0.477
	178.57	1951.20	0.108	0.962
	158.57	1067.98	0.154	0.856
	151.68	658.44	0.189	0.742
	142.72	880.45	0.133	0.812

AUXILIARY INFORMATION

METHOD/APPARATUS/PROCEDURE:	SOURCE AND PURITY OF MATERIALS:
Recirculating vapor flow apparatus with Jerguson liquid level gauge cell and magnetic pump. Nitrogen added to vapor to increase pressure to at least 200 kPa. Vapor analysed by gas chromatography. Partial pressure of carbon dioxide and hydrogen sulfide calculated from pressure and vapor pressure of diethanolamine. Liquid samples passed into sulfuric acid. Carbon dioxide and hydrogen sulfide collected in buret and then analysed by gas chromatography.	1. Liquid Air sample, purity 99.9 mole per cent. 2. Matheson sample, purity 99.5 mole per cent. 3. Purity 99.8 mole per cent. 4. Distilled. Nitrogen used as carrier, purity 99.99 mole per cent.
	ESTIMATED ERROR: $\delta T/K = \pm 0.5$; $\delta P/bar = \pm 1\%$; $\delta\alpha = \pm 3$-5%.
	REFERENCES:

COMPONENTS:	ORIGINAL MEASUREMENTS:

1. Carbon dioxide; CO_2; [124-38-9]
2. Hydrogen sulfide; H_2S; [7783-06-4]
3. 2,2'-Iminobisethanol,
   (Diethanolamine); $C_4H_{11}NO_2$;
   [111-42-2]
4. Water; H_2O; [7732-18-5]
	Lee, J.I.; Otto, F.D.; Mather,
	A.E.
	
	Can. J. Chem. Engng. 1974, *52*,
	125-7.

EXPERIMENTAL VALUES: Solubility of H_2S-CO_2 Mixtures in 2.0 mol dm^{-3} DEA

T/K	P_{H_2S}/kPa	P_{CO_2}/kPa	α H_2S/DEA	α CO_2/DEA
323.15	129.62	5764.65	0.041	1.221
	93.07	7.21	0.653	0.074
	91.35	17.92	0.557	0.146
	88.25	144.78	0.256	0.521
	80.80	71.15	0.385	0.352
	80.66	422.64	0.129	0.742
	79.63	82.94	0.332	0.424
	72.39	33.23	0.434	0.277
	65.70	47.02	0.380	0.345
	49.15	5.23	0.512	0.120
	47.47	5.51	0.525	0.094
	39.71	26.40	0.331	0.328
	38.19	1.47	0.564	0.040
	36.81	152.02	0.129	0.628
	36.26	16.82	0.367	0.238
	34.54	59.36	0.213	0.475
	24.75	423.33	0.045	0.785
	21.71	14.89	0.236	0.323
	14.40	704.63	0.012	0.865
	12.68	11.30	0.197	0.318
	10.79	0.85	0.382	0.078
	10.13	9.43	0.199	0.273
	10.06	0.89	0.370	0.067
	7.65	451.46	0.024	0.815
	7.37	27.09	0.092	0.490
	6.61	0.55	0.304	0.079
	6.41	109.62	0.029	0.662
	4.51	1.39	0.208	0.158
	4.13	9.37	0.092	0.400
	3.85	1.12	0.210	0.155
	3.06	3.47	0.128	0.296
	1.57	0.30	0.178	0.086
	0.88	0.21	0.117	0.076
373.15	1654.72	141.34	1.072	0.033
	1378.94	1678.85	0.713	0.336
	1027.31	1130.73	0.665	0.317
	985.94	277.85	0.820	0.106
	983.87	2772.35	0.504	0.516
	916.99	482.62	0.726	0.176
	889.41	2488.98	0.475	0.516
	872.17	4709.08	0.360	0.700
	848.04	290.95	0.749	0.115
	686.71	4863.52	0.300	0.800
	618.45	98.59	0.695	0.071
	537.78	4907.64	0.220	0.820
	479.87	444.01	0.525	0.250
	417.12	1534.07	0.327	0.545
	399.89	916.99	0.362	0.443
	395.06	1011.45	0.356	0.439
	348.18	3151.56	0.216	0.733
	344.73	37.57	0.634	0.039
	339.21	162.71	0.554	0.146
	317.84	539.85	0.405	0.356
	293.71	5129.65	0.126	0.885
	283.37	760.48	0.318	0.412
	278.54	382.65	0.406	0.317
	237.17	2009.11	0.094	0.782
	217.87	375.07	0.328	0.328
	202.70	1500.28	0.188	0.628

COMPONENTS:	ORIGINAL MEASUREMENTS:

1. Carbon dioxide; CO_2; [124-38-9]
2. Hydrogen sulfide; H_2S; [7783-06-4]
3. 2,2'-Iminobisethanol,
   (Diethanolamine); $C_4H_{11}NO_2$;
   [111-42-2]
4. Water; H_2O; [7732-18-5]
	Lee, J.I.; Otto F.D.; Mather,
	A.E.
	
	Can. J. Chem. Engng. 1974, *52*,
	
	125-7.

EXPERIMENTAL VALUES: Solubility of H_2S-CO_2Mixtures in 2.0 mol dm^{-3} DEA

T/K	P_{H_2S}/kPa	P_{CO_2}/kPa	α H_2S/DEA	α CO_2/DEA
373.15	181.33	48.67	0.451	0.088
	179.26	3160.53	0.107	0.788
	177.88	119.96	0.424	0.170
	170.29	35.16	0.429	0.090
	159.95	5112.42	0.060	0.940
	153.75	577.08	0.226	0.461
	135.82	1571.99	0.123	0.643
	128.24	235.79	0.273	0.321
	115.14	5543.33	0.046	0.897
	113.76	777.72	0.143	0.526
	105.48	2491.05	0.058	0.782
	97.21	502.62	0.143	0.473
	81.08	35.43	0.287	0.098
	79.97	5453.70	0.029	0.962
	75.84	990.76	0.111	0.585
	63.15	65.98	0.233	0.192
	62.05	390.92	0.114	0.462
	56.74	39.23	0.215	0.127
	48.53	186.84	0.140	0.364
	35.71	17.99	0.200	0.092
	27.57	7.51	0.199	0.050
	20.68	5650.20	0.003	1.000
	19.30	9.03	0.151	0.067
	18.27	1545.10	0.015	0.682
	14.75	22.40	0.102	0.152
	13.51	529.51	0.022	0.565
	11.44	170.98	0.038	0.419
	10.13	14.96	0.091	0.117
	9.65	57.70	0.060	0.269
	4.66	5.87	0.049	0.079
	3.26	1.29	0.044	0.017
	0.56	13.61	0.010	0.148

Solubility of H_2S-CO_2 Mixtures in 3.5 mol dm^{-3} DEA

T/K	P_{H_2S}/kPa	P_{CO_2}/kPa	α H_2S/DEA	α CO_2/DEA
323.13	1509.93	1606.46	0.652	0.448
	1450.64	59.01	1.028	0.026
	1071.43	1134.17	0.545	0.465
	1026.62	455.05	0.706	0.264
	952.84	4578.08	0.284	0.824
	625.34	1114.87	0.396	0.550
	413.68	279.92	0.525	0.327
	392.30	59.22	0.727	0.110
	338.52	97.21	0.632	0.199
	307.50	21.02	0.757	0.064
	291.64	153.06	0.511	0.309
	281.30	3069.52	0.095	0.925
	279.23	1141.07	0.157	0.728
	259.93	5228.25	0.060	1.000
	259.24	175.12	0.435	0.357
	244.07	70.32	0.592	0.205
	233.73	62.46	0.486	0.281
	225.45	229.59	0.358	0.419
	187.53	521.92	0.102	0.704
	141.34	293.02	0.200	0.563
	129.48	16.54	0.554	0.151

COMPONENTS:	ORIGINAL MEASUREMENTS:

1. Carbon dioxide; CO_2; [124-38-9]
2. Hydrogen sulfide; H_2S; [7783-06-4]
3. 2,2'-Iminobisethanol,
   (Diethanolamine); $C_4H_{11}NO_2$;
   [111-42-2]
4. Water; H_2O; [7732-18-5]
	Lee, J.I.; Otto, F.D.; Mather, A.E.
	
	Can. J. Chem. Engng. <u>1974</u>,*52*,125-7

EXPERIMENTAL VALUES: Solubility of H_2S-CO_2 Mixtures in 3.5 mol dm^{-3} DEA

T/K	P_{H_2S}/kPa	P_{CO_2}/kPa	α H_2S/DEA	α CO_2/DEA
323.15	125.13	14.06	0.56	0.130
	119.96	1434.78	0.070	0.815
	111.14	35.43	0.451	0.238
	95.14	477.11	0.096	0.684
	89.28	5.13	0.585	0.074
	84.80	79.28	0.364	0.346
	73.77	1159.68	0.053	0.799
	68.60	258.20	0.107	0.602
	63.43	33.23	0.308	0.302
	60.81	87.01	0.200	0.454
	50.81	55.70	0.196	0.425
	45.50	430.22	0.057	0.693
	39.71	12.54	0.317	0.253
	36.33	18.82	0.227	0.338
	33.78	438.50	0.048	0.708
	30.68	46.53	0.163	0.446
	25.99	32.68	0.171	0.392
	21.71	20.33	0.150	0.385
	17.99	3.92	0.310	0.140
	17.09	128.37	0.053	0.595
	16.96	74.53	0.071	0.520
	14.75	22.47	0.125	0.400
	12.27	5.90	0.191	0.259
	9.37	1.34	0.283	0.083
	8.54	31.85	0.066	0.455
	7.79	8.61	0.106	0.317
	6.09	6.83	0.127	0.272
	5.97	11.03	0.072	0.368
	5.03	6.37	0.074	0.354
	4.70	1.51	0.151	0.179
	4.55	0.82	0.159	0.142
	4.51	60.74	0.021	0.541
	2.39	3.48	0.078	0.270
	0.90	0.60	0.085	0.148
373.15	1622.32	1566.47	0.604	0.243
	1585.78	250.96	0.801	0.059
	1416.17	1868.46	0.590	0.293
	1380.31	415.75	0.728	0.083
	1316.88	1533.38	0.566	0.279
	1275.51	361.28	0.716	0.106
	1130.73	4498.79	0.314	0.537
	1064.54	4591.87	0.296	0.596
	1061.78	4525.68	0.356	0.506
	993.52	287.50	0.689	0.099
	754.96	1840.88	0.350	0.405
	703.25	4919.36	0.216	0.594
	670.16	48.95	0.703	0.027
	665.33	923.88	0.410	0.318
	501.93	102.04	0.591	0.071
	495.03	5024.16	0.153	0.681
	485.38	683.26	0.410	0.304
	474.35	1235.53	0.322	0.398
	466.77	340.59	0.480	0.181
	458.49	1875.35	0.243	0.487
	383.69	94.11	0.515	0.086
	375.07	230.97	0.433	0.173

COMPONENTS:	ORIGINAL MEASUREMENTS:

1. Carbon dioxide; CO_2; [124-38-9]
2. Hydrogen sulfide; H_2S; [7783-06-4]
3. 2,2'-Iminobisethanol,
   (Diethanolamine); $C_4H_{11}NO_2$;
   [111-42-2]
4. Water; H_2O; [7732-18-5]
	Lee, J.I.; Otto, F.D.;Mather, A.E.
	
	Can. J. Chem. Engng. 1974, *52*,125-7
	(Complete data in the Centre for
	Unpublished Data, National Science
	NRC, Ottawa, Ontario, K1A 0S2,
	Canada)

EXPERIMENTAL VALUES: Solubility of H_2S-CO_2 Mixtures in 3.5 mol dm^{-3} DEA

T/K	P_{H_2S}/kPa	P_{CO_2}/kPa	α H_2S/DEA	α CO_2/DEA
373.15	337.84	5388.20	0.067	0.755
	320.60	654.99	0.278	0.332
	307.50	6.17	0.545	0.005
	285.44	82.04	0.445	0.091
	258.55	2571.72	0.127	0.626
	255.10	2369.01	0.132	0.639
	222.00	4607.03	0.082	0.725
	203.39	723.94	0.204	0.416
	199.94	575.70	0.221	0.374
	195.12	891.48	0.160	0.472
	131.68	5.53	0.386	0.016
	117.89	151.68	0.227	0.238
	111.55	1.65	0.342	0.006
	108.93	48.67	0.278	0.124
	96.52	49.09	0.228	0.129
	85.90	81.15	0.190	0.192
	66.25	11.58	0.239	0.045
	61.43	31.23	0.210	0.118
	59.29	1625.77	0.043	0.602
	56.19	619.14	0.072	0.470
	52.88	919.75	0.056	0.550
	49.64	181.33	0.103	0.325
	43.78	266.13	0.078	0.387
	37.71	32.12	0.148	0.136
	30.61	193.74	0.070	0.364
	24.20	0.82	0.142	0.006
	15.03	53.77	0.072	0.226
	11.85	1.44	0.102	0.007
	9.23	9.28	0.065	0.088
	6.41	3.61	0.066	0.040

α = Mole ratio in liquid phase.

COMPONENTS:	ORIGINAL MEASUREMENTS:

1. Carbon dioxide; CO_2; [124-38-9]
 2. Hydrogen sulfide; H_2S;
    [7783-06-4]
 3. 2,2'-iminobisethanol;
    (Diethanolamine); $C_4H_{11}NO_2$;
    [111-42-2]
 4. Water; H_2O; [7732-18-5]
	Lee, J.I.; Otto, F.D.; Mather, A.E.
	
	 Can.J.Chem.Eng., <u>1974</u>, *52*, 125 - 127.
	 (Complete data in the Centre for
	 Unpublished Data, National Science
	 Library, National Research Council,
	 Ottawa, Ontario K1A 0S2, Canada)

EXPERIMENTAL VALUES:

The tables give smoothed values of the mole ratio H_2S/DEA in the liquid phase for various partial pressures of H_2S in the gas phase and mole ratios CO_2/DEA in the liquid phase.

Concentration of diethanolamine (DEA) = 2.0 mol dm^{-3}

Mole ratio CO_2/DEA

T/K	P_{H_2S} /psia	P_{H_2S} /kPa**	0.000	0.100	0.200	0.300	0.400	0.500	0.600
323.2	0.0100	0.0689	0.028	0.021*	0.011*	0.0054*	0.0021*	0.0009*	0.00042*
	0.0316	0.218	0.068	0.050*	0.028*	0.016*	0.0073*	0.0034*	0.0016*
	0.100	0.689	0.136	0.097*	0.062*	0.036*	0.021*	0.011*	0.0056*
	0.316	2.18	0.230	0.172	0.120	0.079	0.052	0.033*	0.019*
	1.00	6.89	0.382	0.290	0.225	0.153	0.120	0.080	0.050
	3.16	21.8	0.562	0.455	0.353	0.266	0.208	0.151	0.107
	10.0	68.9	0.730	0.612	0.501	0.406	0.327	0.254	0.195
	31.6	218	0.875	0.760	0.654	0.555	0.464	0.378	0.310
	100	689	1.002	0.906	0.815	0.707	0.625	0.530	0.450
	200	1379	1.215	1.005	0.910	0.808	0.720	0.632	0.546

Mole ratio CO_2/DEA

T/K	P_{H_2S} /psia	P_{H_2S} /kPa**	0.700	0.800	0.900	1.000	1.100	1.200
323.2	0.0100	0.0689	0.00021*					
	0.0316	0.218	0.00088*					
	0.100	0.689	0.00275*	0.0013*				
	0.316	2.18	0.0090 *	0.0044*	0.0019*			
	1.00	6.89	0.028	0.015	0.0070*			
	3.16	21.8	0.065	0.040	0.023			
	10.0	68.9	0.145	0.096	0.061	0.041	0.029	0.021
	31.6	218	0.246	0.189	0.136	0.100	0.073	0.056
	100	689	0.380	0.315	0.250	0.198	0.154	0.118*
	200	1379	0.473	0.410	0.340*	0.280*	0.223*	0.179*

Mole ratio CO_2/DEA

T/K	P_{H_2S} /psia	P_{H_2S} /kPa**	0.000	0.100	0.200	0.300	0.400	0.500	0.600
373.2	0.0100	0.0689	0.0026*	0.0017*	0.0011*				
	0.0316	0.218	0.0084*	0.0067*	0.0037*	0.023*	0.0014*	0.0008*	
	0.100	0.689	0.023	0.0165*	0.0108*	0.0069*	0.0044*	0.0025*	0.0014*
	0.316	2.18	0.055	0.038*	0.027*	0.0176*	0.0117*	0.0068*	0.0040*
	1.00	6.89	0.120	0.082	0.060	0.042	0.029	0.018	0.011
	3.16	21.8	0.208	0.158	0.122	0.092	0.068	0.047	0.029
	10.0	68.9	0.348	0.286	0.237	0.197	0.158	0.122	0.088
	31.6	218	0.535	0.471	0.412	0.358	0.300	0.250	0.200
	100	689	0.808	0.726	0.650	0.579	0.507	0.443	0.380
	316	2180	1.200	1.040	0.950	0.870	0.792	0.725	0.650

* extrapolated values given by the authors.

** calculated by the compiler.

COMPONENTS:	ORIGINAL MEASUREMENTS:

1. Carbon dioxide; CO_2; [124-38-9]
2. Hydrogen sulfide; H_2S;
   [7783-06-4]
3. 2,2'-Iminobisethanol;
   (Diethanolamine); $C_4H_{11}NO_2$;
   [111-42-2]
4. Water; H_2O; [7732-18-5]
	Lee, J.I.; Otto, F.D.; Mather, A.E.
	
	Can.J.Chem.Eng., <u>1974</u>, *52*, 125 - 127.
	(Complete data in the Centre for
	Unpublished Data, National Science
	Library, National Research Council,
	Ottawa, Ontario K1A 0S2, Canada)

EXPERIMENTAL VALUES:
The tables give smoothed values of the mole ratio H_2S/DEA in the liquid phase for various partial pressures of H_2S in the gas phase and mole ratios CO_2/DEA in the liquid phase.

Concentration of diethanolamine (DEA) = 2.0 mol dm^{-3}

Mole ratio CO_2/DEA	0.700	0.800	0.900	1.000	1.100	1.200

T/K	P_{H_2S}/psia	P_{H_2S}/kPa**					
373.2	0.100	0.0689	0.0007*				
	0.316	2.18	0.0021*	0.0011*			
	1.00	6.89	0.0060	0.0034			
	3.16	21.8	0.017	0.011	0.006	0.003*	
	10.0	68.9	0.057	0.039	0.028	0.017*	0.0108*	0.0065*
	31.6	218	0.150	0.115	0.086	0.060*	0.044*	0.032*
	100	689	0.322	0.270	0.220	0.171*	0.139*	0.109*
	316	2180	0.573	0.511	0.450*	0.383*	0.323*	0.276*

Concentration of diethanolamine (DEA) = 3.5 mol dm^{-3}

Mole ratio CO_2/DEA	0.000	0.100	0.200	0.300	0.400	0.500	0.600

T/K	P_{H_2S}/psia	P_{H_2S}/kPa**						
323.2	0.0100	0.0689	0.012*	0.0072*	0.0040*	0.0021*		
	0.0316	0.218	0.040*	0.023	0.0135*	0.0078*	0.0044*	0.0024*	0.00102*
	0.100	0.689	0.087	0.058	0.036	0.024	0.013*	0.0073*	0.0037*
	0.316	2.18	0.160	0.123	0.088	0.054	0.035	0.0190	0.0105
	1.00	6.89	0.278	0.218	0.164	0.110	0.073	0.045	0.029
	3.16	21.8	0.422	0.348	0.272	0.205	0.145	0.095	0.060
	10.0	68.9	0.520	0.510	0.420	0.328	0.245	0.170	0.120
	31.6	218	0.782	0.673	0.572	0.472	0.376	0.288	0.215
	100	689	0.920	0.825	0.726	0.632	0.532	0.434	0.345
	316	2180	1.144	1.007	0.907	0.804	0.710	0.605	0.520

Mole ratio CO_2/DEA	0.700	0.800	0.900	1.000	1.100	1.200

T/K	P_{H_2S}/psia	P_{H_2S}/kPa**					
323.2	0.0316	0.218	0.0006*	0.00034*				
	0.100	0.689	0.0019*	0.0010*				
	0.316	2.18	0.0056*	0.0030*	0.00145*			
	1.00	6.89	0.0155	0.0086*	0.0041*	0.0021*		
	3.16	21.8	0.038	0.023	0.0120	0.0065*	0.0035*	
	10.0	68.9	0.080	0.055	0.0335	0.0195	0.0120*	0.0070*
	31.6	218	0.150	0.110	0.076	0.050	0.0340*	0.0230*
	100	689	0.273	0.210	0.155	0.120*	0.090*	0.065*
	316	2180	0.440	0.360	0.296*	0.244*	0.200*	0.160*

** * extrapolated values given by the authors.
   calculated by the compiler.

COMPONENTS:	ORIGINAL MEASUREMENTS:
1. Carbon dioxide; CO_2; [124-38-9]	Lee, J.I.; Otto, F.D.; Mather, A.E.
2. Hydrogen sulfide; H_2S; [7783-06-4]	$Can.J.Chem.Eng.$, 1974, 52, 125 - 127.
3. 2,2'-Iminobisethanol; (Diethanolamine); $C_4H_{11}NO_2$; [111-42-2]	(Complete data in the Centre for Unpublished Data, National Science Library, National Research Council,
4. Water; H_2O; [7732-18-5]	Ottawa, Ontario K1A 0S2, Canada)

EXPERIMENTAL VALUES:
The tables give smoothed values of the mole ratio H_2S/DEA in the liquid phase for various partial pressures of H_2S in the gas phase and mole ratios CO_2/DEA in the liquid phase.

Concentration of diethanolamine (DEA) = 3.5 mol dm^{-3}

Mole ratio CO_2/DEA	0.000	0.100	0.200	0.300	0.400	0.500	0.600

T/K	P_{H_2S} /psia	P_{H_2S} /kPa**						
373.2	0.0100	0.0689	0.0019*	0.0011*	0.0007*			
	0.0316	0.218	0.0062*	0.0039*	0.0024*	0.0014*	0.0007*	0.0003*
	0.100	0.689	0.020	0.0110*	0.0070*	0.0040*	0.0022*	0.0010*
	0.316	2.18	0.038	0.0275*	0.0185	0.0113*	0.0066*	0.0034* 0.0016*
	1.00	6.89	0.078	0.060	0.045	0.032	0.019*	0.010* 0.0052*
	3.16	21.8	0.155	0.120	0.092	0.070	0.049	0.034 0.017
	10.0	68.9	0.282	0.230	0.185	0.144	0.109	0.081 0.053
	31.6	218	0.475	0.393	0.324	0.262	0.212	0.161 0.119
	100	689	0.712	0.605	0.518	0.434	0.363	0.297 0.226
	316	2180	0.942	0.815	0.711	0.630	0.547	0.470 0.388

Mole ratio CO_2/DEA	0.700	0.800	0.900	1.000	1.100	1.200

T/K	P_{H_2S} /psia	P_{H_2S} /kPa**					
373.3	1.00	6.89	0.0025*	0.0010*			
	3.16	21.8	0.0093*	0.0037*			
	10.0	68.9	0.031	0.013	0.0080*	0.0040*	0.0022* 0.0011*
	31.6	218	0.082	0.046	0.0290*	0.0185*	0.0118* 0.0066*
	100	689	0.165	0.116	0.082*	0.060*	0.043* 0.030*
	316	2180	0.313*	0.248*	0.195*	0.150*	0.122* 0.096*

** * extrapolated values given by the authors.
calculated by the compiler.

COMPONENTS:	ORIGINAL MEASUREMENTS:

1. Carbon dioxide; CO_2; [124-38-9]
2. Hydrogen sulfide; H_2S; [7783-06-4]
3. Water; H_2O; [7732-18-5]
4. 2,2'-Iminobisethanol,
   (Diethanolamine); $C_4H_{11}NO_2$;
   [111-42-2]
	Lawson, J.D. and Garst, A.W.
	
	J. Chem. Engng. Data, 1976, *21*,
	20-30.

VARIABLES:	PREPARED BY:
Temperature, pressure	C.L. Young

EXPERIMENTAL VALUES:

T/K	$P^+_{H_2S}$/bar	$P^+_{CO_2}$/bar	Conc.of DEA /wt%	Liquid comp mol/mol amine H_2S	CO_2
310.93	0.00013	–	25	0.0042	0.0676
	0.00017	–		0.0045	0.0990
	0.00025	–		0.0074	0.130
	0.00027	–		0.0077	0.081
	0.00073	–		0.0095	0.212
	0.00052	–		0.0125	0.113
	0.0013	–		0.0122	0.230
	0.0071	–		0.058	0.221
	0.0055	–		0.059	0.214
	0.0033	–		0.062	0.109
	1.29	3.24		0.106	0.734
	3.56	1.72		0.111	0.715
	0.084	0.136		0.120	0.478
	0.031	–		0.124	0.211
	0.021	0.023		0.124	0.227
	0.013	–		0.127	0.118
	0.217	0.541		0.152	0.605
	3.33	12.66		0.220	0.930
	0.092	0.033		0.237	0.310
	0.373	0.493		0.239	0.527
	0.036	–		0.251	0.160
	2.31	3.18		0.384	0.630
	0.115	–		0.442	0.112
	0.222	0.077		0.442	0.216
	0.885	0.613		0.441	0.410

AUXILIARY INFORMATION

METHOD/APPARATUS/PROCEDURE:	SOURCE AND PURITY OF MATERIALS:
Rocking equilibrium cell fitted with liquid and vapour sampling valves. Pressure measured with Bourdon gauge. Cell charged with amine then gases and methane added as an inert gas to achieve the desired total pressure. Vapour phases analysed by mass spectrometry. Liquid samples analysed by electrometric titration, details in source. Additional analytical methods were used for some samples.	1. Purity 99.99 mole per cent. 2. Purity 99.5 mole per cent. 3. Distilled. 4. Commercial sample purity better than 99 mole per cent as determined by acid titration.

ESTIMATED ERROR:

$\delta T/K = \pm 0.15$ at 310 K increasing to ± 0.6 at 422 K; δP/bar $= \pm 0.5\%$; $\delta x_{CO_2} = \pm 3\%$.

REFERENCES:

COMPONENTS:	ORIGINAL MEASUREMENTS:

1. Carbon dioxide; CO_2; [124-38-9]
2. Hydrogen sulfide; H_2S; [7783-06-4]
3. Water; H_2O; [7732-18-5]
4. 2,2'-Iminobisethanol,
   (Diethanolamine); $C_4H_{11}NO_2$;
   [111-42-2]
	Lawson, J.D. and Garst, A.W.
	
	J.Chem.Engng. Data. <u>1976</u>, *21*,
	20-30.

EXPERIMENTAL VALUES:

T/K	$P^+_{H_2S}$/bar	$P^+_{CO_2}$/bar	Conc. of DEA /wt%	Liquid comp mol/mol amine H_2S	CO_2
310.93	3.80	4.20	25	0.523	0.600
	2.12	1.08		0.585	0.381
	0.813	0.220		0.607	0.281
	1.41	0.200		0.631	0.258
	0.440	0.0346		0.635	0.114
	12.18	4.010		0.861	0.264
	19.98	15.32		0.966	0.688
	8.53	1.35		1.020	0.119
324.82	3.40	14.12		0.250	0.890
	11.32	5.20		0.997	0.277
	9.99	1.93		1.050	0.161
338.71	0.00065	–		0.0040	0.065
	0.00087	–		0.0041	0.080
	0.0021	–		0.0097	0.109
	0.0015	–		0.0099	0.062
	0.0012	–		0.0126	0.107
	0.0043	–		0.0128	0.238
	0.0045	–		0.0291	0.074
	0.0051	–		0.0294	0.127
	0.0160	0.040		0.0512	0.233
	0.0065	–		0.0594	0.115
	0.480	3.930		0.105	0.654
	0.193	0.600		0.113	0.457
	0.164	0.500		0.114	0.495
	0.063	0.067		0.118	0.215
	0.043	0.0147		0.120	0.108
	0.413	1.57		0.147	0.596
	0.600	1.03		0.209	0.452
	1.57	6.05		0.214	0.666
	0.285	0.157		0.252	0.285
	3.33	18.65		0.256	0.842
	0.143	0.0031		0.257	0.129
	2.398	4.796		0.375	0.567
	1.359	1.33		0.413	0.410
	0.586	0.244		0.437	0.207
	0.428	0.060		0.443	0.116
	3.797	5.063		0.510	0.506
	1.665	0.746		0.526	0.252
	2.145	1.226		0.576	0.298
	0.973	0.115		0.624	0.114
	1.372	0.426		0.636	0.218
	8.660	4.450		0.855	0.156
	21.98	21.18		0.924	0.547
	11.78	2.585		0.990	0.118
	13.59	5.462		0.976	0.272
	11.99	2.651		1.010	0.140
352.59	3.064	1.865		0.255	0.770
	14.00	6.928		0.920	0.238
	12.66	2.731		0.920	0.139
366.48	0.0081	0.229		0.0113	0.214
	0.0061	–		0.0125	0.109
	0.0035	–		0.0314	0.054
	0.0095	0.040		0.0327	0.085
	0.0091	–		0.0334	0.063
	0.053	0.085		0.0710	0.122

COMPONENTS:	ORIGINAL MEASUREMENTS:

1.Carbon dioxide; CO_2; [124-38-9]
 2.Hydrogen sulfide; H_2S; [7783-06-4]
 3.Water; H_2O; [7732-18-5]
 4.2,2'-Iminobisethanol,
   (Diethanolamine); $C_4H_{11}NO_2$;
   [111-42-2]
	Lawson, J.D. and Garst, A.W.
	
	 J. Chem. Engng. Data, 1976, 21,
	 20-30.

EXPERIMENTAL VALUES:

T/K	$P^+_{H_2S}$/bar	$P^+_{CO_2}$/bar	Conc. of DEA /wt%	Liquid comp mol/mol amine	
				H_2S	CO_2
366.48	0.593	5.262	25	0.107	0.543
	0.293	1.172		0.114	0.353
	0.135	0.156		0.129	0.123
	0.560	3.784		0.129	0.533
	0.746	3.664		0.156	0.474
	0.608	1.625		0.183	0.354
	0.382	0.520		0.196	0.245
	0.306	0.203		0.196	0.117
	2.931	17.99		0.255	0.710
	2.664	1.972		0.496	0.207
	21.85	21.98		0.881	0.395
	13.86	3.504		0.915	0.123
	15.05	7.594		0.935	0.230
380.37	2.000	19.32		0.172	0.730
	14.25	4.276		0.870	0.127
	16.39	8.393		0.922	0.210
	0.0085	0.906		0.0056	0.185
394.26	0.0135	–		0.0135	0.107
	0.310	4.130		0.0391	0.349
	0.228	4.503		0.0472	0.378
	0.373	6.328		0.0532	0.448
	0.280	1.223		0.0627	0.197
	0.246	1.892		0.0736	0.238
	0.866	19.58		0.0874	0.650
	0.441	1.199		0.124	0.282
	0.786	3.144		0.126	0.273
	0.693	0.999		0.157	0.132
	0.759	0.426		0.176	0.070
	4.54	4.25		0.488	0.218
	14.65	4.66		0.825	0.117
	16.52	9.59		0.840	0.210
	20.12	22.91		0.822	0.405
310.93	0.00016	–	50	0.0026	0.111
	0.693	0.600		0.260	0.438
338.71	0.0011	–		0.0030	0.114
366.48	0.0035	0.085		0.0028	0.116
	2.398	5.595		0.233	0.390

+ partial pressure calculated by compiler using 1 bar = 750.062 mmHg

COMPONENTS:	ORIGINAL MEASUREMENTS:
1. Hydrogen sulfide; H_2S; [7783-06-4]	Lal, D.; Otto, F.D.; Mather, A.E.
2. Carbon dioxide; CO_2; [124-38-9]	Can.J.Chem.Eng. 1985, 63, 681 - 685.
3. Water; H_2O; [7732-18-5]	
4. Ethanol, 2,2'-iminobis- (diethanolamine); $C_4H_{11}NO_2$; [111-42-2]	

VARIABLES:	PREPARED BY:
Temperature, pressure	P.G.T. Fogg

EXPERIMENTAL VALUES:

Concentration of diethanolamine = 2.0 kmol m^{-3} (2.0 mol dm^{-3})

T/K	P_{H_2S}/kPa	P_{CO_2}/kPa	Mole ratio in liquid, H_2S/DEA	Mole ratio in liquid, CO_2/DEA
313.2	0.060	0.165	0.0182	0.173
	0.069	0.068	0.0352	0.103
	0.127	0.057	0.053	0.083
	0.329	0.224	0.055	0.178
	0.431	0.250	0.061	0.182
	0.187	0.070	0.064	0.093
	0.188	0.075	0.073	0.101
	0.410	0.070	0.083	0.101
	0.573	0.133	0.090	0.130
	0.784	0.470	0.090	0.226
	0.625	0.252	0.091	0.180
	0.782	0.369	0.092	0.207
	0.387	0.033	0.101	0.058
	0.524	0.057	0.105	0.077
	1.057	0.270	0.119	0.165

AUXILIARY INFORMATION

METHOD/APPARATUS/PROCEDURE:	SOURCE AND PURITY OF MATERIALS:
Apparatus described in ref. (1) was used. Nitrogen was passed through three 500 cm^3 stainless steel vessels in series. These contained amine solution, dissolved H_2S and CO_2. Temperatures were controlled to ± 0.5 K by an oil-bath. The gas emerging from the last vessel was analysed by gas chromatography. Samples of liquid from this vessel were analysed for H_2S by iodine-thiosulfate titration and for CO_2 by precipitation as barium carbonate and then titration with hydrochloric acid.	4. supplied by Dow chemicals.

	ESTIMATED ERROR:
	$\delta T/K$ = ± 0.5 (authors) $\delta P/P$ = ± 0.15 at P_{H_2S} > 0.05 kPa; larger at lower partial pressures (authors).

REFERENCES:

1. Isaacs, E.E.; Otto, F.D.; Mather, A.E. J.Chem.Eng.Data 1980, 25, 118.

COMPONENTS:	ORIGINAL MEASUREMENTS:
1. Hydrogen sulfide; H_2S; [7783-06-4]	Lal, D.; Otto, F.D.; Mather, A.E.
2. Carbon dioxide; CO_2; [124-38-9]	$Can.J.Chem.Eng.$ 1985, 63, 681 - 685.
3. Water; H_2O; [7732-18-5]	
4. Ethanol, 2,2'-iminobis-($diethanolamine$); $C_4H_{11}NO_2$; [111-42-2]	

EXPERIMENTAL VALUES:

Concentration of diethanolamine = 2.0 kmol m^{-3} (2.0 mol dm^{-3})

T/K	P_{H_2S}/kPa	P_{CO_2}/kPa	Mole ratio in liquid, H_2S/DEA	Mole ratio in liquid, CO_2/DEA
313.2	0.633	0.099	0.120	0.095
	0.700	0.048	0.131	0.062
	0.881	0.090	0.134	0.093
	1.452	0.292	0.141	0.072
	0.930	0.062	0.153	0.170
	0.897	0.114	0.154	0.107
	1.001	0.041	0.155	0.045
	1.039	0.215	0.172	0.146
	1.390	0.283	0.172	0.148
	1.190	0.254	0.174	0.149
	1.108	0.116	0.177	0.090
	1.344	0.063	0.185	0.068
	1.243	0.068	0.186	0.067
	2.231	0.450	0.191	0.175
	1.405	0.365	0.198	0.153
	1.004	0.141	0.200	0.087
	3.392	0.549	0.203	0.182
	1.347	0.120	0.208	0.100
	2.286	0.525	0.209	0.188
	3.185	0.332	0.247	0.141
373.2	0.015	0.749	0.0064	0.042
	0.022	4.077	0.0064	0.099
	0.078	3.266	0.0095	0.098
	0.132	4.571	0.0100	0.123
	0.197	1.018	0.0142	0.041
	0.263	1.472	0.0155	0.062
	0.324	1.423	0.0168	0.059
	0.467	1.918	0.020	0.064
	0.651	4.015	0.022	0.104
	0.787	5.015	0.025	0.105
	0.304	0.0057	0.025	0.0048
	0.762	1.998	0.026	0.060
	0.680	4.668	0.027	0.128
	0.685	1.294	0.030	0.050
	1.272	6.026	0.032	0.137
	0.614	4.548	0.033	0.090
	0.541	4.970	0.036	0.100
	1.319	3.409	0.037	0.091
	0.614	0.426	0.037	0.024
	1.460	3.739	0.041	0.087

COMPONENTS:	ORIGINAL MEASUREMENTS:
1. Hydrogen sulfide; H_2S; [7783-06-4] 2. Carbon dioxide; CO_2; [124-38-9] 3. Water; H_2O; [7732-18-5] 4. Ethanol, 2,2'-iminobis-(*diethanolamine*); $C_4H_{11}NO_2$; [111-42-2]	Lal, D.; Otto, F.D.; Mather, A.E. *Can.J.Chem.Eng.* 1985, *63*, 681 - 685.

EXPERIMENTAL VALUES:

Concentration of diethanolamine = 2.0 kmol m^{-3} (2.0 mol dm^{-3})

T/K	P_{H_2S}/kPa	P_{CO_2}/kPa	Mole ratio in liquid, H_2S/DEA	Mole ratio in liquid, CO_2/DEA
373.2	1.265	1.638	0.043	0.051
	2.101	6.469	0.045	0.106
	1.189	0.752	0.045	0.032
	1.607	1.838	0.047	0.060
	1.914	3.140	0.047	0.065
	1.859	3.568	0.047	0.077
	2.213	4.647	0.047	0.118
	2.295	5.253	0.048	0.115
	2.489	0.393	0.049	0.019
	1.567	1.305	0.050	0.050
	1.740	1.919	0.050	0.051
	1.942	3.526	0.051	0.083
	2.177	1.699	0.056	0.040
	1.623	0.701	0.059	0.027
	3.153	3.547	0.060	0.075
	2.292	4.376	0.062	0.100
	3.203	4.519	0.065	0.085
	2.116	0.312	0.067	0.013
	3.216	4.873	0.069	0.103
	1.940	0.145	0.069	0.0082
	3.035	2.667	0.072	0.072
	2.273	1.843	0.072	0.053
	3.221	2.405	0.074	0.064
	3.598	2.007	0.083	0.055
	3.000	2.874	0.084	0.070
	3.649	1.568	0.084	0.035
	3.646	2.692	0.086	0.064
	3.091	0.243	0.088	0.0076
	3.555	0.105	0.093	0.0048
	3.492	0.121	0.094	0.0077
	4.264	2.048	0.094	0.053
	3.208	0.738	0.096	0.024
	4.793	1.731	0.097	0.041
	4.102	2.093	0.103	0.039
	3.675	0.087	0.103	0.0042
	5.732	5.234	0.103	0.082
	3.868	0.180	0.106	0.0080
	4.832	1.039	0.117	0.031
	5.067	0.834	0.119	0.018
	4.726	0.318	0.124	0.0092

COMPONENTS:	ORIGINAL MEASUREMENTS:
1. Hydrogen sulfide; H_2S; [7783-06-4] 2. Carbon dioxide; CO_2; [124-38-9] 3. Water; H_2O; [7732-18-5] 4. 2-(2-Aminoethoxy)ethanol, (*diglycolamine*); $C_4H_{11}NO_2$; [929-06-6]	Dingman, J.C.; Jackson, J.L.; Moore, T.F.; Branson, J.A. *Proc. 62nd Annual Convention of the Gas Processors Association*, <u>1983</u>, 256-268.
VARIABLES:	PREPARED BY:
Temperature, composition of liquid and gas phases.	P.G.T. Fogg

EXPERIMENTAL VALUES:

Concentration of diglycolamine (DGA) in aqueous solution before addition of gas = 65 wt %

T/°F	T/K*	Mole ratio in liquid phase CO_2/DGA	H_2S/DGA	Partial pressure of CO_2 /psia	/kPa*	Partial pressure of H_2S /psia	/kPa*
100	310.9	0.0026	0	0.0000290	0.000200	0.0	0
		0	0.0027	0.0	0	0.0000480	0.000331
		0	0.0029	0.0	0	0.000133	0.000917
		0	0.0039	0.0	0	0.0000580	0.000400
		0	0.0050	0.0	0	0.000232	0.00160
		0	0.0104	0.0	0	0.00116	0.00800
		0	0.0114	0.0	0	0.000812	0.00560
		0.0140	0.0014	0.00029	0.00200	0.000062	0.000427
		0.0145	0.0030	0.00029	0.00200	0.000422	0.00291
		0	0.0205	0.0	0	0.00296	0.0204
		0.0213	0.0022	0.000319	0.00220	0.000319	0.00220
		0.0271	0.0012	-	-	0.0000430	0.000296
		0.0306	0.0010	0.000841	0.00580	0.0000810	0.000558
		0	0.0335	0.0	0	0.00723	0.0498
		0	0.0412	0.0	0	0.00952	0.0656
		0.0440	0.0020	0.0000870	0.000600	0.000319	0.00220
		0.0474	0.0008	0.000232	0.00160	0.0000640	0.000441
		0.0503	0.0031	0.000232	0.00160	0.000841	0.00580
		0.0537	0	0.000232	0.00160	0.0	0
		0.0547	0.0023	-	-	0.000435	0.00300
		0.0624	0	0.000174	0.00120	0.0	0
		0.0643	0.0010	0.000290	0.00200	0.000133	0.000917
		0.0630	0.0054	0.000232	0.00160	0.00263	0.0181
		0.0610	0.0112	0.00104	0.00717	0.00422	0.0291
		0	0.0820	0.0	0	0.0191	0.132
		0.0926	0	0.000890	0.00614	0.0	0

* calculated by the compiler.

AUXILIARY INFORMATION

METHOD/APPARATUS/PROCEDURE:	SOURCE AND PURITY OF MATERIALS:
Aqueous solution of DGA (600 cm^3) was placed in a stainless steel equilibrium vessel of capacity 1700 cm^3. The space above the was flushed with methane and H_2S and/or CO_2 passed into the vessel to appropriate partial pressures. The total pressure was maintained at 1500 mmHg by addition of methane. The vessel was rocked in a constant temperature bath for 1 - 2 h. The vapor phase was analysed by gas chromatography. CO_2 in the liquid phase was determined by acidification and measurement of the evolved gas. The H_2S in this phase was determined by titration with standard copper(II) nitrate.	No information.
	ESTIMATED ERROR:
	REFERENCES:

COMPONENTS:	ORIGINAL MEASUREMENTS:
1. Hydrogen sulfide; H_2S; [7783-06-4] 2. Carbon dioxide; CO_2; [124-38-9] 3. Water; H_2O; [7732-18-5] 4. 2-(2-Aminoethoxy)ethanol, (*diglycolamine*); $C_4H_{11}NO_2$; [929-06-6]	Dingman, J.C.; Jackson, J.L.; Moore, T.F.; Branson, J.A. *Proc. 62nd Annual Convention of the Gas Processors Association*, <u>1983</u>, 256-268.

EXPERIMENTAL VALUES:

Concentration of diglycolamine (DGA) in aqueous solution before addition of gas = 65 wt %

T/°F	T/K*	Mole ratio in liquid phase		Partial pressure of CO_2		Partial pressure of H_2S	
		CO_2/DGA	H_2S/DGA	/psia	/kPa*	/psia	/kPa*
100	310.9	0.0935	0.0016	0.000957	0.00660	0.000725	0.00500
		0.0960	0.0009	0.000580	0.00400	0.000120	0.000827
		0.0922	0.0107	0.000754	0.00520	0.00638	0.0440
		0.1030	0.0	0.000638	0.00440	0.0	0
		0.1000	0.0030	0.00551	0.0380	0.00197	0.0136
		0.1110	0.0004	0.000638	0.00440	0.0000410	0.000283
		0.1150	0.0021	–	–	0.00102	0.00703
		0.0953	0.0235	0.000890	0.00614	0.0119	0.0820
		0.1170	0.0062	0.00174	0.0120	–	–
		0	0.1250	0.0	0	0.0638	0.440
		0.1250	0.0053	0.000609	0.00420	0.00395	0.0272
		0.1350	0.0022	0.00113	0.00779	0.00193	0.0133
		0.1540	0.0011	0.00116	0.00780	0.000870	0.00600
		0.1600	0.0004	0.00122	0.00841	0.0000850	0.000586
		0.0470	0.1170	–	–	0.105	0.724
		0.1650	0.0012	0.00145	0.0100	0.000580	0.00400
		0.1680	0.0026	0.00197	0.0136	0.00317	0.0219
		0	0.1760	0.0	0	0.0544	0.375
		0.0920	0.0880	–	–	0.0302	0.208
		0	0.1940	0.0	0	0.142	0.979
		0	0.2190	0.0	0	0.208	1.43
		0.0900	0.1320	–	–	0.115	0.793
		0.2250	0.0023	0.00308	0.0212	0.00354	0.0244
		0.2280	0.0009	0.00337	0.0232	0.000841	0.00580
		0	0.2300	0.0	0	0.186	1.28
		0.0926	0.1380	0.00180	0.0124	0.185	1.28
		0.2340	0.0004	0.00352	0.0243	0.000240	0.00165
		0.1900	0.0470	–	–	0.0841	0.580
		0.2500	0.0004	0.00575	0.0396	0.000422	0.00291
		0.2660	0.0012	0.00275	0.0190	0.00183	0.0126
		0.0460	0.2220	–	–	0.319	2.20
		0.1890	0.0800	–	–	0.116	0.800
		0.2590	0.0102	0.00516	0.0356	0.0148	0.102
		0.1900	0.0800	–	–	0.128	0.883
		0.2700	0.0020	0.00642	0.0443	0.00379	0.0261
		0.2790	0.0050	0.00279	0.0192	0.00809	0.0558
		0.0930	0.1920	–	–	0.242	1.67
		0	0.2920	0.0	0	0.244	1.68
		0.2920	0	0.0108	0.0745	0.0	0
		0.2910	0.0011	0.00747	0.0515	0.00296	0.0204
		0	0.3000	0.0	0	0.263	1.81
		0.1870	0.1210	0.00967	0.0667	0.171	1.18
		0.3120	0.0009	0.0109	0.0752	0.00255	0.0176
		0.2820	0.0422	0.0151	0.104	0.0468	0.320
		0.0920	0.2380	–	–	0.389	2.68
		0	0.3390	0.0	0	0.422	2.91
		0.0920	0.2550	–	–	0.545	3.76
		0.1840	0.1680	0.0116	0.0800	0.267	1.84
		0.3590	0	0.0511	0.352	0.0	0
		0.3620	0.0003	0.0209	0.144	0.00139	0.00958
		0.2750	0.0882	0.0213	0.147	0.217	1.50
		0.0930	0.2980	–	–	0.696	4.80
		0.0480	0.3450	–	–	0.584	4.03

* calculated by the compiler.

COMPONENTS:	ORIGINAL MEASUREMENTS:
1. Hydrogen sulfide; H_2S; [7783-06-4] 2. Carbon dioxide; CO_2; [124-38-9] 3. Water; H_2O; [7732-18-5] 4. 2-(2-Aminoethoxy)ethanol, (*diglycolamine*); $C_4H_{11}NO_2$; [929-06-6]	Dingman, J.C.; Jackson, J.L.; Moore, T.F.; Branson, J.A. *Proc. 62nd Annual Convention of the Gas Processors Association*, <u>1983</u>, 256-268.

EXPERIMENTAL VALUES:

Concentration of diglycolamine (DGA) in aqueous solution before addition of gas = 65 wt %

T/°F	T/K*	Mole ratio in liquid phase		Partial pressure of CO_2		Partial pressure of H_2S	
		CO_2/DGA	H_2S/DGA	/psia	/kPa*	/psia	/kPa*
100	310.9	0.1740	0.2240	0.0139	0.0958	0.503	3.47
		0.4030	0	0.0607	0.419	0.0	0
		0.2670	0.1360	0.0344	0.237	0.557	3.84
		0.1660	0.2370	-	-	0.754	5.20
		0.3940	0.0109	0.0464	0.320	0.0503	0.347
		0.3880	0.0180	0.0749	0.516	-	-
		0.3900	0.0340	0.0812	0.560	0.166	1.14
		0.0930	0.3370	-	-	0.917	6.32
		0.0830	0.3470	-	-	1.00	6.89
		0	0.4340	0.0	0	0.911	6.28
		0.3730	0.0610	0.125	0.862	0.406	2.80
		0.2450	0.1910	0.0574	0.396	-	-
		0.2720	0.1680	0.0841	0.580	0.737	5.08
		0.0873	0.3580	0.00604	0.0416	1.06	7.31
		0.4500	0.0064	0.348	2.40	0.0656	0.452
		0.4580	0	0.286	1.97	0.0	0
		0.1700	0.2910	-	-	1.31	9.93
		0.2670	0.2000	0.0580	0.340	1.22	8.41
		0.4450	0.0280	0.596	4.11	0.445	3.07
		0.0900	0.3900	-	-	1.08	7.45
		0.3760	0.1180	0.412	2.84	1.64	11.3
		0.5020	0	1.83	12.6	0.0	0
		0.5060	0	1.77	12.2	0.0	0
		0.4990	0.0080	2.03	14.0	0.232	1.60
		0.3670	0.1420	0.758	5.23	2.98	20.5
		0.4850	0.0270	3.75	25.9	1.19	8.20
		0.5000	0.0160	2.36	16.3	0.449	3.10
		0.2640	0.2580	0.145	1.00	2.51	17.3
		0.4310	0.0917	11.4	78.6	8.27	57.0
		0.4600	0.0690	7.95	54.8	4.76	32.8
		0.0890	0.4400	-	-	1.80	12.4
		0.3610	0.1780	2.03	14.0	6.81	47.0
		0.3520	0.1960	4.14	28.5	11.4	78.6
		0.1680	0.3820	-	-	3.09	21.3
		0	0.5520	0.0	0	2.24	15.4
		0.1630	0.3920	0.0841	0.580	3.31	22.8
		0.2490	0.3260	0.828	5.71	9.28	64.0
		0.1680	0.4160	0.0812	0.560	4.80	33.1
		0.0880	0.5030	-	-	3.21	22.1
		0.3480	0.2420	6.73	46.4	17.6	121.3
		0.5900	0	27.0	186.2	0.0	0
		0	0.5920	0.0	0	2.75	19.0
		0.2510	0.3570	1.44	9.93	14.3	98.6
		0.1620	0.4500	0.131	0.903	6.79	46.8
		0.2510	0.3680	2.07	14.3	18.1	124.8
		0.1540	0.5080	0.717	4.94	19.0	131.0
		0.0880	0.5770	-	-	6.42	44.3
		0.0865	0.5830	0.0429	0.296	6.67	46.0
		0	0.6830	0.0	0	5.51	38.0
		0.1800	0.5270	0.841	5.80	21.6	148.9
		0	0.7020	0.0	0	5.80	40.0
		0.0900	0.6130	-	-	8.76	60.4
		0.1610	0.5500	1.31	9.03	27.1	186.8

* calculated by the compiler.

COMPONENTS:	ORIGINAL MEASUREMENTS:
1. Hydrogen sulfide; H_2S; [7783-06-4] 2. Carbon dioxide; CO_2; [124-38-9] 3. Water; H_2O; [7732-18-5] 4. 2-(2-Aminoethoxy)ethanol, (*diglycolamine*); $C_4H_{11}NO_2$; [929-06-6]	Dingman, J.C.; Jackson, J.L.; Moore, T.F.; Branson, J.A. *Proc. 62nd Annual Convention of the Gas Processors Association*, <u>1983</u>, 256-268.

EXPERIMENTAL VALUES:

Concentration of diglycolamine (DGA) in aqueous solution before addition of gas = 65 wt %

T/°F	T/K*	Mole ratio in liquid phase CO_2/DGA	H_2S/DGA	Partial pressure of CO_2 /psia	/kPa*	Partial pressure of H_2S /psia	/kPa*
100	310.9	0	0.7130	0.0	0	6.32	43.6
		0.0890	0.6300	–	–	12.7	87.6
		0	0.7410	0.0	0	7.76	53.5
		0.0630	0.6830	0.116	0.800	10.3	71.0
		0.0460	0.7050	0.0377	0.260	10.4	71.7
		0.0860	0.6690	0.232	1.60	15.8	108.9
		0.0490	0.7440	0.107	0.738	21.2	146.2
		0	0.8300	0.0	0	20.0	137.9
		0	0.8370	0.0	0	22.2	153.1
		0	0.8510	0.0	0	22.5	155.1
140	333.2	0.0443	0.0466	0.00230	0.0159	0.0754	0.520
		0.1200	0.0095	0.0106	0.0731	0.0117	0.0807
		0.1530	0.0091	0.0253	0.174	0.0263	0.181
		0.1200	0.0878	0.0155	0.107	–	–
		0.2030	0.0094	0.0331	0.228	0.0396	0.273
		0.0443	0.2130	0.00644	0.0444	0.956	6.59
		0.2600	0.0077	0.0557	0.384	0.0451	0.311
		0.0989	0.1950	0.0251	0.173	1.16	8.00
		0.0944	0.2410	0.0391	0.270	1.80	12.4
		0.2500	0.0861	0.0903	0.623	0.621	4.28
		0.0419	0.3040	0.00870	0.0600	2.22	15.3
		0.1420	0.2110	0.0437	0.301	1.73	11.9
		0.1073	0.2850	0.0468	0.323	2.80	19.3
		0.2330	0.1940	0.271	1.87	3.17	21.9
		0.3980	0.0413	1.33	9.17	1.25	8.62
		0.3650	0.0816	1.21	8.34	2.37	16.3
		0.0370	0.4170	–	–	4.41	30.4
		0.1530	0.2970	–	–	4.84	33.3
		0.2910	0.2000	1.13	7.79	6.96	48.0
		0.0228	0.4750	0.0124	0.0855	5.80	40.0
		0.5020	0.0069	12.1	83.4	0.716	4.94
		0.4900	0.0325	23.0	158.6	4.22	29.1
		0.1210	0.4050	0.154	1.06	8.18	56.4
		0.0164	0.5200	0.0173	0.119	8.51	58.7
		0.2830	0.2700	3.71	25.6	19.4	133.8
		0.2040	0.3680	1.46	10.1	19.5	134.4
		0.1080	0.5190	0.321	2.21	17.6	121.3
		0.1030	0.5360	0.627	4.32	26.8	184.8
160	344.3	0.0830	0.2530	0.0609	0.420	3.13	21.6
		0.1780	0.2220	2.46	17.0	22.2	153.1
		0.1020	0.3640	0.201	1.39	8.86	61.1
		0.4200	0.0620	7.14	49.2	4.62	31.9
180	355.4	0	0.0109	0.0	0	0.0110	0.0758
		0	0.0157	0.0	0	0.0464	0.320
		0.0203	0	0.00437	0.0301	0.0	0
		0.0367	0	0.0197	0.136	0.0	0
		0	0.0623	0.0	0	0.271	1.87
		0.0616	0.0145	0.0435	0.300	0.0870	0.600
		0.0689	0.0109	0.0435	0.300	0.0377	0.260
		0.0423	0.0407	0.0250	0.172	0.296	2.04

* calculated by the compiler.

COMPONENTS:	ORIGINAL MEASUREMENTS:
1. Hydrogen sulfide; H_2S; [7783-06-4] 2. Carbon dioxide; CO_2; [124-38-9] 3. Water; H_2O; [7732-18-5] 4. 2-(2-Aminoethoxy)ethanol, (*diglycolamine*); $C_4H_{11}NO_2$; [929-06-6]	Dingman, J.C.; Jackson, J.L.; Moore, T.F.; Branson, J.A. *Proc. 62nd Annual Convention of the Gas Processors Association*, <u>1983</u>, 256-268.

EXPERIMENTAL VALUES:

Concentration of diglycolamine (DGA) in aqueous solution before addition of gas = 65 wt %

T/°F	T/K*	Mole ratio in liquid phase CO_2/DGA	H_2S/DGA	Partial pressure of CO_2 /psia	/kPa*	Partial pressure of H_2S /psia	/kPa*
180	355.4	0.0976	0.0147	0.0897	0.618	0.125	0.862
		0.1160	0	0.0874	0.603	0.0	0
		0.0372	0.0859	0.0331	0.228	0.824	5.68
		0.1180	0.0113	0.0878	0.605	0.0899	0.620
		0	0.1450	0.0	0	1.57	10.8
		0.1640	0	0.147	1.01	0.0	0
		0.0873	0.0850	0.102	0.703	1.07	7.38
		0.0402	0.1760	0.0449	0.310	3.08	21.2
		0.1480	0.0701	0.230	1.59	1.18	8.14
		0.2230	0.0151	0.418	2.88	0.255	1.76
		0.0820	0.1160	0.126	0.869	3.13	21.6
		0	0.2590	0.0	0	4.53	31.2
		0.2590	0	0.588	4.05	0.0	0
		0.2260	0.0417	0.557	3.84	0.882	6.08
		0.2870	0.0103	0.855	5.90	0.250	1.72
		0.0930	0.2090	0.186	1.28	5.22	36.0
		0.1660	0.1570	0.534	3.68	5.26	36.3
		0	0.3240	0.0	0	7.58	52.3
		0.0368	0.2890	0.0859	0.592	7.54	52.0
		0.3330	0	1.44	9.93	0.0	0
		0.0880	0.2480	0.186	1.28	5.92	40.8
		0.2160	0.1250	0.944	6.51	4.53	31.2
		0.2090	0.1350	0.952	6.56	5.26	36.3
		0.3250	0.0417	1.91	13.2	1.53	10.5
		0.3550	0.0151	2.55	17.6	0.592	4.08
		0.1020	0.3530	0.631	4.35	16.4	113.1
		0.3520	0.1160	12.2	84.1	15.3	105.5
		0.4140	0.0593	14.6	100.7	7.20	49.6
		0.1550	0.3210	2.09	14.4	23.4	161.3
		0.2170	0.2600	3.40	23.4	22.2	153.1
		0.4370	0.0401	19.3	133.1	4.80	33.1
		0.4740	0.0113	23.8	164.1	1.38	9.51
		0.4880	0	26.5	182.7	0.0	0
		0.0966	0.4010	0.807	5.56	24.8	171.0
		0	0.5080	0.0	0	16.4	113.1
		0.0639	0.4610	0.391	2.70	26.5	182.7
		0	0.5270	0.0	0	26.5	182.7

Concentration of DGA before addition of gas = 70 wt %

T/°F	T/K*	CO_2/DGA	H_2S/DGA	/psia	/kPa*	/psia	/kPa*
140	333.2	0.3230	0.0230	0.184	1.27	0.263	1.81
160	344.3	0.3190	0.0220	0.623	4.30	0.596	4.11
180	355.4	0.3180	0.0230	1.75	12.1	0.971	6.69

* calculated by the compiler.

COMPONENTS:	ORIGINAL MEASUREMENTS:
1. Hydrogen sulfide; H_2S; [7783-06-4] 2. Carbon dioxide; CO_2; [124-38-9] 3. Water; H_2O; [7732-18-5] 4. 1,1'-Iminobis-2-propanol, (*diisopropanolamine*); $C_6H_{15}NO_2$; [110-97-4]	Isaacs, E.E.; Otto, F.D.; Mather,A.E. *Can. J. Chem. Eng.* <u>1977</u>, *55*, 210-212. (Complete data in the Centre for Unpublished Data, National Science Library, National Research Council, Ottawa, Ontario K1A 0S2, Canada.)
VARIABLES: Temperature, pressure, concentration of components in liquid phase.	PREPARED BY: 　P.G.T. Fogg

EXPERIMENTAL VALUES:

Concentration of DIPA (diisopropanolamine) = 2.5 kmol m^{-3} (2.5 mol dm^{-3})

T/K	P_{H_2S}/kPa	P_{CO_2}/kPa	Mole ratios in liquid phase	
			H_2S/DIPA	CO_2/DIPA
313.2	2231.1	201.3	1.316	0.051
	1569.2	104.1	1.213	0.040
	1597.5	188.9	1.174	0.078
	1969.1	481.2	1.130	0.138
	1687.8	403.3	1.128	0.166
	1638.8	328.8	1.079	0.143
	1090.0	85.4	1.057	0.045
	1307.9	268.2	1.021	0.129
	1677.4	758.4	1.010	0.246
	1374.8	406.0	1.001	0.166
	2091.1	432.2	0.994	0.118
	2138.0	1396.8	0.976	0.370
	1119.0	247.5	0.975	0.142
	1855.3	1263.1	0.956	0.392
	683.2	63.4	0.933	0.059
	1113.4	404.7	0.899	0.223
	1300.3	646.7	0.891	0.307
	1654.0	1411.3	0.877	0.399
	776.3	197.1	0.861	0.142

AUXILIARY INFORMATION

METHOD/APPARATUS/PROCEDURE:	SOURCE AND PURITY OF MATERIALS:
The equilibrium cell consisted of a Jerguson gauge with a 250 cm^3 tubular gas reservoir mounted at the top. Gas was circulated with a magnetic pump (1). Temperatures were measured by thermopiles and controlled by an air-bath. Pressures were measured by a Heise bourdon tube gauge. The cell was charged with about 150 cm^3 of DIPA solution. H_2S & CO_2 were added to give appropriate pressures. Nitrogen was added when necessary to ensure that the total pressure was above 600 kPa. Gases were circulated for at least 4 h. The gas phase was analysed by gas chromatography and the liquid phase by treating samples with H_2SO_4 (7 mol dm^{-3}), recording P-V-T data for the gases evolved and analysing them by gas chromatography.	No information
	ESTIMATED ERROR: $\delta T/K$ = ± 0.5 $\delta(CO_2/DIPA)$; $\delta(H_2S/DIPA)$ = ± 0.02 or 4%, whichever is the larger (authors)
	REFERENCES: 1. Ruska, W.E.A.; Hurt, L.J.; Kobayashi, R. 　*Rev. Sci. Inst.* <u>1979</u>, *41*, 1444.

COMPONENTS:	ORIGINAL MEASUREMENTS:
1. Hydrogen sulfide; H_2S; [7783-06-4] 2. Carbon dioxide; CO_2; [124-38-9] 3. Water; H_2O; [7732-18-5] 4. 1,1'-Iminobis-2-propanol, (*diisopropanolamine*); $C_6H_{15}NO_2$; [110-97-4]	Isaacs, E.E.; Otto, F.D.; Mather,A.E. *Can. J. Chem. Eng.* <u>1977</u>, *55*, 210-212. (Complete data in the Centre for Unpublished Data, National Science Library, National Research Council, Ottawa, Ontario K1A 0S2, Canada.)

EXPERIMENTAL VALUES:

Concentration of DIPA (diisopropanolamine) = 2.5 kmol m^{-3} (2.5 mol dm^{-3})

T/K	P_{H_2S}/kPa	P_{CO_2}/kPa	Mole ratios in liquid phase H_2S/DIPA	CO_2/DIPA
313.2	499.8	48.9	0.859	0.063
	1072.1	512.2	0.856	0.291
	914.9	427.4	0.821	0.262
	349.5	35.1	0.811	0.073
	2140.8	2332.4	0.790	0.531
	447.4	121.3	0.722	0.168
	206.1	19.9	0.719	0.079
	505.3	461.2	0.655	0.352
	379.2	192.3	0.651	0.257
	595.7	110.3	0.650	0.091
	888.7	669.4	0.598	0.364
	255.7	109.6	0.594	0.257
	186.1	62.0	0.593	0.190
	85.4	13.0	0.583	0.094
	1958.0	3834.1	0.554	0.756
	144.7	62.0	0.535	0.247
	117.8	32.4	0.500	0.197
	710.1	843.2	0.498	0.536
	44.1	5.5	0.498	0.126
	113.0	28.2	0.498	0.210
	80.6	23.4	0.470	0.198
	2496.5	3810.0	0.437	0.841
	1891.2	4177.4	0.432	0.851
	150.9	133.0	0.418	0.365
	734.9	1227.9	0.417	0.655
	106.1	96.5	0.411	0.322
	40.6	15.1	0.397	0.219
	106.1	84.1	0.387	0.354
	70.3	34.4	0.381	0.291
	41.3	13.0	0.379	0.230
	488.1	922.5	0.377	0.653
	59.2	32.4	0.376	0.283
	99.2	85.4	0.370	0.355
	230.2	336.4	0.367	0.524
	24.1	4.8	0.366	0.168
	39.2	16.5	0.342	0.253
	15.8	2.0	0.341	0.097
	17.9	6.2	0.339	0.168
	1313.4	4489.8	0.329	0.903
	15.8	4.1	0.327	0.175
	275.0	663.2	0.258	0.665
	86.1	164.7	0.254	0.504
	58.6	93.0	0.253	0.429
	46.1	48.2	0.251	0.364
	48.9	82.7	0.251	0.444
	15.8	9.6	0.239	0.261
	118.5	286.8	0.239	0.536
	84.1	159.2	0.236	0.498
	311.6	1027.9	0.233	0.737
	68.2	111.6	0.231	0.464

COMPONENTS:	ORIGINAL MEASUREMENTS:
1. Hydrogen sulfide; H_2S; [7783-06-4] 2. Carbon dioxide; CO_2; [124-38-9] 3. Water; H_2O; [7732-18-5] 4. 1,1'-Iminobis-2-propanol, (*diisopropanolamine*); $C_6H_{15}NO_2$; [110-97-4]	Isaacs, E.E.; Otto, F.D.; Mather,A.E. *Can. J. Chem. Eng.* 1977, *55*, 210-212. (Complete data in the Centre for Unpublished Data, National Science Library, National Research Council, Ottawa, Ontario K1A 0S2, Canada.)

EXPERIMENTAL VALUES:

Concentration of DIPA (diisopropanolamine) = 2.5 kmol m^{-3} (2.5 mol dm^{-3})

T/K	P_{H_2S}/kPa	P_{CO_2}/kPa	Mole ratios in liquid phase H_2S/DIPA	CO_2/DIPA
313.2	894.9	4896.6	0.229	0.970
	99.2	172.3	0.228	0.494
	273.0	1489.9	0.221	0.802
	273.7	911.4	0.216	0.730
	894.9	4833.1	0.191	1.020
	47.5	106.1	0.182	0.492
	204.0	1047.3	0.165	0.777
	28.2	22.7	0.165	0.351
	326.1	3109.5	0.150	0.940
	237.1	1540.9	0.145	0.869
	15.1	42.7	0.124	0.454
	10.3	26.8	0.122	0.410
	31.0	101.3	0.122	0.531
	13.0	25.5	0.121	0.409
	216.4	5282.0	0.115	1.030
	517.7	5276.5	0.115	1.052
	438.5	4555.3	0.111	1.053
	226.8	3224.6	0.098	0.999
	430.9	5391.6	0.076	1.089
	158.5	3678.3	0.071	1.067
	128.2	2625.5	0.047	1.011
	1.3	8.9	0.045	0.365
	1.3	11.0	0.042	0.392
	3.4	22.0	0.041	0.465
	3.4	22.7	0.041	0.452
	6.2	70.3	0.035	0.554
	17.2	318.5	0.031	0.731
	22.0	549.5	0.028	0.791
373.2	4126.4	574.3	1.254	0.059
	1797.4	523.9	0.903	0.079
	1819.5	754.9	0.829	0.117
	2748.9	2475.1	0.824	0.273
	1851.9	1392.7	0.809	0.176
	1107.9	237.1	0.796	0.065
	2137.3	1375.4	0.765	0.234
	1103.1	324.7	0.757	0.083
	1475.4	999.7	0.707	0.202
	2151.1	2273.8	0.653	0.337
	2363.5	3261.1	0.629	0.381
	606.0	212.3	0.617	0.095
	504.6	56.5	0.602	0.039
	2080.1	4123.0	0.591	0.456
	875.6	373.6	0.579	0.150
	617.7	236.4	0.575	0.098
	2002.9	4029.9	0.559	0.476
	2220.0	3727.9	0.551	0.474
	591.5	191.6	0.551	0.084
	774.9	346.1	0.547	0.161
	810.1	894.9	0.516	0.255
	753.5	951.4	0.480	0.278

COMPONENTS:	ORIGINAL MEASUREMENTS:
1. Hydrogen sulfide; H_2S; [7783-06-4] 2. Carbon dioxide; CO_2; [124-38-9] 3. Water; H_2O; [7732-18-5] 4. 1,1'-Iminobis-2-propanol, (*diisopropanolamine*); $C_6H_{15}NO_2$; [110-97-4]	Isaacs, E.E.; Otto, F.D.; Mather,A.E. *Can. J. Chem. Eng.* <u>1977</u>, *55*, 210-212. (Complete data in the Centre for Unpublished Data, National Science Library, National Research Council, Ottawa, Ontario K1A 0S2, Canada.)

EXPERIMENTAL VALUES:

Concentration of DIPA (diisopropanolamine) = 2.5 kmol m^{-3} (2.5 mol dm^{-3})

T/K	P_{H_2S}/kPa	P_{CO_2}/kPa	Mole ratios in liquid phase H_2S/DIPA	CO_2/DIPA
373.2	313.7	62.0	0.467	0.060
	395.7	170.9	0.459	0.115
	332.3	55.8	0.452	0.053
	890.1	1650.5	0.445	0.358
	868.7	1613.3	0.436	0.353
	244.7	35.1	0.435	0.049
	1567.8	4374.6	0.428	0.537
	1109.3	1851.9	0.427	0.380
	708.0	1233.4	0.421	0.309
	434.3	385.4	0.407	0.179
	185.4	48.2	0.385	0.081
	1296.8	3230.8	0.384	0.465
	162.0	37.2	0.371	0.060
	211.6	140.6	0.369	0.139
	218.5	157.1	0.357	0.151
	725.3	1838.8	0.353	0.411
	253.0	261.9	0.347	0.217
	1005.2	2676.5	0.336	0.501
	146.8	26.8	0.335	0.044
	384.0	677.7	0.322	0.285
	219.9	154.4	0.310	0.151
	302.6	427.4	0.306	0.275
	1213.4	4767.6	0.295	0.610
	129.6	68.9	0.293	0.113
	128.2	36.5	0.289	0.079
	150.9	117.2	0.284	0.145
	392.9	1094.8	0.274	0.395
	222.0	288.8	0.257	0.247
	447.4	1272.0	0.256	0.410
	79.2	29.6	0.253	0.074
	80.6	17.2	0.250	0.043
	133.0	175.1	0.241	0.192
	472.9	1990.4	0.239	0.513
	318.5	1074.1	0.230	0.403
	172.3	144.7	0.227	0.195
	203.3	532.9	0.226	0.312
	155.8	257.8	0.216	0.228
	230.9	854.9	0.212	0.361
	53.7	16.5	0.212	0.050
	187.5	451.6	0.210	0.286
	224.7	743.2	0.210	0.387
	48.9	19.3	0.207	0.068
	319.2	1349.2	0.205	0.458
	215.1	741.8	0.204	0.334
	36.5	9.6	0.185	0.050
	68.9	54.4	0.184	0.125
	717.0	5274.4	0.181	0.683
	31.0	11.0	0.156	0.066
	344.7	2031.8	0.154	0.539
	374.3	3219.8	0.145	0.600

COMPONENTS:	ORIGINAL MEASUREMENTS:
1. Hydrogen sulfide; H_2S; [7783-06-4] 2. Carbon dioxide; CO_2; [124-38-9] 3. Water; H_2O; [7732-18-5] 4. 1,1'-Iminobis-2-propanol, (*diisopropanolamine*); $C_6H_{15}NO_2$; [110-97-4]	Isaacs, E.E.; Otto, F.D.; Mather,A.E. *Can. J. Chem. Eng.* <u>1977</u>, *55*, 210-212. (Complete data in the Centre for Unpublished Data, National Science Library, National Research Council, Ottawa, Ontario K1A 0S2, Canada.)

EXPERIMENTAL VALUES:

Concentration of DIPA (diisopropanolamine) = 2.5 kmol m^{-3} (2.5 mol dm^{-3})

T/K	P_{H_2S}/kPa	P_{CO_2}/kPa	Mole ratios in liquid phase	
			H_2S/DIPA	CO_2/DIPA
373.2	794.9	4421.5	0.134	0.715
	43.4	63.4	0.133	0.161
	53.7	230.9	0.133	0.258
	25.5	24.1	0.132	0.097
	103.4	437.8	0.131	0.320
	44.8	88.2	0.130	0.186
	61.3	204.7	0.130	0.264
	421.9	3872.0	0.129	0.704
	361.9	3854.1	0.125	0.707
	32.7	44.5	0.124	0.121
	92.3	530.2	0.120	0.387
	94.4	661.2	0.119	0.404
	66.8	276.4	0.117	0.300
	28.9	35.8	0.106	0.106
	107.5	1119.0	0.105	0.467
	196.4	4474.6	0.105	0.699
	24.1	22.0	0.103	0.089
	42.7	88.9	0.101	0.193
	46.8	177.8	0.098	0.258
	144.0	3454.2	0.097	0.609
	312.3	5991.4	0.094	0.762
	42.0	63.4	0.090	0.198
	70.3	424.7	0.090	0.355
	173.0	5239.9	0.087	0.975
	278.5	5543.3	0.082	0.775
	323.3	7391.1	0.080	0.825
	8.2	4.8	0.064	0.036
	6.8	4.1	0.063	0.036
	9.9	22.5	0.054	0.102
	5.5	4.1	0.053	0.037
	11.3	33.8	0.047	0.134
	95.8	5885.3	0.025	0.783
	19.9	5619.1	0.018	0.747
	152.3	5591.6	0.005	0.754
	11.0	3157.7	0.002	0.644

COMPONENTS:	ORIGINAL MEASUREMENTS:
1. Hydrogen sulfide; H_2S; [7783-06-4] 2. Carbon dioxide; CO_2; [124-38-9] 3. Water; H_2O; [7732-18-5] 4. 1,1'-Iminobis-2-propanol, (*diisopropanolamine*); $C_6H_{15}NO_2$; [110-97-4]	Isaacs, E.E.; Otto, F.D.; Mather, A.E. *Can. J. Chem. Eng.* <u>1977</u>, *55*, 210-212. (Complete data in the Centre for Unpublished Data, National Science Library, National Research Council, Ottawa, Ontario K1A 0S2, Canada.)

EXPERIMENTAL VALUES:

The tables give values of the mole ratio H_2S/DIPA in the liquid phase for various partial pressures of H_2S in the gas phase and mole ratios CO_2/DIPA in the liquid phase.

Concentration of DIPA = 2.5 kmol m^{-3} (2.5 mol dm^{-3})

Mole ratio CO_2/DIPA	0.000	0.100	0.200	0.300	0.400	0.500	0.600

T/K	P_{H_2S}/kPa							
313.2	1.0	0.063	0.045*	0.028*	0.015*	0.006*		
	3.16	0.185	0.158*	0.136*	0.108	0.069	0.029*	0.010*
	10.0	0.335	0.290	0.253	0.205	0.133	0.075	0.045
	31.6	0.565	0.458	0.383	0.310	0.213	0.143	0.092
	100	0.752	0.612	0.510	0.429	0.338	0.248	0.170
	316	0.902	0.759	0.670	0.606	0.527	0.440	0.347
	1000	1.126	0.983	0.907	0.838	0.772	0.677	0.570
	3000	1.620	1.458	1.255*	1.165*	1.080*	0.960*	0.850*

Mole ratio CO_2/DIPA	0.700	0.800	0.900	1.000	1.100	1.200

T/K	P_{H_2S}/kPa						
313.2	10.0	0.024					
	31.6	0.063	0.039	0.018	0.005*		
	100	0.133	0.102	0.072	0.047	0.029*	0.018*
	316	0.263	0.202	0.160	0.123	0.088*	0.067*
	1000	0.472	0.370	0.293	0.232	0.185	0.155
	3000	0.720*	0.622*	0.488*	0.405*	0.340*	0.288*

Mole ratio CO_2/DIPA	0.000	0.100	0.200	0.300	0.400	0.500

T/K	P_{H_2S}/kPa						
373.2	3.16	0.025	0.018*	0.010*			
	10.0	0.086	0.052	0.036	0.028	0.020	0.015*
	31.6	0.178	0.138	0.092	0.065	0.053	0.041
	100	0.311	0.245	0.200	0.150	0.120	0.093
	316	0.541	0.423	0.346	0.286	0.239	0.189
	1000	0.880	0.705	0.580	0.490	0.412	0.337
	3000	1.200	1.042	0.880	0.770	0.665	0.552*

Mole ratio CO_2/DIPA	0.600	0.700	0.800	0.900	1.000

T/K	P_{H_2S}/kPa					
373.2	10.0	0.010*				
	31.6	0.023	0.016*			
	100.0	0.073	0.058	0.037	0.024*	0.018*
	316	0.142	0.111	0.080*	0.058*	0.039*
	1000	0.272	0.212	0.163*	0.130*	0.107*
	3000	0.470*	0.397*	0.331*	0.286*	0.250*

* extrapolated values given by the authors.

COMPONENTS:	EVALUATOR:
1. Hydrogen sulfide; H_2S; [7783-06-4] 2. Non-aqueous solvents	Peter G.T. Fogg, Department of Applied Chemistry and Life Sciences, Polytechnic of North London, Holloway, London N7 8DB, U.K. July 1987

CRITICAL EVALUATION:

The solubilities of hydrogen sulfide in a wide range of organic solvents have been measured at pressures above and below barometric pressure. These measurements are of varying reliability. Error limits are likely to vary from about ± 2% to about ± 10% but it is not always possible to quantify these error limits. This must be borne in mind when attempts are made to deduce a general pattern of solubilities of hydrogen sulfide from the experimental data which have been reported.

Non-aromatic hydrocarbons

Propane; C_3H_8; [74-98-6]   Butane; C_4H_{10}; [106-97-8]
2-Methylpropane; C_4H_{10}; [75-28-5]   Pentane; C_5H_{12}; [109-66-0]
Hexane; C_3H_{14}; [110-54-3]   Heptane; C_7H_{16}; [142-82-5]
Octane; C_8H_{18}; [111-65-9]   Nonane; C_9H_{20}; [111-84-2]
Decane; $C_{10}H_{22}$; [124-18-5]   Undecane; $C_{11}H_{24}$; [1120-21-4]
Dodecane; $C_{12}H_{26}$; [112-40-3]   Tridecane; $C_{13}H_{28}$; [629-50-5]
Tetradecane; $C_{14}H_{30}$; [629-59-4]   Pentadecane; $C_{15}H_{32}$; [629-62-9]
Hexadecane; $C_{16}H_{34}$; [544-76-3]   Cyclohexane; C_6H_{12}; [110-82-7]
Methylcyclohexane; C_7H_{14}; [108-87-2]   Ethylcyclohexane; C_8H_{16}; [1678-91-7]
Propylcyclohexane; C_9H_{18}; [1678-92-8]   (1-Methylethyl)-cyclohexane;
Decahydronaphthalene; $C_{10}H_{18}$; [91-17-8] C_9H_{18}; [696-29-7]
1,1'-Bicyclohexyl; $C_{12}H_{22}$; [92-51-3]   Liquid paraffin

The solubilities of hydrogen sulfide at a partial pressure of 1.013 bar and temperatures in the range 288 - 343 K in hexane, octane, decane, dodecane, tetradecane, hexadecane have been reported by King & Al-Najjar (1). Makranczy et al. (2) reported solubilities in the straight chain alkanes from pentane to hexadecane also as at a partial pressure of 1.013 bar at temperatures of 298.15 K and 313.15 K. Mole fraction solubilities reported by Makranczy et al. for 298.15 K are all within 4% of values reported by King & Al-Najjar. The mole fraction solubility at 313.15 K in hexadecane reported by Makranczy differs from the interpolated value from King & Al-Najjar by 7%. Measurements of solubilities in other alkanes at 313.15 by Makranczy are within 5% of interpolated values from King & Al-Najjar. Measurements by the two groups indicate that mole fraction solubilities at a partial pressure of hydrogen sulfide of 1.013 bar increase with the length of the carbon chain, in the temperature range 288 - 303 K. This pattern was found by Bell (3) at 293 K for hexane, octane, dodecane and hexadecane. Bell, however, used rather simpler apparatus and individual values are likely to be less reliable than those of King & Al-Najjar or of Makranczy. This pattern of solubilities is not substantiated by measurements of the solubility in hexane by Hayduk & Pahlevanzadeh (4), nor that in heptane measured by Ng et al. (5), nor that in decane measured by Gerrard (6) and by Reamer et al. (7), nor that in hexadecane measured by Tremper & Prausnitz (8). Mole fraction solubilities at 298.15 K from measurements by the different authors are shown in fig. 2. The biggest discrepancy is between the value of the mole fraction solubility in hexadecane of 0.040 by extrapolation of measurements by Tremper & Prausnitz and the value of 0.0573 given by Makranczy et al. and also by King & Al-Najjar. There are similar discrepancies between mole fraction solubilities of sulfur dioxide in alkanes from measurements by different authors. These differences have been discussed by Young (9).

Phase equilibria between propane and hydrogen sulfide from 341 K to 367 K and total pressures from 27.6 bar to 41.4 bar were investigated by Gilliland and Scheeline (10). Data may be extrapolated to give approximate values of mole fraction solubilities for partial pressures of hydrogen sulfide of 1.013 bar in this temperature range. These are consistent with the pattern shown by the higher alkanes mentioned above.

The butane-hydrogen sulfide system was investigated by Robinson et al. (11) in the temperature range 311 K to 394 K and pressure range 4.2 to 78.5 bar. An approximate value of the mole fraction solubility at 298.15 K and a partial pressure of 1.013 bar may be estimated by extrapolation of these measurements. The value of 0.041 is consistent with the pattern of mole fraction solubilities for higher straight chain alkanes given by Makranczy (2).

COMPONENTS:	EVALUATOR:
1. Hydrogen sulfide; H₂S; [7783-06-4] 2. Non-aqueous solvents	Peter G.T. Fogg, Department of Applied Chemistry and Life Sciences, Polytechnic of North London, Holloway, London N7 8DB, U.K. July 1987

CRITICAL EVALUATION:

The 2-methylpropane-hydrogen sulfide system has been studied by Besserer & Robinson (12) over the temperature range 278 K to 378 K and pressure range 1.8 to 61.7 bar. Approximate values for mole fraction solubilities for a partial pressure of 1.013 bar may be estimated. The value for 298.15 K is 0.038, apparently slightly less than the value for the straight chain isomer.

Phase equilibria between pentane and hydrogen sulfide at high pressures have also been investigated. Reamer, Sage and Lacey (13) made measurements on this system between 278 K and 444 K and total pressures from 1.4 bar to 89.6 bar. These measurements are consistent with the data presented by Makranczy et al. (2). Extrapolation of mole fraction compositions of the liquid phase at 298.15 K to a partial pressure of hydrogen sulfide of 1.013 bar gives a value differing from the Makranczy value by less than 1%. The extrapolated value for 313.15 K differs from the corresponding Makranczy value by 5%.

The mole fraction solubility in hexane of 0.0372 at a partial pressure of 1.013 bar and 298.15 reported by Hayduk & Pahlevanzadeh (4) is significantly lower than the Makranczy value of 0.0429. These recent measurements by Hayduk were carried out using apparatus of high precision but the relative reliability of the two values cannot be judged. The solubility in hexane was also measured by Bell (3). Bell's value at 293.15 is lower than the value by extrapolation of Makranczy's measurements. Bell used very simple apparatus and more recent measurements are likely to be of greater reliability.

The heptane-hydrogen sulfide system has been investigated by Ng, Kalra, Robinson & Kubota (5) from 311 - 478 K and 1.6 to 8.4 bar. In this case extrapolated values of mole fraction solubilities for partial pressures of hydrogen sulfide of 1.013 bar are considerably lower than found by Makranczy (2). At 298.15 the value from Ng et al. is 0.037 whereas the Makranczy value is 0.0439. The corresponding values for 313.15 K are 0.028 and 0.0325 respectively. Measurements reported by Ng et al. were made over a pressure range and are internally self-consistent. They cast further doubt on the pattern indicated by Makranczy et al. (2) and by King & Al-Najjar (1).

The nonane-hydrogen sulfide system was investigated by Eakin & DeVaney (14) at temperatures from 311 K to 478 K and pressures from 1.37 bar to 27.6 bar. Measurements are self-consistent. The data may be used to estimate mole fraction solubilities at a partial pressure of 1.013 bar. Extrapolation of these estimated values to 298.15 K gives a mole fraction solubility at this temperature of 0.037 which is appreciably below the Makranczy (2) value of 0.0465.

The decane-hydrogen sulfide system was studied by Reamer et al. (7) in the temperature range 278 K to 444 K and pressure range 1.4 bar to 124 bar. Measurements are self-consistent. Extrapolation of mole fraction solubilities to a partial pressure of hydrogen sulfide of 1.013 bar and a temperature of 298.15 K gives a value of 0.040, close to the value of 0.039 by extrapolation of measurements by Gerrard (6) but appreciably lower than the figures of 0.0481 and 0.0465 from measurements by Makranczy (2) and by King (1) respectively.

The solubilities in hexadecane reported by Tremper & Prausnitz (8) were given as Henry's constants based on measurements made at an unspecified pressure or pressures below 1.33 bar. Mole fraction solubilities may be estimated on the assumption that there is a linear variation of mole fraction solubility with pressure to 1.013 bar. The interpolated value of the mole fraction solubility at 298.15 is 0.040 compared with the value of 0.0573 from measurements by Makranczy (2) and also by King (1). The higher value is also supported by the estimation of a value of 0.0535 from gas chromatographic measurements by Lenoir et al. (15) and by the value of 0.058 at 293.15 K published by Bell (3).

As may be seen in fig 2, data for the solubility of hydrogen sulfide in alkanes fall into two groups. Some data indicate that the mole fraction solubility does not change appreciably with the length of the hydrocarbon chain. Other data indicate a marked increase with increase in chain length. Further experimental measurements are needed to clarify the situation. When solubility data for these systems are used the discrepancies between different measurements should be borne in mind.

COMPONENTS:	EVALUATOR:
1. Hydrogen sulfide; H_2S; [7783-06-4] 2. Non-aqueous solvents	Peter G.T. Fogg, Department of Applied Chemistry and Life Sciences, Polytechnic of North London, Holloway, London N7 8DB, U.K. July 1987

CRITICAL EVALUATION:

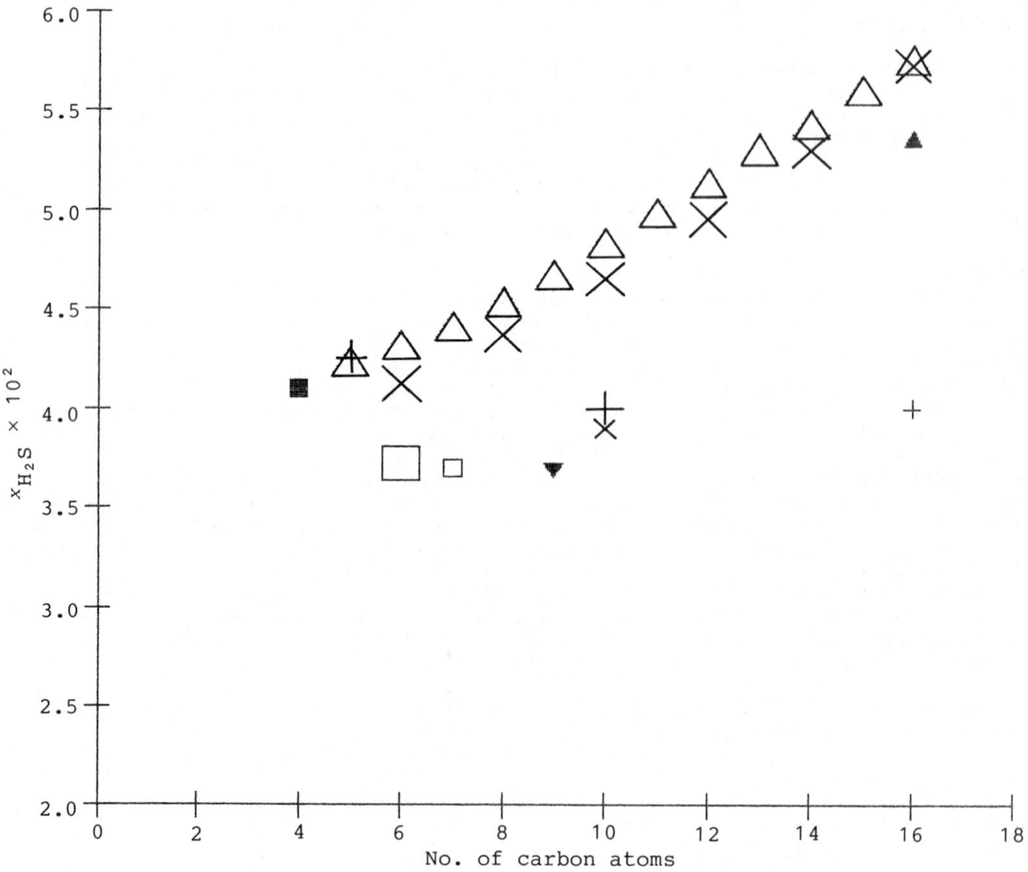

Fig. 2.   Values of the mole fraction solubility of hydrogen sulfide in
straight chain alkanes at 298.15 K and a partial pressure of
1.013 bar from measurements reported by different authors

✕	King & Al-Najjar (1)	✛	Reamer *et al.* (7, 13)
△	Makranczy *et al.* (2)	+	Tremper & Prausnitz (8)
☐	Hayduk & Pahlevanzadeh (4)	◼	Robinson *et al.* (11)
☐	Ng *et al.* (5)	▼	Eakin & DeVaney (14)
✕	Gerrard (6)	▲	Lenoir *et al.* (15)

COMPONENTS:	EVALUATOR:
1. Hydrogen sulfide; H₂S; [7783-06-4] 2. Non-aqueous solvents	Peter G.T. Fogg, Department of Applied Chemistry and Life Sciences, Polytechnic of North London, Holloway, London N7 8DB, U.K. July 1987

CRITICAL EVALUATION:

The solubility of hydrogen sulfide in octane in the presence of methane at various pressures to 103 bar in the temperature range 233 K to 293 K has been studied by Asano *et al.* (16) using a chromatographic method. K-values or ratios of the mole fraction of hydrogen sulfide in the vapor phase to that in the liquid phase are reported. These correspond to very low partial pressures of hydrogen sulfide and may be taken to be equivalent to values for infinite dilution of hydrogen sulfide. These ratios may be used to estimate limiting values of Henry's law constants which are subject to the uncertainties associated with Henry's law constants determined by chromatographic methods. These Henry's law constants can then be used to estimate approximate mole fraction solubilities for a partial pressure of hydrogen sulfide of 1.013 bar. Values indicate that the solubility of hydrogen sulfide decreases with increasing concentration of methane in the liquid phase which is in accord with the apparent decrease of solubility of hydrogen sulfide with decrease in chain length of alkane as discussed above. These apparent Henry's law constants may be extrapolated to zero concentration of methane and the extrapolated value used to estimate solubility of pure hydrogen sulfide in octane to enable direct comparison with data discussed above. At a temperature of 293.15 K the extrapolated value, calculated from the experimental values of K rather than the smoothed values, indicates a mole fraction solubility of about 0.06 for a partial pressure of 1.013 bar. This may be compared with the value of 0.0474 given by King & Al-Najjar (1) at this temperature. In the opinion of the compiler the measurements presented by Asano *et al.* (16) are likely to correspond to the general pattern of behaviour of the hydrogen sulfide-methane-octane system. It must be borne is mind that errors can arise from surface effects when the gas-chromatographic method is used to determine solubilities of gases in liquids.

Solubilities in cyclohexane were measured at pressures from 0.133 to 1.067 bar and temperatures from 283.2 to 313.2 K by Tsiklis & Svetlova (17). These are a self-consistent set of measurements. Bell (3) reported the mole fraction solubility at 293.2 K and a partial pressure of 1.013 bar to be 0.0338. This is appreciably different from the interpolated value of 0.0422 from measurements by Tsiklis & Svetlova. The measurements by Tsiklis & Svetlova are probably the more reliable because they were carried out over a temperature and pressure range although they do not fit into the pattern shown by alkyl substituted cyclohexanes.

Solubilities in methyl, ethyl and propylcyclohexane were measured by Robinson and co-workers (18,19) and in (1-methylethyl)-cyclohexane by Eakin & DeVaney (14) at pressures above 1.013 bar and temperatures in the range 311 K to 478 K. These measurements are consistent with one another and may be accepted on a tentative basis. Mole fraction solubilities for a partial pressure of 1.013 bar may be found by extrapolation. There is an increase in these mole fraction solubilities as the number of carbon atoms in the side chain is increased with the values for (1-methylethyl)-cyclohexane close to those for propylcyclohexane. Values for cyclohexane itself, from measurements by Tsiklis (17), do not fall into the pattern.

Smoothed and extrapolated values of solubilities of cyclohexanes estimated by the evaluator are shown below:

Solvent	mole fraction solubility, x_{H_2S} (1.013 bar)		
	300 K	350 K	400 K
Cyclohexane	0.038		
Methylcyclohexane	0.034	0.015	0.009
Ethylcyclohexane	0.036	0.018	0.011
Propylcyclohexane	0.043	0.023	0.014
(1-Methylethyl)-cyclohexane	0.041	0.023	0.015

Solubilities in 1,1'-bicyclohexyl, published by Tremper & Prausnitz (8), and those in decalin, found by Lenoir *et al.* (15) from chromatographic measurements, are consistent with solubilities reported for cyclohexanes.

Solubilities in liquid paraffin, reported by Devyatykh *et al.* (20), are subject to the uncertainties associated with measurements by chromatography.

COMPONENTS:	EVALUATOR:
1. Hydrogen sulfide; H₂S; [7783-06-4] 2. Non-aqueous solvents	Peter G.T. Fogg, Department of Applied Chemistry and Life Sciences, Polytechnic of North London, Holloway, London N7 8DB, U.K. July 1987

CRITICAL EVALUATION:

Hannaert et al. (21) have reported the solubilities, in a form equivalent to Henry's constants, in kerosene A-1 (distillation range 423 - 553 K; average relative molecular mass 170). Values are similar to those for pure dodecane which also has a relative molecular mass of 170.

Aromatic hydrocarbons

Benzene; C_6H_6; [71-43-2]   Methylbenzene; C_7H_8; [108-88-3]
1,2-Dimethylbenzene; C_8H_{10}; [95-47-6]   1,3-Dimethylbenzene; C_8H_{10}; [108-38-3]
Ethylbenzene; C_8H_{10}; [100-41-4]   1,3,5-Trimethylbenzene; C_9H_{12}; [108-67-8]
Propylbenzene; C_9H_{12}; [103-65-1]   (1-Methylethyl)-benzene; C_9H_{12}: [98-82-8]
1-Methylnaphthalene; $C_{11}H_{10}$; [1321-94-4]   1,1'-Methylenebisbenzene; $C_{13}H_{12}$;
9-Methylanthracene; $C_{15}H_{12}$; [779-02-2]   [101-81-5]

Solubilities in methylbenzene have been measured by Gerrard (6) and by Bell (3) at barometric pressure from 265 K to 298 K. Ng, Kalra, Robinson and Kubota (5) investigated phase equilibria at total pressures from 2 bar to 116 bar in the temperature range 311 K to 478 K. Mole fraction solubilities at partial pressures of 1.013 bar may be found by extrapolating mole fraction solubilities at higher pressures calculated from data given by Ng et al. These extrapolated values are consistent with mole fraction solubilities at lower temperatures calculated from data presented by Gerrard and by Bell. All mole fraction solubilities at a partial pressure of 1.013 bar lie above the reference line defined by the Raoult's law relationship and may be fitted to the equation :

$$\log_{10} x_{H_2S} = - 3.811 + 768 / (T/K)$$

The standard deviation in values of x_{H_2S} is ± 0.005. The equation in based upon data for the range 265 K - 478 K.

Phase equilibria between hydrogen sulfide and 1,3,5-trimethylbenzene have been investigated by Eakin & DeVaney (14) at total pressures from 1.3 bar to 32.6 bar at 311, 366 and 478 K. The solubility at a partial pressure of 1.013 bar and 298 K has been reported by Patyi et al. (22). This value is consistent with mole fraction solubilities estimated by extrapolation of data provided by Eakin and DeVaney to a partial pressure of 1.013 bar. The four values of mole fraction solubilities at a partial pressure of 1.013 bar lie above the reference line corresponding to the Raoult's law equation but below values for methylbenzene . They fit the equation :

$$\log_{10} x_{H_2S} = - 3.495 + 648 / (T/K)$$

The standard deviation in values of x_{H_2S} is ± 0.001. The equation is based upon data for the range 298 K to 478 K.

The mole fraction solubility in benzene at a partial pressure of 1.013 bar was reported by Bell (3) to be 0.0563 at 293.15 K. Gerrard (6) made measurements of solubility at barometric pressure in the range 265.15 K to 293.15 K. The mole fraction solubility at 293.15 K from Gerrard's measurements, corrected to a partial pressure of 1.013 bar is 0.062. Measurements by Bell and by Gerrard are incompatible with the value of 0.0358 at 298.15 K calculated from a solubility measurement by Patyi et al. (22). Gerrard's measurements at three temperatures are consistent with each other and may be more reliable than the single measurements reported by Patyi.

Gerrard measured solubilities in 1,2-dimethylbenzene at barometric pressure in the temperature range 265 K to 293 K. Huang & Robinson (23) measured solubilities in 1,3-dimethylbenzene at various total pressures from 1.5 to 131 bar in the temperature range 311 K to 478 K. These data for the xylenes may be accepted on a tentative basis.

The mole fraction solubilities at a partial pressure of 1.013 bar and 298 K of ethylbenzene, propylbenzene and (1-methylethyl)-benzene may also be calculated from solubility data given by Patyi et al. (22). The values for propylbenzene and (1-methylethyl)-benzene are close to values for methylbenzene and 1,3,5-trimethylbenzene. The value for ethylbenzene is about 20% lower and

COMPONENTS:	EVALUATOR:
1. Hydrogen sulfide; H_2S; [7783-06-4] 2. Non-aqueous solvents	Peter G.T. Fogg, Department of Applied Chemistry and Life Sciences, Polytechnic of North London, Holloway, London N7 8DB, U.K. July 1987

CRITICAL EVALUATION:

closer to the value for benzene, discussed above, from measurements by these
authors. The merit of these measurements cannot be judged.

Tremper & Prausnitz (8) reported Henry's constant for solubilities in
1-methylnapthalene and in 1,1'-methylenebisbenzene for the temperature range
300 K to 475 K. No other measurements on these systems are available for
comparison but there is no reason to doubt the reliability of these
measurements. Mole fraction solubilities for a partial pressure oa 1.013 bar
may be calculated on the assumption that mole fraction solubility varies
linearly with pressure at pressures to 1.013 bar.

Kragas & Kobayashi (24) measured limiting values of Henry's constants for
dissolution of hydrogen sulfide in 9-methylanthracene. A chromatographic method
was used with hydrogen at pressures of 13 bar to 58 bar as the carrier gas.
Temperatures were in the range 373 K to 423 K. Values of Henry's constant in
the absence of hydrogen were found by extrapolation. These Henry's constants
should be used with caution in the estimation of solubilities for finite
pressures of hydrogen sulfide because of the uncertainties associated with the
chromatographic method of measuring solubilities.

Apparent values of mole fraction solubilities in aromatic hydrocarbons at a
partial pressure of 1.013 bar and 298.15 K based on the measurements discussed
above, are given below.

Hydrocarbon	x_{H_2S} at P_{H_2S} = 1.013 bar and T = 298.15 K	Source
Benzene	0.056	Gerrard (6) (extrapolated)
	0.036	Patyi *et al.* (22)
Methylbenzene	0.058	Eqn. given above
1,2-dimethylbenzene	0.060	Gerrard (6) (extrapolated)
1,3-dimethylbenzene	0.037	Huang & Robinson (23) (extrap.)
1,3,5-trimethylbenzene	0.048	Eqn. given above
Ethylbenzene	0.042	Patyi *et al.* (22)
Propylbenzene	0.052	Patyi *et al.* (22)
(1-methylethyl)-benzene	0.053	Patyi *et al.* (22)
1-methylnapthalene	0.033	Tremper & Prausnitz (8)
1,1'-methylenebisbenzene	0.031	Tremper & Prausnitz (8)
Value from Raoult's law equation	0.049	

Solubilities in alcohols

Methanol; CH_4O; [67-56-1]   Ethanol; C_2H_6O; [64-17-5]
1-Butanol; $C_4H_{10}O$; [71-36-3]   1-Octanol; $C_8H_{18}O$; [111-87-5]
Benzenemethanol; C_7H_8O; [100-51-6]   2-Ethoxyethanol; $C_4H_{10}O_2$; [110-80-5]
1,2-Ethanediol (*ethylene glycol*); $C_2H_6O_2$; [107-21-1]

Solubilities in methanol, ethanol, butanol and octanol have been reported.
Available data indicate that mole fraction solubilities for a partial pressure
of 1.013 bar increase from ethanol to octanol. Below 273 K the mole fraction
solubility in methanol is less than that in ethanol but extrapolation of the
values for methanol indicates that the solubility in methanol would be higher
than that of ethanol at higher temperatures. Methanol shows an apparently
similar anomaly when the mole fraction solubilities of amines in alkanols are
compared (25).

Solubilities in methanol in the range of total pressures from 2.0 to 5.9 bar and
from 248 - 273 K have been measured by Yorizane, Sadamoto, Masuoka and Eto (26).
Short, Sahgal and Hayduk (27) reported mole fraction solubilities at 263 K and
333 K and a partial pressure of hydrogen sulfide of 1.013 bar. The mole
fraction solubilities at high pressures reported by Yorizane *et al.* are
self-consistent. In addition mole fraction solubilities at a partial pressure

COMPONENTS:	EVALUATOR:
1. Hydrogen sulfide; H₂S; [7783-06-4] 2. Non-aqueous solvents	Peter G.T. Fogg, Department of Applied Chemistry and Life Sciences, Polytechnic of North London, Holloway, London N7 8DB, U.K. July 1987

CRITICAL EVALUATION:

of 1.013 bar estimated by extrapolation of these data are consistent with the mole fraction solubilities reported by Short *et al*.

Bezdel & Teodorovich (28) measured solubilities in methanol at partial pressures to 0.080 bar at temperatures 223.2 K, 243.2 K and 303.2 K. Mole fraction solubilities may be calculated from these data. Extrapolation to a partial pressure of 1.013 bar, assuming linear variation of mole fraction solubility with pressure at pressures to 1.013 bar, gives values of 0.35 at 223.2 K, 0.16 at 243.2 K and 0.021 at 303.2 K. The mole fraction solubility, estimated from data given by Short *et al*. and by Yorizane *et al*. is 0.12 at 243.2 K and 0.028 at 303.2 K. The evaluator considers that measurements by Short *et al*. and by Yorizane *et al*. are more reliable than those published by Bezdel & Teodorovitch.

Solubilities in ethanol and octanol were measured by Gerrard (6). Values for ethanol at 283 K and 293 K are within 5% and the value for 273 K is within 10% of values reported many years ago by Fauser (29). Gerrard's data for these two alcohols may be accepted as tentative values.

The solubilities in butanol at a partial pressure of 1,013 bar reported by Short, Sahgal and Hayduk (27) are likely to be reliable and can be accepted as tentative values.

Lenoir, Renault and Renon (15) used a gas-liquid chromatographic technique to measure the Henry's constant for dissolution of hydrogen sulfide in benzenemethanol 298.2 K and very low partial pressures. This measurement may be unreliable because of surface adsorption of gas. The estimated mole fraction solubility at 298 K and 1.013 bar, assuming a linear relationship between mole fraction solubility and pressure to 1.013 bar, is close to the value for 1-octanol mentioned above and about 85% of the Raoult's law value.

Solubility data for 2-ethoxyethanol, obtained using a chromatographic method by Devyatykh *et al*. (20) should also be considered to be of semi-quantitative significance until they can be independently confirmed.

Measurements of solubilities in 1,2-ethanediol made by Short *et al*. (27) agree closely with earlier measurements by Gerrard (6). This may be seen in the following table:

T/K	Mole fraction solubility for P_{H_2S} = 1.013 bar	
	Gerrard	Short *et al*. (interpolated)
265.15	0.0352	0.0309
267.15	0.032	0.0288
273.15	0.023	0.0235
283.15	0.017	0.0175
293.15	0.013	0.0136

Lenoir *et al*. (15) measured the Henry's constant at 298.15 by a chromatographic method. The corresponding mole fraction for a partial pressure of gas of 1.013 bar is 0.0189 compared with a value of 0.0122 given by Short *et al*. (27) Since solubilities measured by chromatographic methods are prone to errors due to surface effects this high value may be disregarded. Measurements of solubility at 297.1 K have been reported by Byeseda *et al*. (30) The Ostwald coefficient given by these workers corresponds to a mole fraction solubility at a partial pressure of 1.013 bar of 0.016 compared with the value of 0.0125. The apparatus used by Short *et al*. is likely to have yielded the more reliable solubility data and the higher value should be disregarded. Data published by Short *et al*. are recommended by the evaluator.

Triethylene glycol and polyethylene glycols are discussed below.

COMPONENTS:	EVALUATOR:
1. Hydrogen sulfide; H_2S; [7783-06-4] 2. Non-aqueous solvents	Peter G.T. Fogg, Department of Applied Chemistry and Life Sciences, Polytechnic of North London, Holloway, London N7 8DB, U.K. July 1987

CRITICAL EVALUATION:

Compounds containing carbon-oxygen-carbon links

1,1'-Oxybisethane; (diethyl ether); $C_4H_{10}O$; [60-29-7]
1,1'-Oxybisoctane; (dioctyl ether); $C_{16}H_{34}O$; [629-82-3]
Ethoxybenzene; $C_8H_{10}O$; [103-73-1]
Tetrahydrofuran; C_4H_8O; [109-99-9]   1,4-Dioxane; $C_4H_8O_2$; [123-91-1]
Oxybispropanol (dipropylene glycol); $C_6H_{14}O_3$; [25265-71-8]
2,2'[1,2-Ethanediylbis(oxy)]bisethanol (triethylene glycol); $C_6H_{14}O_4$;
                                                                    [112-27-6]
α-hydro-ω-hydroxy-poly(oxy-1,2-ethanediyl) (polyethylene glycol);
                                        $(C_2H_4O)_nH_2O$; [25322-68-3]
2-(2-Methoxyethoxy)ethanol (diethylene glycol monomethyl ether); $C_5H_{12}O_3$;
                                                                    [111-77-3]
3,6,9,12-Tetraoxahexadecan-1-ol (tetraethylene glycol monobutyl ether);
                                        $C_{12}H_{26}O_5$; [1559-34-8]
1,1'-Oxybis(2-methoxyethane) (diethylene glycol dimethyl ether);
                                        $C_6H_{14}O_3$; [111-96-6]
2,5,8,11-Tetraoxadodecane (triethylene glycol dimethyl ether);
                                        $C_8H_{18}O_4$; [112-49-2]
2,5,8,11,14-Pentaoxapentadecane (tetraethylene glycol dimethyl ether);
                                        $C_{10}H_{22}O_5$; [143-24-8]
2-Methyl-3,6,9,12,15,18-hexaoxanonadecane (pentaethylene glycol methyl
isopropyl ether); $C_{14}H_{30}O_6$;
®Sepasolv MPE (A mixture of oligo methyl isopropyl ethers developed by BASF)
1,4,7,10-Tetraoxacyclododecane (12-crown-4); $C_8H_{16}O_4$; [294-93-9]
1,4,7,10,13-Pentaoxacyclopentadecane (15-crown-5); $C_{10}H_{20}O_5$; [33100-27-5]
4-Methyl-1,3-dioxolan-2-one (propylene carbonate); $C_4H_6O_3$; [108-32-7]

The solubility in diethyl ether at 299.15 K and a total pressure of 0.986 bar
was measured by Parsons (31). The average concentration of hydrogen sulfide in
the liquid corresponds to a mole fraction of 0.026. The vapor pressure of pure
diethyl ether is about 0.074 bar at this temperature so the partial pressure of
solvent was much greater than the partial pressure of hydrogen sulfide in these
experiments. The data indicate that the mole fraction solubility for a partial
pressure of gas of 1.013 bar is about 0.10, more than double the Raoult's law
value of 0.047. There are, however, no other measurements of the solubility in
diethyl ether with which to make comparison so the measurements must be
classified as tentative.

Gerrard (6) measured solubilities in 1,1'-oxybisoctane at a total pressure of
1.015 bar and temperatures of 265 K to 293 K. Extrapolation of these
measurements to 298.15 K and correction to a partial pressure of 1.013 bar gives
a mole fraction solubility of 0.088, comparable with the value estimated for
diethyl ether. In the absence of other data for comparison, Gerrard's data for
1,1'-oxybisoctane should be accepted on a tentative basis.

Solubilities in ethoxybenzene and 1,4-dioxane reported by Gerrard (6) are
consistent with the general pattern found for ethers and may be accepted on a
tentative basis. Measurements of the solubility in tetrahydrofuran at 263.15 K
and 298.15 K reported by Short et al. (27) are likely to be reliable but must
also be accepted on a tentative basis in the absence of measurements on this
system by other workers.

Byeseda et al. (30) reported the Ostwald coefficient for dissolution in
2,2'[1,2-ethanediylbis(oxy)]bisethanol (triethylene glycol) at 297.1 K. The
mole fraction solubility for a partial pressure of 1.013 bar corresponding to
this Ostwald coefficient is 0.0618. This is close to the value of 0.065
estimated from graphs of solubilities from 273.15 K to 373.15 K and partial
pressures of gas from 1.013 to 11.355 bar published by Blake (32). The data at
other temperatures and pressures given by Blake cannot be confirmed by
comparison with other measurements. Blake gave no information on the method
used and data can only be considered to be tentative.

Solubility data for four polyethylene glycols of different average relative
masses have been published by Gestrich & Reinke (33). This work appears to be
self consistent and the measurements fit into a general pattern discussed below.

COMPONENTS:	EVALUATOR:
1. Hydrogen sulfide; H₂S; [7783-06-4] 2. Non-aqueous solvents	Peter G.T. Fogg, Department of Applied Chemistry and Life Sciences, Polytechnic of North London, Holloway, London N7 8DB, U.K. July 1987

CRITICAL EVALUATION:

Solubilities in diethylene glycol monomethyl ether and in tetraethylene glycol monobutyl ether have been measured by Sciamanna (34). These measurements appear to be of high accuracy but must be considered to be tentative until other measurements on these systems are reported.

Measurements of solubilities in polyglycol ethers have been made by Sciamanna (34), by Härtel (35) and by Sweeney (36). Each of these workers measured the solubility in tetraethylene glycol dimethyl ether. Solubilities were reported as Henry's constants, H. Sciamanna made measurements in the temperature range 288 K to 373 K and reported limiting values for zero partial pressure in the form of equations of the form:

$$\ln (H/\text{kPa}) \quad = \quad A/T + B$$

Härtel based his values on measurements with the mole fraction of hydrogen sulfide in the liquid phase not greater than 0.16. In this range the variation of mole fraction solubility was found to be almost linear with a correlation coefficient greater than 0.99. The Henry's law constant published by Sweeney is based upon gas chromatographic measurements. Henry's law constants in tetraglyme from measurements by the different authors are as follows:

T/K	Sciamanna	H/bar Härtel	Sweeney
293.15	3.71	3.390	
298.15	4.22	3.97*	3.91
313.15	6.07	6.179	
323.15	7.58	7.20*	7.46
333.15	9.35	8.310	
353.15	13.73	11.321	
373.15	19.34	16.901	

* interpolated values.
(Henry's law constant = partial pressure of gas / mole fraction solubility)

The evaluator recommends that, for the temperature range 298.15 to 323.15, the Henry's constant is taken to be the mean of values from measurements by Sciamanna and Härtel. Outside this temperature range the divergence of measurements is too great for values to be recommended.

Sciamanna (34) also measured solubilities in diethylene glycol dimethyl ether and triethylene glycol dimethyl ether.

The available experimental solubility data for glycols, polyglycols and ethers derived from these compounds can be correlated with the number of carbon-oxygen -carbon links in these molecules. The evaluator has estimated mole ratio solubilities (H₂S/solvent) at 298.15 K for a partial pressure of 1.013 bar. If the value for ethylene glycol, propylene glycol triethylene glycol and the polyethylene glycols are plotted against the number of carbon-oxygen-carbon links in the molecule then the points lie close to a straight line (see fig.3). In the case of the polyethylene glycols the number of links is taken as the average number calculated from the average relative molecular masses of the samples.

Mole ratio solubilities for five of the glycol diethers fall close to a similar straight line if a value for ®Sepasolv MBE based upon measurements by Wolfer et al. (37) is included. The mole ratio solubility of dioctyl ether also lies close to this line. The estimated value for pentaethylene glycol methyl isopropyl ether does not fit into the pattern. Härtel (35) published Henry's law constants for this compound and has shown that the mole fraction solubility is proportional to pressure to a mole fraction of 0.16, with a correlation coefficient of better than 0.99. The evaluator has estimated the mole ratio solubility at 298.15 K and 1.013 bar from an apparent mole fraction solubility of 0.323 calculated from an interpolated value of Henry's law constant. This may be an over-estimate because it is outside the range in which a linear relationship between mole fraction solubility and pressure has been experimentally demonstrated.

COMPONENTS:	EVALUATOR:
1. Hydrogen sulfide; H₂S; [7783-06-4] 2. Non-aqueous solvents	Peter G.T. Fogg, Department of Applied Chemistry and Life Sciences, Polytechnic of North London, Holloway, London N7 8DB, U.K. July 1987

CRITICAL EVALUATION:

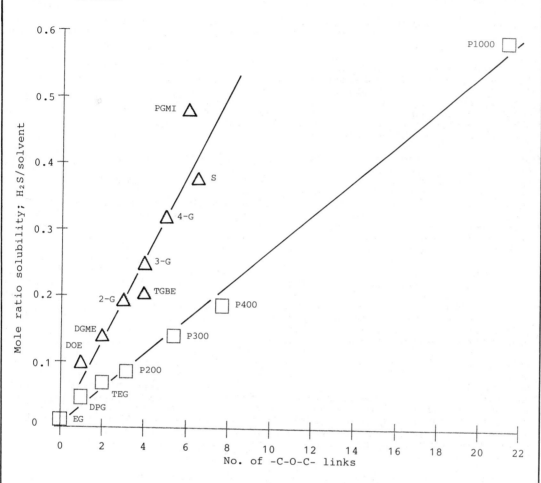

Fig. 3   Mole ratio solubilities at 298.15 K and a partial pressure of 1.013 bar in glycols, glycol ethers and dioctyl ether as a function of the number of -C-O-C- links in a molecule of solvent

glycols

EG   *ethylene glycol* (6,27)

DPG   *dipropylene glycol* (15)

TEG   *triethylene glycol* (30,32)

P200; P300; P400; P1000
      polyethylene glycols (33)

glycol ethers; dioctyl ether

DOE   *dioctyl ether* (6)

DGME *diethylene glycol monomethyl ether* (34)

2-G   *diethylene glycol dimethyl ether* (34)

3-G   *triethylene glycol dimethyl ether* (34)

4-G   *tetraethylene glycol dimethyl ether* (34,35)

TGME *tetraethylene glycol monobutyl ether* (34)

PGMI *pentaethylene glycol methyl isopropyl ether* (35)

S   ®*Sepasolv MBE* (37)

COMPONENTS:	EVALUATOR:
1. Hydrogen sulfide; H₂S; [7783-06-4] 2. Non-aqueous solvents	Peter G.T. Fogg, Department of Applied Chemistry and Life Sciences, Polytechnic of North London, Holloway, London N7 8DB, U.K. July 1987

CRITICAL EVALUATION:

The overall graphical pattern of measurements on the two series of compounds indicates that most of the solubilities in individual solvents are likely to be reliable.

Mole fraction solubilities in two of the crown ethers, 12-crown-4 and 15-crown-5, have been published by Linford & Thornhill (38). The values at 295.2 K for a partial pressure of 1.013 bar are 0.31 and 0.33 respectively and are greater than values for polyglycol ethers having the same number of carbon-oxygen-carbon bonds discussed above (0.20 and 0.24 at 298.15 K respectively). No other solubilities in crown ethers are available for comparison but these values may be accepted as being tentative.

Solubilities in propylene carbonate have been measured by several workers and there is fairly good agreement between different sets of measurement. Lenoir *et al.* (15) and Sweeney (36) reported Henry's constants measured by a chromatographic method. Mole fraction solubilities from these constants are higher than those from other measurements and may be less reliable. Shakhova *et al.* (39) measured solubilities over the temperature range 273.2 K to 313.2 K and pressure range 0.267 to 1.013 bar. The apparent values of mole fraction solubilities do not vary linearly with pressure over this pressure range. Isaacs, Otto & Mather (40) measured solubilities at 313.2 K and 373.2 K over a pressure range of 1.847 to 49.60 bar. The evaluator has estimated the corresponding mole fraction solubilities at 1.013 bar by use of the Krichevskii-Il'inskaya equation. The apparent mole fraction solubility at 313.2 K is 0.023 and at 373.2 K is 0.009. The mole fraction solubility at 313.2 K and 1.013 bar from Shakhova *et al.* is 0.0264. Data given by Isaacs *et al.* appear to be reliable with the exception of the solubility at 373.2 K and 1.847 bar. The solubility which has been reported is less than the experimental error in the measurements and seems to be out of line with other measurements.

Henry's constants at temperatures from 263.15 K to 373.15 K have been reported by Rivas & Prausnitz (41). These may be used to estimate mole fraction solubilities at a partial pressure of 1.013 bar but the apparent deviation from a linear variation of mole fraction solubility with variation of pressure indicated by Shakhova's measurements should be borne in mind. Mole fraction solubilities for a partial pressure of 1.013 bar from measurements by Shakhova *et al.*, by Isaacs *et al.*, and by Rivas & Prausnitz are shown in fig.4. These values are fitted by the equation:

$$\log_{10} x_{H_2S} = -3.889 + 740/(T/K)$$
$$\delta x_{H_2S} = \pm 0.003$$

Other compounds of carbon, hydrogen and oxygen

Acetic acid; C₂H₄O₂; [64-19-7]   Hexanoic acid; C₆H₁₂O₂; [142-62-1]
Acetic anhydride; C₄H₆O₃; [108-24-7]   2-Propanone; C₃H₆O; [67-64-1]
2-Hydroxybenzoic acid, methyl ester (*methyl salicylate*); C₈H₈O₃; [119-36-8]
Phenol; C₆H₆O; [108-95-2]   2-Furancarboxaldehyde; C₅H₄O₂; [98-01-1]
Benzenedicarboxylic acid, didecyl ester; C₂₈H₄₆O₄; [84-77-5]
Dibenzofuran; C₁₂H₈O; [132-64-9]

Solubilities in acetic acid at 298 K and 303 K were reported by Short, Sahgal and Hayduk (27) and in hexanoic acid over the range 265 - 293 K by Gerrard (6). In each case the mole fraction solubilities for a partial pressure of 1.013 bar fall below values corresponding to the Raoult's law equation. Extrapolation of solubilities to common temperatures shows a higher mole fraction solubility for the longer chain acid. Solubilities in either solvent can be accepted as tentative values.

Solubilities in acetic anhydride were also measured by Gerrard (6). Mole fraction solubilities lie above a reference line corresponding to Raoult's law in this case and when extrapolated to 303 K are about double the values for acetic acid. This difference may be due to stronger solvent-solvent hydrogen bonding in the case of acetic acid. There is no reason to doubt the data for acetic anhydride.

COMPONENTS:	EVALUATOR:
1. Hydrogen sulfide; H₂S; [7783-06-4] 2. Non-aqueous solvents	Peter G.T. Fogg, Department of Applied Chemistry and Life Sciences, Polytechnic of North London, Holloway, London N7 8DB, U.K. July 1987

CRITICAL EVALUATION:

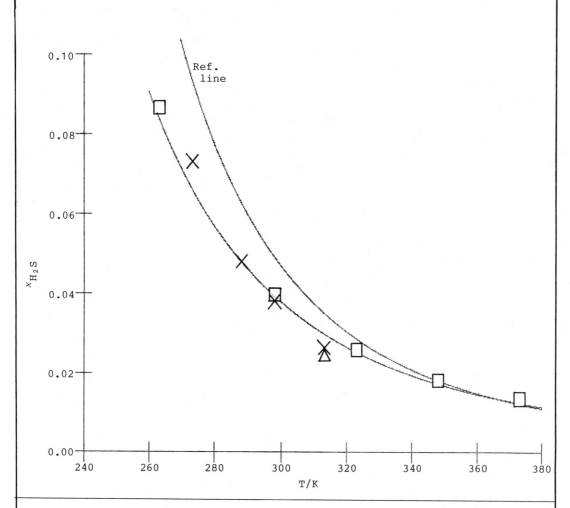

Fig. 4   Variation with temperature of the mole fraction solubility of hydrogen sulfide in propylene carbonate at a partial pressure of 1.013 bar

Points based upon experimental measurements have been superimposed upon a curve corresponding to the equation:

$$\log_{10} x_{H_2S} = -3.889 + 740/(T/K)$$

A reference line corresponding to values from the Raoult's law equation is also shown.

X Shakhova *et al*. (39)   ☐ Rivas & Prausnitz (41)
△ Isaacs *et al*. (40)

COMPONENTS:	EVALUATOR:
1. Hydrogen sulfide; H$_2$S; [7783-06-4] 2. Non-aqueous solvents	Peter G.T. Fogg, Department of Applied Chemistry and Life Sciences, Polytechnic of North London, Holloway, London N7 8DB, U.K. July 1987

CRITICAL EVALUATION:

Short *et al*. (27) reported solubilities at a partial pressure of 1.013 bar in 2-propanone at 263 K and 298 K. These measurements are likely to be reliable but must be accepted on a tentative basis in the absence of similar systems for comparison. Mole fraction solubilities, in this case, are greater than those in acetic anhydride under the same conditions.

Solubilities in methyl salicylate at a total pressure of 1.004 bar in the temperature range 265.15 K to 293.15 K have been measured by Gerrard (6). The values are self-consistent and may be accepted as tentative values until other measurements on this or similar systems are available for comparison.

Lenoir *et al*. (15) measured the Henry's law constant for phenol at 323.2 K by a chromatographic method. The reliability of this method cannot be judged because of uncertainties associated with the method used. No other measurements on this system are available for comparison. Devyatykh *et al*. (20) also used a similar chromatographic method to measure solubilities in 2-furancarboxaldehyde and in 2-benzenedicarboxylic acid, didecyl ester. Again there are no similar systems for comparison and no judgement of reliability can be made.

Kragas & Kobayashi (24) used a high pressure chromatographic method to measure limiting values of Henry's constants for dissolution of hydrogen sulfide in dibenzofuran in the presence of hydrogen. Hydrogen pressures ranged from 7 bar to 59 bar. Temperatures were 373.2 K and 398.2 K. At each temperature a plot of the reciprocal of the Henry's law constant against hydrogen pressure is close to a straight line enabling approximate values of Henry's constants at zero hydrogen pressure to be estimated. No other measurements on this system are available for comparison and the reliability of these Henry's constants, for estimation of solubilities at finite partial pressure, cannot be judged.

Halogenated compounds

Bromoethane; C$_2$H$_5$Br; [74-96-4]   1-Bromobutane; C$_4$H$_9$Br; [109-65-9]
1-Bromooctane; C$_8$H$_{17}$Br; [111-83-1]   1,2-Dibromoethane; C$_2$H$_4$Br$_2$; [106-93-4]
Tribromomethane; CHBr$_3$; [75-25-2]   1,1,2,2-Tetrabromoethane; C$_2$H$_2$Br$_4$;
Trichloromethane; CHCl$_3$; [67-66-3]   [79-27-6]
Tetrachloromethane; CCl$_4$; [56-23-5]   1,1,2,2-Tetrachloroethane; C$_2$H$_2$Cl$_4$;
Tetrachloroethene; C$_2$Cl$_4$; [127-18-4]   [79-34-5]
Pentachloroethane; C$_2$HCl$_5$; [76-01-7]   2,2,2-Trichloroethanol; C$_2$H$_3$Cl$_3$O;
2,2-Dichloroacetic acid; C$_2$H$_2$Cl$_2$O$_2$; [79-43-6]   [115-20-8]
1,1'-Oxybis(2-chloroethane); C$_4$H$_8$Cl$_2$O; [111-44-4]
Chlorobenzene; C$_6$H$_5$Cl; [108-90-7]   Bromobenzene;   C$_6$H$_5$Br; [108-86-1]
Iodobenzene; C$_6$H$_5$I; [591-50-4]

Solubilities in 1-bromooctane at a total pressure of 1.007 bar in the temperature range 265.15 K to 293.15 K were measured by Gerrard (6). Mole fraction solubilities for a partial pressure of 1.013 bar calculated from these measurements fall on a smooth curve lying above the reference line based on the Raoult's law equation when plotted against temperature. The mole fraction solubility in 1-bromobutane for 265.15 K and 1.013 bar partial pressure estimated from the solubility reported by Gerrard is 0.135 compared with 0.143 in 1-bromooctane under the same conditions. The mole fraction solubility in bromoethane given by Bell (3) for 1.013 bar partial pressure and 293.15 K is 0.126 compared with a value of 0.074 for 1-bromooctane at this temperature. The difference in this case could be due to bromoethane having the greater tendency to form hydrogen bonds with sulfur in hydrogen sulfide.

Gerrard's measurements (6) indicate that mole fraction solubilities in tribromomethane and 1,2-dibromoethane are within about 2% of each other at a partial pressure of 1.013 bar and 283.15 K and 293.15 K. Values at 293.2 K are close to values given by Bell (3). These mole fraction solubilities are about 20% lower than corresponding values for 1-bromooctane. Bell's value for mole fraction solubility in 1,1,2,2-tetrabromoethane at 293.15 K and 1.013 bar is about 60% of the value for 1-bromooctane under the same conditions. Steric factors may play an important role in determining the tendency for hydrogen bonding to occur and hence the solubility.

COMPONENTS:	EVALUATOR:
1. Hydrogen sulfide; H$_2$S; [7783-06-4] 2. Non-aqueous solvents	Peter G.T. Fogg, Department of Applied Chemistry and Life Sciences, Polytechnic of North London, Holloway, London N7 8DB, U.K. July 1987

CRITICAL EVALUATION:

Solubilities in these brominated compounds measured by Bell and by Gerrard should be treated as tentative values.

Bell (3) also measured the mole fraction solubility in trichloromethane at a partial pressure of 1.013 bar and 293.15 K. The value at 265.15 K may be calculated from Gerrard's measurements and is consistent with Bell's value for 293.15 K. The value of 0.103 at 293.15 K may be compared with the value for tribromomethane mentioned above of 0.059 under these conditions. There would be less steric hindrance to hydrogen bond formation with hydrogen sulfide in the case of the chloro-compound and this may account for the apparent large difference in mole fraction solubilities. The corresponding reference value based upon Raoult's law is 0.055.

The mole fraction solubility in 1,1,2,2-tetrachloroethane at 293.15 K and 1.013 bar given by Bell (3) is 0.0702. This is close to the value of 0.0719 for 1,2-dichloroethane. The corresponding tetrabromo compound has a lower value 0.0446. Bell's values for mole fraction solubilities under these conditions in pentachloroethane and in trichloroethene are 0.0514 and 0.0482 respectively. Values for tetrachloromethane and tetrachloroethene, which contain no hydrogen, are even lower; 0.0419 and 0.0372 respectively.

The mole fraction solubility in 2,2,2-trichloroethanol, at 273.15 and 1.013 bar, calculated from Gerrard's data (6), is about 15% greater than the corresponding value for ethanol. The value for 2,2-dichloroacetic acid at 283.15 K from Gerrard's measurements is about 30% greater than the hypothetical value for acetic acid found by extrapolating solubilities measured at temperatures above the melting point. The mole fraction solubility in 1,1'-oxybis(2-chloroethane) at 293.2 K and 1.013 bar may be estimated from the chromatographic data given by Devyatykh *et al.* (20) to be 0.053. This may be compared with an approximate value for 1,1'-oxybisethane (diethyl ether) of 0.1 at 299.2 K and a more reliable value of 0.088 for 1,1'-oxybisoctane (dioctyl ether). However, values of solubilities from chromatographic data should be used with caution.

There is no reason, however, to doubt solubilities in chloro-compounds discussed above and reported by Bell and by Gerrard. Values should therefore be accepted as tentative.

Solubilities in chlorobenzene have been measured by Bell (3), by Gerrard (6), by Patyi *et al.* (20) and by Short *et al.* (27) in the temperature range 263.15 K to 333.15 K and at barometric pressures. Interpolated values of mole fraction solubilities for a partial pressure of 1.013 bar from data published by Short *et al.* are within 4% of values calculated from Gerrard's data for 273.15 K, 283.15 K and 293.15 K but are about 12% lower than Gerrard's values for 265.15 K and 267.15 K. The interpolated value for 293.15 is about 40% higher than Bell's value. The value for 298.15 is about 30% higher than the value given by Patyi *et al.* Despite these differences the evaluator is of the opinion that the measurements by Short *et al.* using modern techniques and apparatus are the most reliable and should be accepted on a tentative basis.

Solubilities in bromobenzene have been measured by Gerrard (6) at barometric pressure over the temperature range 266.15 K to 293.15 K. The mole fraction solubility for a partial pressure of gas of 1.013 bar calculated from Gerrard's data is about 49% greater than the figure given by Bell (3). The extrapolated value for 298.15 K is about 34% greater than the value given by Patyi *et al.* (20). Values from Gerrard's measurements are, however, close to interpolated values for chlorobenzene from the work by Short *et al.* (27) mentioned above. They are also close to values for iodobenzene and for benzene which may also be found from the data which he has reported.

Gerrard (6) measured the solubility of hydrogen sulfide at barometric pressure in iodobenzene over the range 265.15 K to 293.15 K. As indicated above, mole fraction solubilities fit into a general pattern but no other solubility data for this gas in this solvent have been found in the literature.

COMPONENTS:	EVALUATOR:
1. Hydrogen sulfide; H₂S; [7783-06-4] 2. Non-aqueous solvents	Peter G.T. Fogg, Department of Applied Chemistry and Life Sciences, Polytechnic of North London, Holloway, London N7 8DB, U.K. July 1987

CRITICAL EVALUATION:

Compounds containing nitrogen

1-Methyl-2-pyrrolidinone; C_5H_9NO; [872-50-4]   Benzenamine (*aniline*); C_6H_7N;
N,N-Dimethylbenzenamine (*dimethylaniline*); $C_8H_{11}N$; [121-69-7]   [62-53-3]
N-Methylbenzenamine (*methylaniline*); C_7H_9N; [100-61-8]
N-Ethylbenzenamine (*ethylaniline*); $C_8H_{11}N$; [103-69-5]
N,N-Diethylbenzenamine (*diethylaniline*); $C_{10}H_{15}N$; [91-66-7]
Pyridine; C_5H_5N; [110-86-1]   Quinoline; C_9H_7N; [91-22-5]
3-Methyl-1H-pyrazole; $C_4H_6N_2$; [1453-58-3]
1,3-Dimethyl-1H-pyrazole; $C_5H_8N_2$; [694-48-4]
Nitrobenzene; $C_6H_5NO_2$; [98-95-3]   Acetonitrile; C_2H_3N; [75-05-8]
Benzonitrile; C_7H_5N; [100-47-0]   N,N-Dimethylformamide; C_3H_7NO; [68-12-2]
N,N-Dimethylacetamide; C_4H_9NO; [127-19-5]
Hexahydro-1-methyl-2H-azepin-2-one (*N-methyl-ε-caprolactam*); $C_7H_{13}NO$;
                                                          [2556-73-2]

Mole fraction solubilities of hydrogen sulfide in compounds containing basic nitrogen atoms are relatively high compared with solubilities in other compounds under the same conditions of temperature and pressure.

Various measurements of the solubilities in 1-methyl-2-pyrrolidinone under different conditions of temperature and pressure have been reported. (15, 36, 41-44). There are discrepancies between solubilities reported by different authors as may be seen below:

	Henry's constant/bar	
	298.15 K	323.15 K
Lenoir *et al* (15)	5.56	
Sweeney (36)	5.81	12.3
Rivas & Prausnitz (41,43)	7.6	13.5
Yarym-Agaev *et al* (42)	6.41†	11.6†
Murrieta-Guevara & Rodriguez (44)	7.5	

† calculated by the evaluator from the solubility at 1.013 bar

The chromatographic method used by Lenoir *et al* and by Sweeney can give unreliable results. Murrieta-Guevara & Rodriguez have plotted their measurements at 298.2 K and partial pressures from 0.21 to 1.64 bar to show a linear variation of mole fraction solubility with pressure with a slope corresponding to a Henry's constant of 7.5, close to the value obtained by Rivas & Prausnitz. The line, however, does not pass through the origin. This may be an artefact of the apparatus or may reflect the limitations of Henry's law as applied to this system. The high pressure measurements by Yarym-Agaev *et al*. may be accepted on a tentative basis as reflecting the overall pattern of behaviour over the temperature and pressure range in which measurements were made. Mole fraction solubilities at a partial pressure of 1.013 bar may be estimated from Henry's constants reported by Rivas & Prausnitz but the evaluator cannot judge whether these estimated solubilities are more reliable than solubilities at this pressure reported by Yarym-Agaev *et al*.

Rivas & Prausnitz (43) also measured solubilities in methyl pyrrolidinone in the presence of about 12 mole % of 2-(2-aminoethoxy)-ethanol (diglycolamine). Muerieta-Guevara and Rodriguez (44) measured solubilities in the presence of 5.1 wt% and 14.3 wt% of 2-aminoethanol (monoethanolamine). These measurements may be accepted on a tentative basis.

Solubilities in benzenamine at pressures of about 1.013 bar have been measured by Gerrard (6) over the temperature range 265.15 K to 293.15 K and by Patyi *et al*. (22) at 298.15 K. Bancroft & Belden (45) made measurements over a pressure range from 0.14 to 1.55 bar at 295.15 K. Extrapolation of mole fraction solubilities for a partial pressure of 1.013 bar calculated from Gerrard's measurements give a value which is 24% above the value given by Patyi *et al*. and about 7% above a value from the work by Bancroft & Belden. The value calculated from the Henry's constant given by Lenoir *et al*. (15) is about 24% greater than the value from Gerrard.

COMPONENTS:	EVALUATOR:
1. Hydrogen sulfide; H₂S; [7783-06-4] 2. Non-aqueous solvents	Peter G.T. Fogg, Department of Applied Chemistry and Life Sciences, Polytechnic of North London, Holloway, London N7 8DB, U.K. July 1987

CRITICAL EVALUATION:

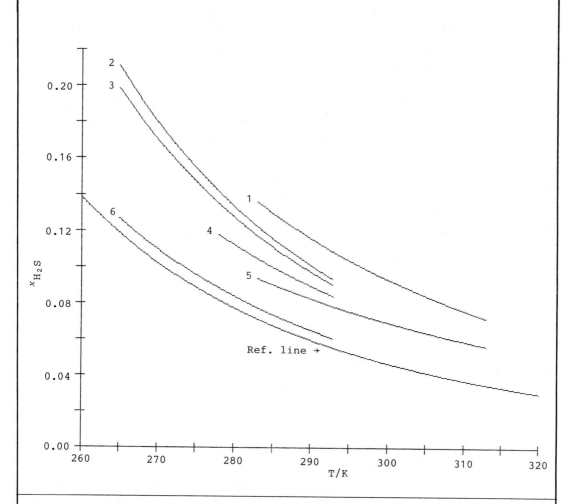

Fig. 5   The general pattern of mole fraction solubilities of hydrogen sulfide
           at a partial pressure of 1.013 bar in nitrogen bases

1. 1,3-dimethyl-1H-pyrazole; (ref.46)
2. pyridine; (ref.6)
3. quinoline; (ref.6)
4. N,N-dimethylbenzenamine (dimethylaniline); (ref.6)
5. 3-methyl-1H-pyrazole; (ref.46)
6. benzenamine (aniline); (ref.6)

The reference line corresponds to values from the Raoult's law equation.

COMPONENTS:	EVALUATOR:
1. Hydrogen sulfide; H$_2$S; [7783-06-4] 2. Non-aqueous solvents	Peter G.T. Fogg, Department of Applied Chemistry and Life Sciences, Polytechnic of North London, Holloway, London N7 8DB, U.K. July 1987

CRITICAL EVALUATION:

The solubility measurement by Lenoir et al. can be rejected because of the uncertainties associated with the chromatographic technique which was used. In the case of other solvents solubilities reported by Patyi et al. tend to be lower than those reported by other workers and the solubility at a single temperature may be less reliable than those reported by Gerrard over a temperature range in this case. Gerrard's measurements should therefore be accepted on a tentative basis.

Solubilities in N,N-dimethylbenzenamine have been measured at barometric pressure by Gerrard (6) at 278.15 K, 283.15K & 293.15 K and by Patyi et al. (22) at 298.15 K. The extrapolated value of the mole fraction solubility at 298.15 K from Gerrard's measurements is again appreciably higher than the value from Patyi et al.; 0.076 compared with 0.056. The evaluator considers the Gerrard value to be the more reliable but the discrepancy should be borne in mind.

Solubilities in N-methylbenzenamine, N-ethylbenzenamine and N,N-diethylbenzenamine have been reported by Patyi et al. (22) but there are no other measurements on these compounds for comparison.

Solubilities in pyridine and in quinoline have been measured by Gerrard (6) at barometric pressure in the temperature range 265.15 K to 293.15 K. The mole fraction solubilities for a partial pressure of 1.013 bar are similar in the two compounds (see fig.5). The solubilities in these compounds can be accepted as tentative values.

Solubilities in 3-methyl-1H-pyrazole and in 1,3-dimethyl-1H-pyrazole were measured by Egorova et al. (46) at pressures below 1.013 bar in the temperature range 283.2 K to 313.2 K. Values of Henry's constants were reported. No other data on these compounds are available for comparison and these measurements should be accepted on a tentative basis.

The general pattern of solubilities in nitrogen bases at a partial pressure of gas of 1.013 bar, on the basis of measurements discussed above, is shown in fig.5.

Solubilities in nitrobenzene at barometric pressure were measured by Gerrard (6) at temperatures from 265.15 K to 293.15 K. Solubilities have also been measured by chromatographic techniques by Devyatykh et al. (20) and by Lenoir et al. (15) The mole fraction solubility at 293.15 K calculated from data given by Devyatkh et al. is 0.053. Although, by itself, this value is of uncertain status because it is based upon chromatographic measurements it does substantiate the data reported by Gerrard. However, mole fraction solubilities at other temperatures, estimated from the distribution coefficient at 293.15 K and the heat of solution given by Devyatykh et al, do not agree with values from Gerrard's measurements. The mole fraction solubility for a partial pressure of 1.013 bar from data given by Lenoir et al. is 0.0535 at 298.15 K compared with a value of 0.046 by extrapolation of values based upon Gerrard's work. The solubilities given by Gerrard may be accepted as tentative values, in preference to solubilities based upon chromatographic methods.

Solubility in acetonitrile has been measured by Hayduk & Pahlevanzadeh (4) at a total pressure close to 1.013 bar and temperatures in the range 268.15 K to 333.15 K. They found the corrected mole fraction solubility at 298.15 for a partial pressure of gas of 1.013 bar to be 0.0476. This value is in sharp contrast to the figure of 0.031 from measurements by Evans & Blount (47). In the opinion of the evaluator the data presented by Hayduk & Pahlevanzadeh are likely to be the more reliable and may be accepted on a tentative basis.

Gerrard (6) measured the solubility in benzonitrile at barometric pressure and temperatures from 265.15 K to 293.15 K. Mole fraction solubilities from this work lie close to a smooth curve above the reference line corresponding to the Raoult's law equation. There is no reason to doubt the reliability of these values which should be accepted on a tentative basis.

COMPONENTS:	EVALUATOR:
1. Hydrogen sulfide; H$_2$S; [7783-06-4] 2. Non-aqueous solvents	Peter G.T. Fogg, Department of Applied Chemistry and Life Sciences, Polytechnic of North London, Holloway, London N7 8DB, U.K. July 1987

CRITICAL EVALUATION:

Solubilities in N,N-dimethylformamide measured at pressures of about 1 bar, have
been reported by the following:

DuPont du Nemours and Co. (Inc.) (48)   (298.15 K)
Gerrard (6)   (265.15 K - 293.15 K)
Byeseda, Deetz & Manning (30)   (297.1 K)
Hayduk & Pahlevanzadeh (4)   (268.15 K - 333.15 K)

When measurements by Gerrard and by Hayduk are corrected to a partial pressure
of 1.013 bar there is very good agreement between the two sets of results over
the whole of the temperature range. The DuPont value is slightly lower and the
Byeseda value appreciably lower than the Gerrard and the Hayduk values. The
corrected mole fraction solubility at 298.15 K from Gerrard's measurements is
0.119 compared with a value of 0.116 given by Hayduk and 0.109 from the
solubility given by DuPont. The value of 0.097 at 297.1 K from measurements by
Byeseda must be disregarded.   The following equation for mole fraction
solubilities at a partial pressure of 1.013 bar is based upon measurements by
Gerrard and by Hayduk & Pahlevanzadeh :

$$\log_{10} x_{H_2S} = -4.053 + 930/(T/K)$$
$$\delta\ x_{H_2S} = \pm\ 0.003 \text{ ; valid for 265.15 K to 333.15 K.}$$

Hayduk & Pahlevanzadeh (4) have also measured solubilities in
N,N-dimethylacetamide over the temperature range 268.15 K to 333.15 K. These
measurements are likely to be reliable but should be accepted on a tentative
basis until other measurements on this system are available for comparison.

Solubility in hexahydro-1-methyl-2H-azepin-2-one (N-methyl-ε-caprolactam) has
been reported by Wehner et al. (49). The evaluator has estimated that the mole
fraction solubility is about 0.18 at 293.2 K and a partial pressure of 1.013
bar. There are no data on comparable systems.

Compounds containing phosphorus

Phosphoric acid, trimethyl ester; C$_3$H$_9$O$_4$P; [512-56-1]
Phosphoric acid, triethyl ester; C$_6$H$_{15}$O$_4$P; [78-40-0]
Phosphoric acid, tripropyl ester; C$_9$H$_{21}$O$_4$P; [513-08-6]
Phosphoric acid, tributyl ester; C$_{12}$H$_{27}$O$_4$P; [126-73-8]
Phosphoric acid, tris(2-methylpropyl) ester; C$_{12}$H$_{27}$O$_4$P; [126-71-6]
Phosphoric triamide, hexamethyl; C$_6$H$_{18}$NO$_3$P; [680-31-9]

The solubility in tributyl phosphate has been measured by Härtel (35) over the
temperature range 293 - 373 K. Measurements were reported as Henry's constants.
based on measurements to a mole fraction concentration not greater than 0.16.
The author found that mole fraction solubility varied approximately linearly
with pressure. Henry's constants at infinite dilution were also reported by Vei
et al. (50) for the temperature range 298 to 383 K and by Lenoir et al. (15) for
298.15 K and by Sweeney (36) for 298.15 & 323.15 K. Sergienko et al. (51)
published small scale graphs showing variation of mole fraction solubility with
pressure to 0.86 bar in the temperature range 223 to 313 K. These may be used
to estimate limiting values of Henry's constant for comparison purposes. Values
of Henry's constants from measurements by various workers are shown in fig.6.

Lenoir et al. (15) also published Henry's law constants, (partial pressure of
gas/mole fraction solubility) for the trimethyl, triethyl, tripropyl and
tris(2-methyl propyl) esters of phosphoric acid and also hexamethyl phosphoric
triamide. These were also measured by gas chromatography at low partial
pressures of hydrogen sulfide. , The evaluator has found no other measurements
of solubilities of hydrogen sulfide in these compounds recorded in the
literature. An estimation of the reliability of these measurements is not
possible. The chromatographic technique can give unreliable solubility
measurements.   There are divergences between values of solubilities in other
solvents published in this paper and values from other sources. The Henry's law
constant reported for hexamethylphosphoric triamide is exceptionally low (1.63
bar or 1.61 atm).   Even if this value is reliable it is incorrect to assume a

COMPONENTS:	EVALUATOR:
1. Hydrogen sulfide; H_2S; [7783-06-4] 2. Non-aqueous solvents	Peter G.T. Fogg, Department of Applied Chemistry and Life Sciences, Polytechnic of North London, Holloway, London N7 8DB, U.K. July 1987

CRITICAL EVALUATION:

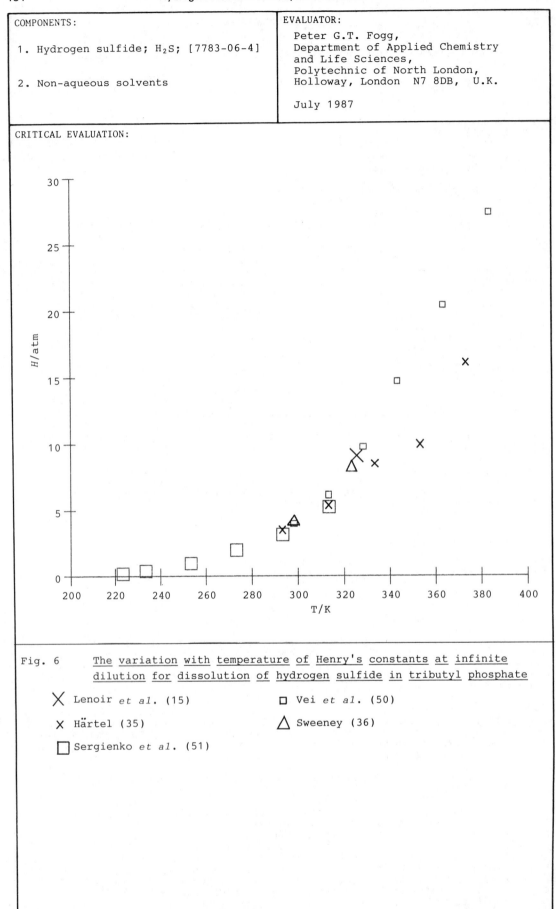

Fig. 6   The variation with temperature of Henry's constants at infinite dilution for dissolution of hydrogen sulfide in tributyl phosphate

✕ Lenoir *et al.* (15)   □ Vei *et al.* (50)

✗ Härtel (35)   △ Sweeney (36)

□ Sergienko *et al.* (51)

COMPONENTS:	EVALUATOR:
1. Hydrogen sulfide; H_2S; [7783-06-4] 2. Non-aqueous solvents	Peter G.T. Fogg, Department of Applied Chemistry and Life Sciences, Polytechnic of North London, Holloway, London N7 8DB, U.K. July 1987

CRITICAL EVALUATION:

linear variation of mole fraction solubility with partial pressure to 1.013 bar.
Linear variation would give a mole fraction of 0.621 but appreciable deviation
from proportionality will occur at a much lower concentration.

Compounds containing sulfur

Sulfinylbismethane (*dimethylsulfoxide*); C_2H_6OS; [67-68-5]
Tetrahydro-1,1-thiophene (®*Sulfolane*); $C_6H_8O_2S$; [126-33-0]
Carbon disulfide; CS_2; [75-15-0]   Sulfur; S; [7704-34-9]

Lenoir *et al.* (15) measured Henry's constant in dimethylsulfoxide by a
chromatographic method. The reliability of the measurement cannot be judged but
it is subject to the same uncertainty as other chromatographic measurements of
solubility.

Henry's constants for solubility in ®Sulfolane have been measured by Rivas &
Prausnitz (41) at temperatures from 303.15 K to 373.15 and unspecified pressures
below 1.013 bar. Mole fraction solubilities for a partial pressure of 1.013 bar
may be calculated on the assumption that there is linear variation of mole
fraction solubility with pressure. Extrapolation of these solubilities to 297.1
K gives a mole fraction solubility of 0.055. This value may be compared with
the value of 0.0686 at this temperature which may be calculated from the data
given by Byeseda *et al.* (30). The Rivas & Prausnitz data are, however, likely
to be the more reliable and may be accepted on a tentative basis.

Vapor pressures of various mixtures of hydrogen sulfide and carbon disulfide
have been measured by Gattow & Krebs (52) in the temperature range 153 - 213 K.
These measurements correspond to a self-consistent pattern of variation of mole
fraction solubility with variation in partial pressure of hydrogen sulfide and
can be accepted on a tentative basis. There are no other measurements on this
system available for comparison.

The solubility of hydrogen sulfide in molten sulphur has been measured by
Fanelli (53) at barometric pressure from 399.2 K to 623.2 K. The reversible
absorption of hydrogen sulfide increases with increase in temperature until it
reaches a maximum at about 644 K. It subsequently decreases as the boiling
point of sulfur is approached. This behaviour is consistent with the reversible
formation of unstable polyhydrogen sulfides by an endothermic process. The
measurements reported by Fanelli form a consistent set but there are no other
comparable measurements in the literature to judge them against. There is no
reason to question their reliability and they can be accepted as tentative
measurements.

Compounds containing silicon

Triethoxysilane; $C_6H_{16}O_3Si$; [998-30-1]
Silicic acid, tetraethyl ester; $C_8H_{28}O_4Si$; [78-10-4]
Silicone oils.

Devyatykh *et al.* (20) measured distribution constants for triethoxysilane,
tetraethylsilicate, and three different silicones by a chromatographic method.
It is uncertain whether these may be equated with Ostwald coefficients. Surface
effects and non-equilibrium conditions can cause errors in the estimation of
solubility by the chromatographic method.

References

1.   King, M.B.; Al-Najjar, H. *Chem. Eng. Sci.* 1977, *32*, 1241-1246.

2.   Makranczy, J.; Megyery-Balog, K.; Rusz, L.; Patyi, L.
     Hung. J. Ind. Chem. 1976, *4*, 269-280.

COMPONENTS:	EVALUATOR:
1. Hydrogen sulfide; H_2S; [7783-06-4] 2. Non-aqueous solvents	Peter G.T. Fogg, Department of Applied Chemistry and Life Sciences, Polytechnic of North London, Holloway, London N7 8DB, U.K. July 1987

CRITICAL EVALUATION:

3.   Bell, R.P. *J. Chem. Soc.* 1931, 1371-1382.

4.   Hayduk, W.; Pahlevanzadeh, H. *Can. J. Chem. Eng.* 1987, *65*, 299-307.

5.   Ng, H.-J.; Kalra, H.; Robinson, D.B.; Kubota, H.
     J. Chem. Eng. Data 1980, *25*, 51-55.

6.   Gerrard, W. *J. Appl. Chem. Biotechnol.* 1972, *22*, 623-650.

7.   Reamer, H.H.; Selleck, F.T.; Sage, B.H.; Lacey, W.N.
     Ind. Eng. Chem. 1953, *45*, 1810-1812.

8.   Tremper, K.K.; Prausnitz, J.M. *J. Chem. Engng. Data* 1976, *21*, 295-298.

9.   Young, C.L.
     Solubility Data Series Volume 12, Pergamon Press, New York, 1983, 116.

10.   Gilliland, E.R.; Scheeline, H.W. *Ind. Eng. Chem.* 1940, *32*, 48-54.

11.   Robinson, D.B.; Hughes, R.E.; Sandercock, J.A.W.
      Can. J. Chem. Eng. 1964, *42(4)*, 143-146.

12.   Besserer; G.J.; Robinson, D.B. *J. Chem. Eng. Japan* 1975, *8*, 11-13.

13.   Reamer, H.H.; Sage, B.H.; Lacey, W.N.
      Ind. Eng. Chem. 1953, *45*, 1805-1809.

14   Eakin, B.E.; DeVaney, W.E.
     Am. Inst. Chem. Engnrs. Symp. Ser. 1974, No. 140, *70*, 80-90.

15.   Lenoir, J-Y.; Renault, P.; Renon, H.
      J. Chem. Eng. Data 1971, *16*, 340-342.

16.   Asano, K.; Nakahara, T.; Kobayashi, R.
      J. Chem. Eng. Data 1971, *16*, 16-18.

17.   Tsiklis, D.S.; Svetlova, G.M. *Zh. Fiz. Khim.* 1958, *32*, 1476-1480.

18.   Ng, H-J.; Robinson, D.B. *Fluid Phase Equilibria* 1979, *2*, 283-292.

19.   Huang, S. S.-S.; Robinson, D.B. *J. Chem. Eng. Data* 1985, *30*, 154-157.

20.   Devyatykh, G.G.; Ezheleva, A.E.; Zorin, A.D.; Zueva, M.V.
      Zh. Neorgan. Khim. 1963, *8(6)*, 1307-1313.
      Russ. J. Inorg. Chem. 1963, *8*, 678-682.

21.   Hannaert, H.; Haccuria, M.; Mathieu, M.P.
      Ind. Chim. Belge 1967, *32*, 156-164.

22.   Patyi, L.; Furmer, I.E.; Makranczy, J.; Sadilenko, A.S.; Stepanova, Z.G.;
      Berengarten, M.G. *Zh. Prikl. Khim.* 1978, *51*, 1296-1300.
      J. Applied Chem. (Russ.) 1978, *51*, 1240-1243.

23.   Huang, S.S.-S.; Robinson, D.B. *Fluid Phase Equilibria* 1984, *17*, 373-382.

24.   Kragas, T.K.; Kobayashi, R. *Fluid Phase Equilibria* 1984, *16*, 215-236.

25.   Fogg, P.G.T.
      Solubility Data Series Vol.21, Pergamon Press, New York, 1985, 85-91.

26.   Yorizane, M.; Sadamoto, S.; Masuoka, H.; Eto, Y.
      Kogyo Kagaku Zasshi 1969, 72, 2174-2177.

27.   Short, I.; Sahgal, A.; Hayduk, W. *J. Chem. Eng. Data* 1983, *28*, 63-66.

28.   Bezdel, L.S.; Teodorovich, V.P. *Gazovaya Prom.* 1958, *No.8*, 38-43.

COMPONENTS:	EVALUATOR:
1. Hydrogen sulfide; H$_2$S; [7783-06-4] 2. Non-aqueous solvents	Peter G.T. Fogg, Department of Applied Chemistry and Life Sciences, Polytechnic of North London, Holloway, London N7 8DB, U.K. July 1987

CRITICAL EVALUATION:

29.   Fauser, G. *Math. Naturwiss. Ber. Ung.* 1888, *6*, 154.

30.   Byeseda, J.J.; Deetz, J.A.; Manning, W.P.
      Proc. Laurance Reid Gas Cond. Conf. 1985.

31.   Parsons, L.B. *J. Amer. Chem. Soc.* 1925, *47*, 1820-1830.

32.   Blake, R.J. *Oil Gas J.* 1967, *65(2)*, 105-108.

33.   Gestrich, W.; Reinke, L. *Chem.-Ing.-Tech.* 1983, *55*, 629-631.

34.   Sciamanna, S.F.
      Ph.D. Dissertation, Department of Chemical Engineering, University of
      California,Berkeley, U.S.A. 1986. (Directed by Lynn, S.)

35.   Härtel, G.H. *J. Chem. Eng. Data* 1985, *30*, 57-61.

36.   Sweeney, C.W. *Chromatographia* 1984, *18*, 663-667.

37.   Wolfer, W.; Schwartz, E.; Vodrazka, W.; Volkamer, K.
      Oil Gas J. 1980, *78(3)*, 66-70.

38.   Linford, R.G.; Thornhill, D.G.T.
      J. Chem. Thermodynamics 1985, *17*, 701-702.

39.   Shakhova, S.F.; Bondareva, T.I.; Zubchenko, Yu.P.
      Khim. Prom. 1966, *(10)*, 753-754.

40.   Isaacs, E.E.; Otto, F.D.; Mather, A.E.
      Can. J. Chem. Eng. 1977, *55*, 751-752.

41.   Rivas, O.R.; Prausnitz, J.M. *Am. Inst. Chem. Engnrs.* 1979, *25*, 975-984.

42.   Yarym-Agaev, V.G.; Matvienko, V.G.; Povalyaeva, N.V.
      Zhur. Prikl. Khim. 1980, *53*, 2456-2461.
      J. Applied Chem. (Russ.) 1980, *53*, 1810 -1814.

43.   Rivas, O.R.; Prausnitz, J.M. *Ind. Eng. Chem. Fundam.* 1979, *18*, 289-292.

44.   Murrieta-Guevara, F.; Rodriguez, A.T.
      J. Chem. Eng. Data 1984, *29*, 456-460.

45.   Bancroft, W.D.; Belden, B.C. *J. Phys. Chem.* 1930, *34*, 2123-2124.

46.   Egorova, V.I.; Grishko, N.I.; Neokladnova, L.N.; Furmanov, A.S.;
      Podvigailova, I.G. *Deposited document* 1976, *VINITI 2907-76.*

47.   Evans, J.F.; Blount, H.N. *Anal. Lett.* 1974, *7(6)*, 445-451.

48.   DuPont de Nemours and Co. (Inc.) *Chem. Eng. News* 1955, *33*, 2366.

49.   Wehner, K; Burk, W.; Kisan, W.
      Chem. Tech. (Leipzig) 1977, *29(8)*, 445-448.

50.   Vei,D.; Furmer, I.E.; Sadilenko, A.S.; Efimova, N.M.; Stepanova, Z.G.;
      Gracheva, N.V. *Gazov. Prom-st.* 1975, *7(7)*, 47-49.

51.   Sergienko, I.D.; Kosyakov, N.E.; Yushko, V.L.; Khokhlov, S.F.;
      Pushkin, A.G. *Vop. Khim. Tekhnol.* 1973, *29*, 57-60.

52.   Gattow, G.; Krebs, B. *Z. anorg. allgem. Chem.* 1963, *325*, 15-25.

53.   Fanelli, R. *Ind. Eng. Chem.* 1949, *41*, 2131-2033.

COMPONENTS:	ORIGINAL MEASUREMENTS:
1. Hydrogen sulfide; H_2S; [7783-06-4] 2. Alkanes, C_5 to C_{10}	Makranczy, J.; Megyery-Balog, K.; Rusz, L.; Patyi, L. *Hung. J. Ind. Chem.* 1976, *4*, 269-280.

VARIABLES:	PREPARED BY:
Temperature	P.G.T. Fogg

EXPERIMENTAL VALUES:

Solvent	T/K	Ostwald coefficient L	Mole fraction of H_2S [*] x_{H_2S} (1.013 bar)
Pentane; C_5H_{12}; [109-66-0]	298.15 313.15	9.147 6.93	0.0421 0.0314
Hexane; C_6H_{14}; [110-54-3]	298.15 313.15	8.230 6.23	0.0429 0.0319
Heptane; C_7H_{16}; [142-82-5]	298.15 313.15	7.520 5.69	0.0439 0.0325
Octane; C_8H_{18}; [111-65-9]	298.15 313.15	6.986 5.28	0.0451 0.0335
Nonane; C_9H_{20}; [111-84-2]	298.15 313.15	6.560 4.96	0.0465 0.0345
Decane; $C_{10}H_{22}$; [124-18-5]	298.15 313.15	6.232 4.71	0.0481 0.0356

$$P_{H_2S} = 1.013 \text{ bar.}$$

[*] estimated by the compiler using densities from the chemical literature.

AUXILIARY INFORMATION

METHOD/APPARATUS/PROCEDURE:	SOURCE AND PURITY OF MATERIALS:
The authors stated that apparatus described previously by Bodor *et al.* was used (ref. (1)). However Bodor *et al.* described apparatus for use below 273 K and referred to another paper (2) in which an apparatus for use above 273 K was described. Bodor *et al.* stated that, in each case, the volume of gas absorbed by a given quantity of liquid at a particular pressure was measured by a gas burette. Bodor *et al.* gave details of a method of calculating gas solubilities applicable to either apparatus, with allowance for the vapor pressure of the solvent.	Analytical grade reagents were used.

ESTIMATED ERROR:

$\delta L/L$ = ± 0.03 (authors)

REFERENCES:

1. Bodor, E; Bor, G.J. Mohai, B.; Sipos, G. Veszpremi. *Vegyip. Egy. Kozl.* 1957, *1*, 55.
2. Schay, G.; Szekely, G.; Racz, Gy. *Periodica Polytechnica Ser. Chem. Eng. (Budapest)* 1958, *2*, 1.

COMPONENTS:	ORIGINAL MEASUREMENTS:
1. Hydrogen sulfide; H_2S; [7783-06-4] 2. Alkanes, C_{11} to C_{16}	Makranczy, J.; Megyery-Balog, K.; Rusz, L.; Patyi, L. *Hung. J. Ind. Chem.* 1976, *4*, 269-280.
VARIABLES: Temperature	PREPARED BY: P.G.T. Fogg

EXPERIMENTAL VALUES:

Solvent	T/K	Ostwald coefficient L	Mole fraction of H_2S* x_{H_2S} (1.013 bar)
Undecane; $C_{11}H_{24}$; [1120-21-4]	298.15 313.15	5.949 4.50	0.0496 0.0368
Dodecane; $C_{12}H_{26}$; [112-40-3]	298.15 313.15	5.698 4.31	0.0511 0.0379
Tridecane; $C_{13}H_{28}$; [629-50-5]	298.15 313.15	5.501 4.15	0.0528 0.0390
Tetradecane; $C_{14}H_{30}$; [629-59-4]	298.15 313.15	5.305 4.02	0.0541 0.0402
Pentadecane; $C_{15}H_{32}$; [629-62-9]	298.15 313.15	5.152 3.89	0.0558 0.0413
Hexadecane; $C_{16}H_{34}$; [544-76-3]	298.15 313.15	4.999 3.80	0.0573 0.0427

$$P_{H_2S} = 1.013 \text{ bar.}$$

* estimated by the compiler using densities from the chemical literature.

AUXILIARY INFORMATION

METHOD/APPARATUS/PROCEDURE:	SOURCE AND PURITY OF MATERIALS:
The authors stated that apparatus described previously by Bodor *et al.* was used (ref. (1)). However Bodor *et al.* described apparatus for use below 273 K and referred to another paper (2) in which an apparatus for use above 273 K was described. Bodor *et al.* stated that, in each case, the volume of gas absorbed by a given quantity of liquid at a particular pressure was measured by a gas burette. Bodor *et al.* gave details of a method of calculating gas solubilities applicable to either apparatus, with allowance for the vapor pressure of the solvent.	Analytical grade reagents were used.
	ESTIMATED ERROR: $\delta L/L = \pm 0.03$ (authors)
	REFERENCES: 1. Bodor, E; Bor, G.J. Mohai, B.; Sipos, G. *Veszpremi. Vegyip. Egy. Kozl.* 1957, *1*, 55. 2. Schay, G.; Szekely, G.; Racz, Gy. *Periodica Polytechnica Ser. Chem. Eng. (Budapest)* 1958, *2*, 1.

COMPONENTS:	ORIGINAL MEASUREMENTS:
1. Hydrogen sulfide; H_2S; [7783-06-4] 2. Alkanes	Bell, R.P. *J. Chem. Soc.* <u>1931</u>, 1371-1382.
VARIABLES:	PREPARED BY: C.L. Young

EXPERIMENTAL VALUES:

Solvent	T/K	Partition coefficient, s^+	Mole fraction[§] of hydrogen sulfide in liquid, x_{H_2S}
Hexane; C_6H_{14}; [110-54-3]	293.15	6.30	0.0341
Octane; C_8H_{18}; [111-65-9]		6.80	0.0440
Dodecane; $C_{12}H_{26}$; [112-40-3]		5.71	0.0513
Hexadecane; $C_{16}H_{34}$; [544-76-3]		5.05	0.0578
Cyclohexane; C_6H_{12}; [110-82-7]		7.50	0.0338

[+] s defined as $s = 22.4 \times \dfrac{293}{273} \times c$ where c is the "solubility in equivalents/litre".

[§] for a partial pressure of 101.325 kPa.

AUXILIARY INFORMATION

METHOD/APPARATUS/PROCEDURE:	SOURCE AND PURITY OF MATERIALS:
Volumetric apparatus consisting of bulb (~50cm³ capacity) extended at the top as a graduated tube and joined at bottom to a capillary u-tube. Liquid saturated with gas at atmospheric pressure. Gas withdrawn in a current of air absorbed in sodium hydroxide and hydrogen peroxide. Excess hydrogen peroxide removed by heating and excess sodium hydroxide titrated.	1. Prepared by reaction of sodium sulfide on hydrochloric acid. 2. Merck and Kahlbaum samples dried over calcium chloride and fractionally distilled.

	ESTIMATED ERROR:
	          $\delta T/K = \pm 0.1$; $\delta x_{H_2S} = \pm 1\%$.
	          (estimated by compiler)

	REFERENCES:

COMPONENTS:	ORIGINAL MEASUREMENTS:
1. Hydrogen sulfide; H_2S; [7783-06-4] 2. Propane, C_3H_8; [74-98-6]	Gilliland, E.R.; Scheeline, H.W. *Ind. Eng. Chem.* 1940, *32*, 48-54.

VARIABLES:
    Temperature, pressure
	PREPARED BY:
	    C.L. Young

EXPERIMENTAL VALUES:

T/K	$P/10^5$Pa	Mole fraction of hydrogen sulfide in liquid, x_{H_2S}	in gas, y_{H_2S}
340.9	27.65	0.037	0.122
338.2	27.58	0.054	0.159
327.0	27.58	0.241	0.379
324.3	27.58	0.332	0.499
355.4	34.54	0.014	0.021
349.8	34.47	0.081	0.130
335.9	34.47	0.395	0.547
367.0	41.44	0.055	0.076
364.8	41.44	0.078	0.101
351.5	41.37	0.342	0.451
343.2	41.37	0.455	0.592

AUXILIARY INFORMATION

METHOD/APPARATUS/PROCEDURE:	SOURCE AND PURITY OF MATERIALS:
Mixture studied in a high pressure glass equilibrium still. Analysis of gas and liquid samples used absorption of hydrogen sulfide in caustic soda solution and subsequent titration with iodine and sodium thiosulfate. Remaining propane measured volumetrically. Details of apparatus in ref. (1).	1. Matheson Co. C.P. grade. 2. Phillips Petroleum C.P. grade purity 99.9 mole per cent.

ESTIMATED ERROR:
$\delta T/K = \pm 0.6$; $\delta P/10^5$Pa $= \pm 0.15$;
$\delta x_{H_2S'} = \delta y_{H_2S} = \pm 0.005$.

REFERENCES:
1. Scheeline, H.W.; Gilliland, D.R.

Ind. Eng. Chem. 1939, *31*, 1050

COMPONENTS:	ORIGINAL MEASUREMENTS:
1. Hydrogen sulfide; H_2S; [7783-06-4] 2. Butane; C_4H_{10}; [106-97-8]	Robinson, D.B.; Hughes, R.E.; Sandercock, J.A.W. *Can. J. Chem. Eng.* <u>1964</u>, *42,(4)*, 143-6.†
VARIABLES: Temperature, pressure	PREPARED BY: P.G.T. Fogg

EXPERIMENTAL VALUES:

T/°F	T/K*	Total pressure /psia	Total pressure /bar*	Mole fraction of H_2S in liquid phase	Mole fraction of H_2S in gas phase
100	310.93	60.4	4.16	0.021	0.133
		69.4	4.78		0.243
		75.3	5.19		0.322
		84.8	5.85	0.066	0.386
		88.0	6.07	0.076	
		111.0	7.65	0.131	0.539
		117.0	8.07		0.564
		146.8	10.12	0.209	0.655
		176.6	12.18		0.713
		203.4	14.02	0.355	0.763
		230.4	15.89		0.805
		239.3	16.50	0.468	0.818
		249.0	17.17	0.497	
		283.8	19.57	0.607	0.870
		315.8	21.77	0.716	0.904
		329.8	22.74		0.924
		343.0	23.65	0.805	
		350.7	24.18		0.945
		358.8	24.74	0.858	

* calculated by compiler;   psia = pound-force per square inch (absolute)

† data in document no. 7952 with the Auxiliary Publications Project, Photoduplication Service, Library of Congress, Washington DC, USA.

AUXILIARY INFORMATION

METHOD/APPARATUS/PROCEDURE:	SOURCE AND PURITY OF MATERIALS:
Mixtures of hydrogen sulfide and butane were confined in a stainless steel cell. This cell was thermostatted and the contents were compressed by a mercury pump. Contents of the cell could be mixed by spraying liquid phase through the vapor phase and observed through a glass window. Pressures were measured by Heise bourdon tube gauges. There was provision for withdrawing either samples of the liquid or the vapor and subsequent analysis by chromatography.	1. commercial sample dried, frozen and evacuated; purity at least 99.9 mol% 2. commercial sample distilled, frozen, evacuated and dried; purity at least 99.9 mol%

ESTIMATED ERROR:

$\delta T/K$ = ± 0.02
$\delta P/psia$ = ±0.5 (1-1000 psia);
= ±5 (> 1000 psia) [authors]

REFERENCES:

COMPONENTS:	ORIGINAL MEASUREMENTS:
1. Hydrogen sulfide; H_2S; [7783-06-4]	Robinson, D.B.; Hughes, R.E.; Sandercock, J.A.W.
2. Butane; C_4H_{10}; [106-97-8]	*Can. J. Chem. Eng.* <u>1964</u>, *42*,(4), 143-6

EXPERIMENTAL VALUES:

T/°F	T/K*	Total pressure /psia	Total pressure /bar*	Mole fraction of H_2S in liquid phase	Mole fraction of H_2S in gas phase
125	324.82	102	7.03	0.035	
		128	8.83	0.096	0.405
		144	9.93	0.118	0.460
		160	11.03	0.150	
		201	13.86	0.247	0.635
		214	14.75	0.262	0.633
		234	16.13		0.688
		246	16.96	0.325	
		247	17.03	0.332	
		253	17.44	0.349	0.725
		275	18.96		0.769
		282	19.44	0.399	
		312	21.51	0.469	0.811
		392	27.03	0.631	0.877
		419	28.89		0.891
		420	28.96	0.716	
		447	30.82	0.778	0.913
		457	31.51	0.791	
		483	33.30		0.945
		487	33.58	0.862	
175	352.59	291	20.06	0.184	0.525
		411	28.34	0.364	0.680
		477	32.89	0.465	0.749
		627	43.23	0.692	0.840
		741	51.09	0.810	0.900
200	366.48	340	23.44		0.431
		343	23.65	0.172	
		435	29.99	0.275	0.582
		437	30.13	0.287	
		537	37.02	0.414	
		554	38.20		0.693
		654	45.09	0.520	0.770
		683	47.09	0.560	
		768	52.95	0.674	
		853	58.81	0.767	0.854
		990	68.26		0.906
		994	68.53	0.861	
		1060	73.1	0.920	
		1100	75.8		0.955
		1105	76.2	0.937	
		1139	78.5		0.985
225	380.37	365	25.17	0.133	0.322
		509	35.09	0.270	0.509
		705	48.61	0.465	0.631
		745	51.37		0.683
		785	54.12	0.540	
		930	64.12	0.709	0.784
		1008	69.5	0.725	0.800
		1050	72.4	0.785	0.812
		1080	74.5	0.809	
250	394.26	494	34.06	0.159	0.319
		537	37.02	0.187	0.380
		628	43.30	0.284	0.454
		751	51.78	0.396	0.523
		820	56.54	0.459	
		818	56.40		0.558
		907	62.54	0.530	0.581
		931	64.19	0.541	0.588

* calculated by compiler;   psia = pound-force per square inch (absolute)

COMPONENTS:	ORIGINAL MEASUREMENTS:
1. Hydrogen sulfide; H$_2$S; [7783-06-4]	Besserer, G.J.; Robinson, D.B.
2. 2-Methylpropane; C$_4$H$_{10}$; [75-28-5]	J.Chem.Eng.Japan 1975, 8, 11 - 13.

VARIABLES:	PREPARED BY:
Temperature, concentration	P.G.T. Fogg

EXPERIMENTAL VALUES:

T/°F	T/K*	P_{total}/psia	P_{total}/bar*	Mole fraction in liquid phase x_{H_2S}	Mole fraction in gas phase y_{H_2S}
40.1	277.65	26.4	1.82	0.0000	0.0000
		30.0	2.07	0.0175	0.1270
		56.9	3.92	0.1231	0.6025
		67.0	4.62	0.1792	0.6700
		92.1	6.35	0.3090	0.7831
		112.8	7.78	0.4460	0.8460
		131.0	9.03	0.6011	0.8819
		147.9	10.20	–	0.9239
		152.7	10.53	0.8287	–
		165.0	11.38	0.9772	0.9846
		168.0	11.58	1.0000	1.0000

* calculated by compiler.

AUXILIARY INFORMATION

METHOD/APPARATUS/PROCEDURE:	SOURCE AND PURITY OF MATERIALS:
The apparatus is described in ref. (1). It consisted of a variable volume metal cell with a piston at each end and an observation window at the centre. Pressures were measured with a pressure transducer attached to the cell. Gas and liquid phases were analysed by gas chromatography.	1. Matheson C.P. grade; purity > 99.8 mol%; distilled before use. 2. Matheson instrument grade; purity > 99.9 mol%.

ESTIMATED ERROR:

REFERENCES:
1. Besserer, G.J.; Robinson, D.B. Can. J. Chem. Eng. 1971, 49, 651.

COMPONENTS:	ORIGINAL MEASUREMENTS:
1. Hydrogen sulfide; H₂S; [7783-06-4]	Besserer, G.J.; Robinson, D.B.
2. 2-Methylpropane; C₄H₁₀; [75-28-5]	J.Chem.Eng.Japan 1975, 8, 11 - 13.

EXPERIMENTAL VALUES:

$T/°F$	T/K[*]	P_{total}/psia	P_{total}/bar[*]	Mole fraction in liquid phase x_{H_2S}	Mole fraction in gas phase y_{H_2S}
100.0	310.93	72.5	5.00	0.0000	0.0000
		107.4	7.40	0.0849	0.3755
		153.2	10.56	0.1985	0.5912
		176.3	12.16	0.2574	0.6559
		224.3	15.46	0.3927	0.7563
		259.1	10.97	0.4895	0.8024
		310.8	21.43	0.6600	0.8621
		343.1	23.66	0.7680	0.8992
		365.8	25.22	0.8646	0.9304
		385.4	26.57	0.9497	0.9713
		394.0	27.17	1.0000	1.0000
160.2	344.37	160.2	11.05	0.000	0.000
		211.2	14.56	0.0774	0.2746
		267.2	18.42	0.1613	0.4353
		344.5	23.75	0.2809	0.5834
		405.2	27.94	0.3824	0.6658
		515.6	35.55	0.5464	0.7662
		620.5	42.78	0.7299	0.8654
		701.4	48.36	0.8548	0.9186
		741.9	51.15	0.9316	0.9576
		773.0	53.30	1.0000	1.0000
220.0	377.59	313.5	21.62	0.000	0.000
		383.1	26.41	0.0663	0.1543
		497.2	34.28	0.1851	0.3485
		618.2	42.62	0.3303	0.5138
		692.2	47.73	0.4047	0.5810
		738.2	50.90	0.4524	0.6043
		787.9	54.32	0.5065	0.6271
		811.8	55.97	0.5205	0.6330
		867.0	59.78	0.5804	0.6635
		872.5	60.16	0.5847	0.6653
		894.6	61.68	0.6023	0.6775

[*] calculated by compiler.

COMPONENTS:	ORIGINAL MEASUREMENTS:
1. Hydrogen sulfide; H_2S; [7783-06-4] 2. Pentane; C_5H_{12}; [109-66-0]	Reamer, H.H.; Sage, B.H.; Lacey, W.N. *Ind. Eng. Chem.*, 1953, *45*, 1805-1812.

VARIABLES:	PREPARED BY:
Temperature, pressure	C. L. Young

EXPERIMENTAL VALUES:

Mole fraction of hydrogen sulfide

T/K	P/bar	in liquid, x_{H_2S}	in gas, y_{H_2S}
277.6	1.38	0.0617	0.7842
	2.76	0.1425	0.8950
	4.14	0.2260	0.9304
	5.52	0.3232	0.9534
	6.89	0.4372	0.9685
	8.62	0.6106	0.9820
	10.34	0.821	0.993
310.9	3.45	0.0788	0.6684
	6.89	0.1951	0.8310
	10.34	0.3151	0.8970
	13.79	0.4380	0.9280
	17.24	0.5662	0.9491
	20.68	0.7080	0.9675
	24.13	0.860	0.985
344.3	6.89	0.0799	0.5124
	13.79	0.2218	0.7355
	20.68	0.3626	0.8279
	27.58	0.4995	0.8840
	34.47	0.6372	0.9277
	41.37	0.7687	0.9553
	48.26	0.900	0.981

AUXILIARY INFORMATION

METHOD/APPARATUS/PROCEDURE:	SOURCE AND PURITY OF MATERIALS:
PVT cell pressure measured with pressure balance, temperature measured with platinum resistance thermometer. Co-existing phases sampled and analysed by gas density measurements. Details in source and ref. (1).	1. Prepared by hydration of pure aluminium sulfide. Fractionated twice. 2. Phillips Petroleum Co. sample, minimum purity 99.5 mole per cent. Dried and fractionated.

ESTIMATED ERROR:

$\delta T/K = \pm 0.01$; $\delta P/bar = \pm 0.02$;

$\delta x_{H_2S} = \pm 0.003$ (Calculated by compiler)

REFERENCES:

1. Sage, B.H.; Lacey, W.N., *Trans. Am. Inst. Mining and Met. Eng.*, 1940, *136*, 136.

| T/K | P/bar | Mole fraction of hydrogen sulfide | |
		in liquid, x_{H_2S}	in gas, y_{H_2S}
377.6	6.89	0.0062	0.0559
	13.79	0.1014	0.4698
	20.68	0.1965	0.6251
	27.58	0.2912	0.7147
	34.47	0.3838	0.7745
	41.37	0.4740	0.8185
	48.26	0.5604	0.8518
	55.16	0.6421	0.8769
	62.05	0.7165	0.8963
	68.95	0.7859	0.9125
	75.84	0.8506	0.9289
	82.74	0.9110	0.9474
	89.63	0.966	0.968
410.9	13.79	0.0118	0.0662
	20.68	0.0897	0.3452
	27.58	0.1630	0.4850
	34.47	0.2326	0.5698
	41.37	0.3003	0.6292
	48.26	0.3655	0.6709
	55.16	0.4294	0.7018
	62.05	0.4910	0.7230
	68.95	0.5510	0.7356
	75.84	0.6108	0.7420
	82.74	0.680	0.749
444.3	27.58	0.0402	0.1385
	34.47	0.0983	0.2732
	41.37	0.1585	0.3689
	48.26	0.2217	0.4420
	55.16	0.2880	0.4990
	62.05	0.3547	0.5352
	68.95	0.428	0.566
	75.84	0.515	0.575

COMPONENTS:	ORIGINAL MEASUREMENTS:
1. Hydrogen sulfide; H_2S; [7783-06-4] 2. Hexane; C_6H_{14}; [110-54-3]	King, M. B.; Al-Najjar, H. *Chem. Eng. Sci.* 1977, *32*, 1241-6.

VARIABLES:	PREPARED BY:
Temperature	C. L. Young

EXPERIMENTAL VALUES:

T/K	Mole fraction of hydrogen sulfide[*] at a partial pressure of 101.3 kPa
288.2	0.0483
293.2	0.0446
298.2	0.0412
303.2	0.0381

[*] allowance was made for the non-ideal gas behaviour of hydrogen sulfide.

AUXILIARY INFORMATION

METHOD/APPARATUS/PROCEDURE:	SOURCE AND PURITY OF MATERIALS:
Solvent degassed by spraying into a continuously evacuated chamber. Solvent flows in a thin film down a glass spiral into a buret system containing the gas to be dissolved. Flow rates may be varied over a wide range without affecting the solubility. Similar to the apparatus of Morrison and Billett.	No details given.

ESTIMATED ERROR:

$\delta T/K = \pm 0.1$; $\delta x_{H_2S} = \pm 2\%$

(estimated by compiler).

REFERENCES:
1. Morrison, T. J.; Billett, F.
   J. Chem. Soc.
   1952, 3819.

COMPONENTS:	ORIGINAL MEASUREMENTS:

COMPONENTS:

1. Hydrogen sulfide; H_2S;
    [7783-06-4]
2. Ethyl Acetate; $C_3H_8O_2$;
    [141-78-6]
    Hexane; C_6H_{14}; [110-54-3]
    Chlorobenzene; C_6H_5Cl;
    [108-90-7]

ORIGINAL MEASUREMENTS:

Hayduk, W.; Pahlevanzadeh, H.

Can. J. Chem. Eng. 1987, *65*,
299-307.

VARIABLES:

   Temperature

PREPARED BY:

   P.G.T. Fogg

EXPERIMENTAL VALUES:

T/K	Total press. /kPa	Mole fraction solubility at total press.	Mole fraction solubility at partial press. of 101.325 kPa.	Ostwald* coeff. L
Ethyl Acetate				
268.15	101.6	0.183	0.186	52.2
298.15	101.3	0.0764	0.0866	23.3
333.15	100.8	0.0187	0.0413	11.3
Hexane				
268.15	100.4	0.0902	0.0905	18.0
298.15	101.0	0.0299	0.0372	7.10
Chlorobenzene				
298.15	100.7	0.0497	0.0508	12.7

* The Oswald coefficient was calculated as the volume of gas
at a gas partial pressure of 1 atm. which will completely dissolve in
one volume of solvent.

AUXILIARY INFORMATION

METHOD/APPARATUS/PROCEDURE:

Solubilities were measured using a
constant solvent flow apparatus
described in (1). Solvent vapor
pressures were calculated from
Antoine constants given in (2).
Densities of solvents were as
reported in (3). The gas molar
volumes were calculated using the
second virial coefficient given in
(4).

SOURCE AND PURITY OF MATERIALS:

1. from Matheson of Canada.

2. from Aldrich Chemicals;
   ethyl acetate & chlorobenzene HPLC
   grade of minimum purity 99.9%;
   hexane of minimum purity 99.0%

ESTIMATED ERROR:

REFERENCES:
1. Asatani, H.; Hayduk, W.
   Can.J.Chem.Eng. 1983, *61*, 227.
2. Reid, R.C.; Prausnitz, J.M.;
   Sherwood, T.K. *The Properties of
   Gases and Liquids*, 1977,
   McGraw-Hill, New York.
3. Zhang, G.; Hayduk, W. *Can.J.Chem.
   Eng.* 1984, *62*, 713.
4. Dymond, J.H.; Smith, E.B. *The
   Virial Coefficients of Pure Gases
   and Mixtures*, 1980, Oxford
   University Press, New York.

COMPONENTS:	ORIGINAL MEASUREMENTS:
1. Hydrogen sulfide; H_2S; [7783-06-4] 2. Heptane; C_7H_{16}; [142-82-5]	Ng, H.-J.; Kalra, H.; Robinson, D. B.; Kubota, H. *J. Chem. Eng. Data* <u>1980</u>, *25*, 51-55.

VARIABLES:	PREPARED BY:
Temperature, pressure	C. L. Young

EXPERIMENTAL VALUES:

T/K	P/psia	P/MPa	Mole fraction of hydrogen sulfide in liquid, x_{H_2S}	in vapor, y_{H_2S}
310.93	23.2	0.160	0.043	0.933
	81.9	0.565	0.166	0.977
	88.0	0.607	0.180	0.978
	148	1.02	0.311	0.988
	149	1.03	0.318	0.989
	194	1.34	0.435	0.992
	234	1.61	0.538	0.992
	236	1.63	0.545	0.991
	293	2.02	0.675	0.993
	294	2.03	0.680	0.994
	332	2.29	0.819	0.996
	335	2.31	0.845	0.996
	351	2.42	0.927	0.997
352.59	40.5	0.279	0.037	0.780
	151	1.04	0.164	0.937
	154	1.06	0.167	0.937
	286	1.97	0.347	0.963
	289	1.99	0.351	0.963
	451	3.11	0.556	0.976
	455	3.14	0.561	0.977
	619	4.27	0.745	0.985
	625	4.31	0.750	0.986

(cont.)

AUXILIARY INFORMATION

METHOD/APPARATUS/PROCEDURE:	SOURCE AND PURITY OF MATERIALS:
Stirred static cell used at highest temperature as described in ref. (1). Variable volume static cell used in measurement at lower temperatures. Details in ref. (2). Temperature measured with thermocouple and pressure measured with Bourdon gauge. After equilibrium established gas and liquid phase sampled and analysed by gas chromatography using a thermal conductivity detector. Details in source.	1. Thio-Pet Chemicals sample, distilled, final purity at least 99.9 mole per cent, as determined by GC. 2. Aldrich Chemicals sample, purity better than 99.9 mole per cent as determined by GC.

ESTIMATED ERROR:
$\delta T/K = \pm 0.06$; $\delta P/MPa = \pm 0.02$; δx_{H_2S}, $\delta y_{H_2S} = \pm 0.003$.

REFERENCES:
1. Ng, H.-J.; Robinson, D. B. *J. Chem. Eng. Data* <u>1978</u>, *23*, 325.

2. Besserer, G. J.; Robinson, D. B. *Can. J. Chem. Eng.* <u>1971</u>, *8*, 334.

COMPONENTS:	ORIGINAL MEASUREMENTS:
1. Hydrogen sulfide; H$_2$S; [7783-06-4] 2. Heptane; C$_7$H$_{16}$; [142-82-5]	Ng, H.-J.; Kalra, H.; Robinson, D. B.; Kubota, H. *J. Chem. Eng. Data* 1980, *25*, 51-55.

EXPERIMENTAL VALUES:

T/K	P/psia	P/MPa	Mole fraction of hydrogen sulfide in liquid, x_{H_2S}	in vapor, y_{H_2S}
352.59	790	5.45	0.896	0.992
	807	5.56	0.908	0.991
394.26	87.8	0.605	0.041	0.702
	159	1.10	0.094	0.812
	162	1.12	0.096	0.810
	338	2.33	0.234	0.916
	509	3.51	0.385	0.933
	509	3.51	0.384	0.935
	729	5.03	0.549	0.948
	909	6.27	0.665	0.953
	912	6.29	0.667	0.955
	1063	7.33	0.749	0.957
	1079	7.44	0.757	0.959
	1198	8.26	0.811	0.959
	1221	8.42	0.821	0.960
	1354	9.34	0.871	0.957
	1385	9.55	0.880	0.954
477.59	182	1.25	0.0177	0.104
	242	1.67	0.0414	0.260
	334	2.30	0.0929	0.443
	488	3.36	0.188	0.589
	652	4.50	0.279	0.652
	856	5.90	0.386	0.677
	964	6.65	0.447	0.760
	1093	7.54	0.514	0.707
	1168	8.05	0.552	0.704
	1213	8.36	0.578	0.676

COMPONENTS:	ORIGINAL MEASUREMENTS:
1. Hydrogen sulfide, H_2S; [7783-06-4] 2. Octane; C_8H_{18}; [111-65-9]	King, M. B.; Al-Najjar, H. *Chem. Eng. Sci.* <u>1977</u>, *32*, 1241-6.

VARIABLES:	PREPARED BY:
Temperature	C. L. Young

EXPERIMENTAL VALUES:

T/K	Mole fraction of hydrogen sulfide[*] at a partial pressure of 101.3 kPa
288.2	0.0513
293.2	0.0474
298.2	0.0437
303.2	0.0402

[*] allowance was made for the non-ideal gas behaviour of hydrogen sulfide.

AUXILIARY INFORMATION

METHOD/APPARATUS/PROCEDURE:	SOURCE AND PURITY OF MATERIALS:
Solvent degassed by spraying into a continuously evacuated chamber. Solvent flows in a thin film down a glass spiral into a buret system containing the gas to be dissolved. Flow rates may be varied over a wide range without affecting the solubility. Similar to the apparatus of Morrison and Billett.	No details given.

ESTIMATED ERROR:
$\delta T/K = \pm 0.1$; $\delta x_{H_2S} = \pm 2\%$

(estimated by compiler).

REFERENCES:
1. Morrison, T. J.; Billett, F.
   J. Chem. Soc.
   <u>1952</u>, 3819.

COMPONENTS:	ORIGINAL MEASUREMENTS:
1. Hydrogen sulfide; H_2S; [7783-06-4] 2. Nonane; C_9H_{20}; [111-84-2]	Eakin, B.E.; DeVaney, W.E. *Am. Inst. Chem. Engnrs. Symp. Ser.* <u>1974</u>, No. 140. *70*, 80-90.
VARIABLES: Temperature, pressure	PREPARED BY: C.L. Young

EXPERIMENTAL VALUES:

T/K	$P/10^5 Pa$	Mole fraction of hydrogen sulfide	
		in liquid, x_{H_2S}	in gas, y_{H_2S}
310.9	1.37	0.0448	0.99131
	2.26	0.0747	0.99436
	3.43	0.1160	0.99666
	4.61	0.1584	0.99693
	5.79	0.2004	0.99749
366.5	2.55	0.0419	0.9052
	4.88	0.0799	0.9670
	7.39	0.1234	0.9760
	9.71	0.1645	0.9824
	12.27	0.2088	0.9864
477.6	8.23	0.0441	0.5449
	12.88	0.0778	0.6635
	17.58	0.1222	0.7314
	22.55	0.1616	0.8054
	27.65	0.2035	0.8048

AUXILIARY INFORMATION

METHOD/APPARATUS/PROCEDURE:	SOURCE AND PURITY OF MATERIALS:
Rocking stainless steel equilibrium vessel, fitted with liquid and gas sampling ports. Pressure measured with Bourdon gauge. Components charged into cell under pressure, equilibrated and samples withdrawn. Samples analysed by G.C. using thermal conductivity detector and Porapak P column.	1. Matheson Gas sample, purity better than 99 mole per cent. 2. Phillips Petroleum sample, purity 99.6 mole per cent.
	ESTIMATED ERROR: $\delta T/K = \pm 0.06$ (at 310.9 K and 366.5 K) ± 0.12 (at 477.6K) $\delta P/MPa$ $= \pm 0.2$ δx_{H_2S}, $\delta y_{H_2S} = \pm 0.001$. (estimated by compiler)
	REFERENCES:

COMPONENTS:	ORIGINAL MEASUREMENTS:
1. Hydrogen sulfide; H_2S; [7783-06-4] 2. Decane, $C_{10}H_{22}$; [124-18-5]	Reamer, H.H.; Selleck, F.T.; Sage, B.H.; Lacey, W.N. *Ind. Eng. Chem.* 1953, 45, 1810-1812

VARIABLES:	PREPARED BY:
Temperature, pressure	C.L. Young

EXPERIMENTAL VALUES:

T/K	$P/10^5$Pa	Mole fraction of hydrogen sulfide in liquid x_{H_2S}	in gas y_{H_2S}
277.6	1.38	0.075	0.999
	2.76	0.154	0.999
	4.14	0.238	0.999
	5.52	0.339	0.999
	6.89	0.446	0.999
	8.62	0.613	0.999
	10.34	0.814	0.999
310.9	3.45	0.1153	0.999
	6.89	0.2332	0.999
	10.34	0.3543	0.999
	13.79	0.4780	0.999
	17.24	0.6045	0.999
	20.68	0.7363	0.999
	24.13	0.8738	0.999
344.3	6.89	0.1572	0.9970
	13.79	0.3051	0.9973
	20.68	0.4444	0.9976
	27.58	0.5760	0.9979
	34.47	0.6971	0.9982
	41.37	0.8107	0.9985
	48.26	0.9192	0.9990

AUXILIARY INFORMATION

METHOD/APPARATUS/PROCEDURE:	SOURCE AND PURITY OF MATERIALS:
PVT static cell used. Pressure measured with pressure balance. Temperature measured using platinum resistance thermometer. Samples of co-existing phases analysed by removing hydrogen sulfide from known weight of sample. Details in source and ref. (1).	1. Prepared by hydration of pure aluminium sulfide. Fractionated twice. 2. Eastman Kodak sample. Fractionated sample impurities mainly isomers.

ESTIMATED ERROR:

$\delta T/K = \pm 0.1$; δP/bar = ± 0.015;
$\delta x_{H_2S} = \pm 0.15\%$

REFERENCES:

1. Sage, B.H.; Lacey, W.N.; *Trans. Am. Inst. Mining.Met. Engnrs.* 1940,136, 136.

COMPONENTS:	ORIGINAL MEASUREMENTS:
1. Hydrogen sulfide; H_2S; [7783-06-4] 2. Decane, $C_{10}H_{22}$; [124-18-5]	Reamer, H.H.; Selleck, F.T. Sage, B.H.; Lacey, W.N. *Ind. Eng. Chem.* <u>1953</u>, *45*, 1810-1812

EXPERIMENTAL VALUES:

T/K	$P/10^5$Pa	Mole fraction of hydrogen sulfide	
		in liquid, x_{H_2S}	in gas, y_{H_2S}
377.6	13.79	0.2157	0.9914
	27.58	0.4044	0.9927
	41.37	0.5681	0.9938
	55.16	0.7086	0.9949
	68.95	0.8308	0.9960
	82.74	0.9324	0.9966
410.9	13.79	0.1660	0.9690
	27.58	0.3149	0.9808
	41.37	0.4439	0.9832
	55.16	0.5558	0.9839
	68.95	0.6552	0.9841
	82.74	0.7423	0.9843
	96.53	0.8201	0.9831
	110.32	0.8970	0.9807
444.3	13.79	0.1352	0.9234
	27.58	0.2612	0.9560
	41.37	0.3703	0.9646
	55.16	0.4662	0.9682
	68.95	0.5530	0.9688
	82.74	0.6297	0.9663
	96.53	0.7010	0.9632
	110.32	0.7668	0.9577
	124.11	0.8307	0.9478

COMPONENTS:	ORIGINAL MEASUREMENTS:
1. Hydrogen sulfide; H$_2$S; [7783-06-4] 2. Decane; C$_{10}$H$_{22}$; [124-18-5]	Gerrard, W. J. Appl. Chem. Biotechnol. 1972, 22, 623-650.

VARIABLES:	PREPARED BY:
Temperature	P.G.T. Fogg

EXPERIMENTAL VALUES:

T/K	Mole ratio	Mole fraction of H$_2$S[*] x_{H_2S}
267.15	0.096	0.088
273.15	0.080	0.074

The total pressure was equal to barometric pressure (not stated).

[*] calculated by the compiler for the stated total pressure.

AUXILIARY INFORMATION

METHOD/APPARATUS/PROCEDURE:	SOURCE AND PURITY OF MATERIALS:
Hydrogen sulfide was bubbled into a weighed amount of component 2 in a bubbler tube as described in detail in the source. The amount of gas absorbed at equilibrium for the observed temperature was found by weighing.	It was stated that "All materials purified and attested by conventional methods.
	ESTIMATED ERROR: $\delta\, x_{H_2S}$ = ± 4% (author)
	REFERENCES:

COMPONENTS:	ORIGINAL MEASUREMENTS:
1. Hydrogen sulfide; H_2S; [7783-06-4] 2. Decane; $C_{10}H_{22}$; [124-18-5]	King, M. B.; Al-Najjar, H. *Chem. Eng. Sci.* <u>1977</u>, *32*, 1241-6.

VARIABLES:	PREPARED BY:
Temperature	C. L. Young

EXPERIMENTAL VALUES:

T/K	Mole fraction of hydrogen sulfide[*] at a partial pressure of 101.3 kPa
288.2	0.0541
293.2	0.0502
298.2	0.0465
303.2	0.0428
323.2	0.0325
343.2	0.0252

[*] allowance was made for the non-ideal gas behaviour of hydrogen sulfide.

AUXILIARY INFORMATION

METHOD/APPARATUS/PROCEDURE:	SOURCE AND PURITY OF MATERIALS:
Solvent degassed by spraying into a continuously evacuated chamber. Solvent flows in a thin film down a glass spiral into a buret system containing the gas to be dissolved. Flow rates may be varied over a wide range without affecting the solubility. Similar to the apparatus of Morrison and Billett.	No details given.

ESTIMATED ERROR:

$\delta T/K = \pm 0.1$; $\delta x_{H_2S} = \pm 2\%$

(estimated by compiler).

REFERENCES:

1. Morrison, T. J.; Billett, F. *J. Chem. Soc.* <u>1952</u>, 3819.

COMPONENTS:	ORIGINAL MEASUREMENTS:
1. Hydrogen sulfide; H_2S; [7783-06-4] 2. Dodecane; $C_{12}H_{26}$; [112-40-3]	King, M. B.; Al-Najjar, H. *Chem. Eng. Sci.* <u>1977</u>, *32*, 1241-6.

VARIABLES:	PREPARED BY:
Temperature	C. L. Young

EXPERIMENTAL VALUES:

T/K	Mole fraction of hydrogen sulfide[*] at a partial pressure of 101.3 kPa
288.2	0.0572
293.2	0.0533
298.2	0.0495
303.2	0.0475
323.2	0.0347
343.2	0.0268

[*] allowance was made for the non-ideal gas behaviour of hydrogen sulfide.

AUXILIARY INFORMATION

METHOD/APPARATUS/PROCEDURE:	SOURCE AND PURITY OF MATERIALS:
Solvent degassed by spraying into a continuously evacuated chamber. Solvent flows in a thin film down a glass spiral into a buret system containing the gas to be dissolved. Flow rates may be varied over a wide range without affecting the solubility. Similar to the apparatus of Morrison and Billett.	No details given.

ESTIMATED ERROR:
$\delta T/K = \pm 0.1$; $\delta x_{H_2S} = \pm 2\%$ (estimated by compiler).

REFERENCES:
1. Morrison, T. J.; Billett, F.   *J. Chem. Soc.*   <u>1952</u>, 3819.

COMPONENTS:	ORIGINAL MEASUREMENTS:
1. Hydrogen sulfide; H_2S; [7783-06-4] 2. Tetradecane; $C_{14}H_{30}$; [629-59-4]	King, M. B.; Al-Najjar, H. *Chem. Eng. Sci.* <u>1977</u>, *32*, 1241-6.

VARIABLES:	PREPARED BY:
Temperature	C. L. Young

EXPERIMENTAL VALUES:

T/K	Mole fraction of hydrogen sulfide[*] at a partial pressure of 101.3 kPa
288.2	0.0615
293.2	0.0571
298.2	0.0530
303.2	0.0490
323.2	0.0370
343.2	0.0284

[*] allowance was made for the non-ideal gas behaviour of hydrogen sulfide.

AUXILIARY INFORMATION

METHOD/APPARATUS/PROCEDURE:	SOURCE AND PURITY OF MATERIALS:
Solvent degassed by spraying into a continuously evacuated chamber. Solvent flows in a thin film down a glass spiral into a buret system containing the gas to be dissolved. Flow rates may be varied over a wide range without affecting the solubility. Similar to the apparatus of Morrison and Billett.	No details given.

ESTIMATED ERROR:

$\delta T/K = \pm 0.1$; $\delta x_{H_2S} = \pm 2\%$

(estimated by compiler).

REFERENCES:
1. Morrison, T. J.; Billett, F.
   J. Chem. Soc.
   <u>1952</u>, 3819.

HSDS—H

COMPONENTS:	ORIGINAL MEASUREMENTS:
1. Hydrogen sulfide; H_2S; [7783-06-4] 2. Hexadecane; $C_{16}H_{34}$; [544-76-3]	Lenoir, J-Y.; Renault, P.; Renon, H. J. Chem. Eng. Data, <u>1971</u>, 16, 340-2

VARIABLES:	PREPARED BY:
	C. L. Young

EXPERIMENTAL VALUES:

T/K	Henry's constant H_{H_2S}/atm	Mole fraction at 1 atm* x_{H_2S}
298.2	18.7	0.0535

* Calculated by compiler assuming a linear function of p_{H_2S} vs x_{H_2S}, i.e., x_{H_2S}(1 atm) = $1/H_{H_2S}$

AUXILIARY INFORMATION

METHOD/APPARATUS/PROCEDURE:	SOURCE AND PURITY OF MATERIALS:
A conventional gas-liquid chromatographic unit fitted with a thermal conductivity detector was used. The carrier gas was helium. The value of Henry's law constant was calculated from the retention time. The value applies to very low partial pressures of gas and there may be a substantial difference from that measured at 1 atm. pressure. There is also considerable uncertainty in the value of Henry's constant since surface adsorption was not allowed for although its possible existence was noted.	(1) L'Air Liquide sample, minimum purity 99.9 mole per cent. (2) Touzart and Matignon or Serlabo sample, purity 99 mole per cent.
	ESTIMATED ERROR: $\delta T/K = \pm0.1$; δH/atm = $\pm6\%$ (estimated by compiler).
	REFERENCES:

COMPONENTS:	ORIGINAL MEASUREMENTS:
1. Hydrogen sulfide; H_2S; [7783-06-4] 2. Hexadecane; $C_{16}H_{34}$; [544-76-3]	Tremper, K.K.; Prausnitz, J.M. *J. Chem. Engng. Data* <u>1976</u>, *21*, 295-9
VARIABLES: Temperature	PREPARED BY: C.L. Young

EXPERIMENTAL VALUES:

T/K	Henry's Constant[a] /atm	Mole fraction[b] of hydrogen sulfide at 1atm partial pressure, x_{H_2S}
300	25.2	0.0397
325	32.6	0.0307
350	40.9	0.0244
375	49.7	0.0201
400	58.9	0.0170
425	68.0	0.0147
450	76.6	0.0131
475	82.7	0.0121

a. Authors stated measurements were made at several pressures and values of solubility used were all within the Henry's Law region.

b. Calculated by compiler assuming linear relationship between mole fraction and pressure.

AUXILIARY INFORMATION

METHOD/APPARATUS/PROCEDURE:	SOURCE AND PURITY OF MATERIALS:
Volumetric apparatus similar to that described by Dymond and Hildebrand (1). Pressure measured with a null detector and precision gauge. Details in ref. (2).	Solvent degassed. No other details given.

ESTIMATED ERROR:

$\delta T/K = \pm 0.1$; $\delta x_{H_2S} = \pm 1\%$

REFERENCES:
1. Dymond, J.; Hildebrand, J.H. *Ind. Chem. Eng. Fundam.* <u>1967</u>,*6*,130.
2. Cukor, P.M.; Prausnitz, J.M. *Ind. Chem. Eng. Fundam.* <u>1971</u>,*10*,638.

COMPONENTS:	ORIGINAL MEASUREMENTS:
1. Hydrogen sulfide; H_2S; [7783-06-4] 2. Hexadecane; $C_{16}H_{34}$; [544-76-3]	King, M. B.; Al-Najjar, H. *Chem. Eng. Sci.* <u>1977</u>, *32*, 1241-6.

VARIABLES:	PREPARED BY:
Temperature	C. L. Young

EXPERIMENTAL VALUES:

T/K	Mole fraction of hydrogen sulfide[*] at a partial pressure of 101.3 kPa
298.2	0.0573
303.2	0.0529
323.2	0.0401
343.2	0.0308

[*] allowance was made for the non-ideal gas behaviour of hydrogen sulfide.

AUXILIARY INFORMATION

METHOD/APPARATUS/PROCEDURE:	SOURCE AND PURITY OF MATERIALS:
Solvent degassed by spraying into a continuously evacuated chamber. Solvent flows in a thin film down a glass spiral into a buret system containing the gas to be dissolved. Flow rates may be varied over a wide range without affecting the solubility. Similar to the apparatus of Morrison and Billett.	No details given.

ESTIMATED ERROR:
$\delta T/K = \pm 0.1$; $\delta x_{H_2S} = \pm 2\%$

(estimated by compiler).

REFERENCES:
1. Morrison, T. J.; Billett, F.
   J. Chem. Soc.
   <u>1952</u>, 3819.

COMPONENTS:	ORIGINAL MEASUREMENTS:
1. Hydrogen sulfide; H_2S; [7783-06-4] 2. Octane; C_8H_{18}; [111-65-9] 3. Methane; CH_4; [74-82-8]	Asano.K.; Nakahara, T.; Kobayashi, R. *J.Chem.Eng.Data* <u>1971</u>, *16*, 16-18.
VARIABLES:	PREPARED BY:
Temperature, Pressure of methane	P.G.T. Fogg

EXPERIMENTAL VALUES:

T/K	P_{CH_4}/psia	P_{CH_4}/bar	Mole fraction of CH_4 in liquid phase*	K-value for H_2S Exptl.	Smoothed**
233.15	20	1.38	0.0070	2.77	2.20
	100	6.89	0.0336	0.649	0.650
	200	13.79	0.0658	0.364	0.390
	400	27.58	0.1282	0.230	0.230
	600	41.37	0.1869	0.184	0.184
	800	55.16	0.2410	0.181	0.181
	1000	68.95	0.2882	0.186	0.186
	1250	86.18	0.3367	0.217	0.217
	1500	103.42	0.3774	0.275	0.265

The K-value for H_2S is defined as:

 mole fraction of H_2S in vapor phase/mole fraction of H_2S in liquid phase

These K-values correspond to essentially infinite dilution.

* calculated by the compiler from K-values for methane in octane from
  ref.(1).

** given by the authors

AUXILIARY INFORMATION

METHOD/APPARATUS/PROCEDURE:	SOURCE AND PURITY OF MATERIALS:
Retention volumes for small samples of hydrogen sulfide were measured by gas chromatography with octane as the stationary phase and methane the carrier gas. Temperatures were maintained to ± 0.1 K and the pressure of methane measured to ± 0.2%. Experimental details are given in refs. (2) - (4).	1. 99.6% pure. 2. 99.88% pure. 3. 99.99% pure.
	ESTIMATED ERROR:
	δ T/K = ± 0.1; δP_{CH_4} = ± 2% (authors)
	REFERENCES:
	1. Kohn, J.P.; Bradish, W.F. *J.Chem.Eng.Data* <u>1964</u>, *9*, 5. 2. Van Horn, L.D.; Kobayashi, R. *J.Chem.Eng.Data* <u>1967</u>, *12*, 294. 3. Masukawa, S.; Kobayashi, R. *J.Gas Chromatogr.* <u>1968</u>, *6*, 257. 4. Koonce, K.T.; Kobayashi, R. *J.Chem.Eng.Data* <u>1964</u>, *9*, 494.

COMPONENTS:	ORIGINAL MEASUREMENTS:
1. Hydrogen sulfide; H_2S; [7783-06-4] 2. Octane; C_8H_{18}; [111-65-9] 3. Methane; CH_4; [74-82-8]	Asano.K.; Nakahara, T.; Kobayashi, R. *J.Chem.Eng.Data* 1971, *16*, 16-18.

EXPERIMENTAL VALUES:

T/K	P_{CH_4}/psia	P_{CH_4}/bar	Mole fraction of CH_4 in liquid phase*	K-value for H_2S Exptl.	Smoothed**
253.15	20	1.38	0.0070	5.81	4.00
	100	6.89	0.0336	1.26	1.16
	200	13.79	0.0658	0.670	0.670
	400	27.58	0.1282	0.407	0.400
	600	41.37	0.1869	0.306	0.310
	800	55.16	0.2410	0.290	0.284
	1000	68.95	0.2882	0.275	0.280
	1250	86.18	0.3367	0.298	0.298
	1500	103.42	0.3774	0.318	0.318
273.15	20	1.38	0.0070	9.37	6.70
	100	6.89	0.0336	2.05	1.90
	200	13.79	0.0658	1.10	1.10
	400	27.58	0.1282	0.626	0.650
	600	41.37	0.1869	0.488	0.488
	800	55.16	0.2410	0.421	0.421
	1000	68.95	0.2882	0.399	0.400
	1250	86.18	0.3367	0.378	0.390
	1500	103.42	0.3774	0.392	0.318
293.15	20	1.38	0.0070	12.5	10.5
	100	6.89	0.0336	3.17	2.99
	200	13.79	0.0658	1.67	1.70
	400	27.58	0.1282	0.954	0.970
	600	41.37	0.1869	0.714	0.714
	800	55.16	0.2410	0.589	0.589
	1000	68.95	0.2882	0.517	0.540
	1250	86.18	0.3367	0.525	0.520
	1500	103.42	0.3774	0.518	0.518

The K-value for H_2S is defined as:

mole fraction of H_2S in vapor phase/mole fraction of H_2S in liquid phase

These K-values correspond to essentially infinite dilution.

* calculated by the compiler from K-values for methane in octane from ref.(1).

COMPONENTS:	ORIGINAL MEASUREMENTS:
(1) Hydrogen sulfide; H_2S; [7783-06-4] (2) Kerosene A-1	Hannaert, H.; Haccuria, M.; Mathieu, M. P. *Ind. Chim. Belge* 1967, *32*, 156-164.

VARIABLES:	PREPARED BY:
$T/K = 233.15 - 273.15$	E. L. Boozer H. L. Clever

EXPERIMENTAL VALUES:

Temperature Interval of Measurements T/K	Hydrogen Sulfide Mol % Range $10^2 x_1$/mol %	$K\pi\nu$/atm[1] at 293.15 K	Enthalpy of Dissolution ΔH/kcal mol^{-1}	Constant A
233.15-273.15	1 - 6	(18)	3.17	3.63

[1] $\log (K\pi\nu/\text{atm}) = A - (\Delta H/\text{cal mol}^{-1})/(2.3\ R(T/K))$

The author's definitions are:

$$K = y_1/x_1 = \frac{\text{mole fraction gas in gas phase}}{\text{mole fraction gas in liquid phase}}$$

π/atm = total pressure,

ν = coefficient of fugacity.

The function, $K\pi\nu$/atm, is equivalent to a Henry's constant in the form $H_{1,2}$/atm $= (f_1/\text{atm})/x_1$ where f_1 is the fugacity.

AUXILIARY INFORMATION

METHOD/APPARATUS/PROCEDURE:	SOURCE AND PURITY OF MATERIALS:
The authors describe three methods: 1.A. [Saturat. n°1]. A measure of the static pressure of saturation in an apparatus which gave a precision of 10 - 15 %. 1.B. [Saturat. n°2]. A measure of the static pressure of saturation in an apparatus which gave a precision of 2 - 5 %. 2. [Chromato]. A Gas liquid chromatographic method estimated to have a precision of 2 - 5 %. 3. [Anal. directe]. Direct analysis of the gaseous and liquid phases. Method 1.B. was used for this system.	(1) Hydrogen sulfide. Matheson Co., Inc. Purity 99.5 per cent. (2) Kerosene A-1

(2) Kerosene A-1

Distillation range, °C	density gcm^{-3},20°C	mol wt
A-1 150-280	0.7805	170

ESTIMATED ERROR:

REFERENCES:

COMPONENTS:	ORIGINAL MEASUREMENTS:
1. Hydrogen sulfide; H_2S; [7783-06-4] 2. Cyclohexane; C_6H_{12}; [110-82-7]	Tsiklis, D.S.; Svetlova, G.M. *Zh. Fiz. Khim.* 1958, *32*, 1476-1480.

VARIABLES:
 	
    Temperature, pressure
	PREPARED BY:
	
	    P.G.T. Fogg

EXPERIMENTAL VALUES:

P_{H_2S}/mmHg	P_{H_2S}/bar	Mole fraction, x_{H_2S}		
		283.2 K	293.2 K	313.2 K
100	0.133	0.0034	0.0024	0.0020
200	0.267	0.0076	0.0058	0.0046
300	0.400	0.0126	0.0100	0.0080
400	0.533	0.0186	0.0154	0.0122
500	0.667	0.0265	0.0218	0.0170
600	0.800	0.0334	0.0290	0.0224
700	0.933	0.0424	0.0370	0.0288
800	1.067	0.0524	0.0458	0.0360

AUXILIARY INFORMATION

METHOD/APPARATUS/PROCEDURE:	SOURCE AND PURITY OF MATERIALS:
A known weight of hydrogen sulfide was dissolved in a known weight of cyclohexane and the resulting total pressure was measured by a manometer. The pressure of hydrogen sulfide was equal to the total pressure minus the vapor pressure of cyclohexane. Temperatures were controlled to ± 0.1 K.	Not stated.
	ESTIMATED ERROR: $\delta T/K$ = ± 0.1 (authors)
	REFERENCES:

COMPONENTS:	ORIGINAL MEASUREMENTS:
1. Hydrogen sulfide; H_2S; [7783-06-4] 2. Methylcyclohexane; C_7H_{14}; [108-87-2]	Ng, H-J.; Robinson, D. B. *Fluid Phase Equilibria* <u>1979</u>, *2*, 283-292.

VARIABLES:	PREPARED BY:
Temperature, pressure	C. L. Young

EXPERIMENTAL VALUES:

T/K	P/MPa	Mole fraction of hydrogen sulfide in liquid, x_{H_2S}	in gas, y_{H_2S}	T/K	P/MPa	Mole fraction of hydrogen sulfide in liquid, x_{H_2S}	in gas, y_{H_2S}
310.9	0.265	0.065	0.963	394.3	0.470	0.030	0.630
	0.682	0.204	0.988		1.303	0.127	0.860
	1.372	0.449	0.993		2.841	0.307	0.932
	2.061	0.732	0.995		4.406	0.470	0.950
	2.434	0.913	0.999		5.681	0.600	0.958
	2.599	0.970	0.999		7.433	0.762	0.963
352.6	0.250	0.029	0.800		8.701	0.858	0.967
	0.752	0.114	0.934		9.322	0.900	0.963
	1.600	0.265	0.969	477.6	1.820	0.071	0.411
	2.689	0.462	0.981		3.199	0.184	0.644
	3.757	0.659	0.983		4.316	0.256	0.740
	4.847	0.844	0.991		5.426	0.341	0.787
	5.654	0.943	0.994		5.722	0.351	0.799
					7.329	0.460	0.833
					8.536	0.549	0.816
					9.453	0.612	0.801

AUXILIARY INFORMATION

METHOD/APPARATUS/PROCEDURE:	SOURCE AND PURITY OF MATERIALS:
Stirred static cell fitted with glass window. Temperature measured with thermocouple and pressure measured with Bourdon gauge. After equilibrium established gas and liquid phases sampled and analysed by gas chromatography using a thermal conductivity detector. Details in ref. (1).	1. Thio-Pet Chemicals sample, distilled, final purity about 99.9 mole per cent. 2. Baker Analyzed Reagent grade, purity better than 99.9 mole per cent.

ESTIMATED ERROR:
$\delta T/K = \pm 0.06$; $\delta P/MPa = \pm 0.2$; δx_{H_2S}, $\delta y_{H_2S} = \pm 0.005$.

REFERENCES:
1. Ng, H-J.; Robinson, D. B.   *J. Chem. Engng. Data*   <u>1978</u>, *23*, 325.

COMPONENTS:	ORIGINAL MEASUREMENTS:
1. Hydrogen sulfide; H_2S; [7783-06-4] 2. Ethylcyclohexane; C_8H_{16}; [1678-91-7]	Huang, S. S.-S.; Robinson, D. B. *J. Chem. Eng. Data* 1985, *30*, 154-157.

VARIABLES:	PREPARED BY:
Temperature, pressure	C. L. Young

EXPERIMENTAL VALUES:		Mole fraction of hydrogen sulfide		K-values	
T/K	P/MPa	in liquid, x_{H_2S}	in vapor, y_{H_2S}	K_{H_2S}	$K_{C_8H_{16}}$
310.0	0.170	0.0490	0.9811	20.0	0.0199
	0.465	0.1495	0.9936	6.65	0.00753
	0.972	0.3060	0.9964	3.26	0.00519
	1.54	0.5362	0.9975	1.86	0.00539
	2.21	0.8178	0.9988	1.22	0.00659
	2.52	0.9447	0.9995	1.06	0.00904
	2.66	0.9840	0.9998	1.02	0.0125
352.6	0.273	0.0465	0.9324	20.1	0.0709
	0.848	0.1518	0.9770	6.44	0.0271
	1.72	0.3157	0.9876	3.13	0.0181
	3.25	0.5822	0.9921	1.70	0.0189
	4.59	0.8014	0.9947	1.24	0.0267
	5.67	0.9443	0.9971	1.06	0.0521
	6.10	0.9841	0.9988	1.01	0.0755
394.3	0.383	0.0362	0.8067	22.3	0.201
	1.14	0.1273	0.9332	7.33	0.0765
	2.36	0.2767	0.9644	3.49	0.0492
	4.42	0.5045	0.9769	1.94	0.0466
	6.39	0.6820	0.9801	1.44	0.0626
	8.68	0.8604	0.9801	1.14	0.143
	9.65	0.9198	0.9787	1.06	0.266
477.6	0.820	0.0163	0.3444	21.1	0.666
	1.65	0.0735	0.6593	8.97	0.368
	3.92	0.2368	0.8325	3.52	0.219
	6.63	0.4109	0.8742	2.13	0.214
	8.82	0.5437	0.8814	1.62	0.260
	11.13	0.6702	0.8698	1.30	0.395
	12.50	0.8313	0.8416	1.01	0.939

AUXILIARY INFORMATION

METHOD/APPARATUS/PROCEDURE:	SOURCE AND PURITY OF MATERIALS:
Stirred static cell fitted with glass window. Temperature measured with thermocouple and pressure with Bourdon gauge. After equilibrium established gas and liquid phases sampled and analysed using gas chromatography using a thermal conductivity detector. Details in source and ref. (1).	1. Matheson sample, purity 99.9 mole per cent. Distilled. 2. Obtained from Aldrich Chemical; purity 99 mole per cent or better; no impurity detected by chromatography.

ESTIMATED ERROR:
$\delta T/K = \pm 0.06$; $\delta P/MPa = \pm 0.007$; (up to 6.9 MPa); ± 0.02 (above 6.9 MPa).

REFERENCES:

1. Ng, H.-J.; Robinson, D. B.
   J. Chem. Eng. Data
   1978, *23*, 325.

COMPONENTS:	ORIGINAL MEASUREMENTS:
1. Hydrogen sulfide; H_2S; [7783-06-4] 2. Propylcyclohexane; C_9H_{18}; [1678-92-8]	Huang, S. S.-S.; Robinson, D. B. *J. Chem. Eng. Data* <u>1985</u>, *30*, 154-157.
VARIABLES: Temperature, pressure	PREPARED BY: C. L. Young

EXPERIMENTAL VALUES:

T/K	P/MPa	Mole fraction of hydrogen sulfide in liquid, x_{H_2S}	in vapor, y_{H_2S}	K-values K_{H_2S}	$K_{C_9H_{18}}$
310.9	0.177	0.0642	0.9913	15.5	0.0093
	0.585	0.2096	0.9981	4.76	0.0024
	1.01	0.3810	0.9988	2.62	0.00194
	1.54	0.5670	0.9992	1.76	0.00185
	2.03	0.7804	0.9995	1.28	0.00228
	2.28	0.8503	0.9996	1.18	0.00267
	2.58	0.9598	0.9998	1.04	0.00498
352.6	0.220	0.0455	0.9651	21.2	0.0366
	0.752	0.1543	0.9892	6.41	0.0128
	1.80	0.3623	0.9947	2.75	0.00831
	3.40	0.6455	0.9966	1.54	0.00959
	4.93	0.8676	0.9974	1.15	0.0196
	5.65	0.9494	0.9983	1.05	0.0336
	6.02	0.9832	0.9993	1.02	0.0417
394.3	0.342	0.0418	0.8980	21.5	0.106
	1.11	0.1505	0.9668	6.42	0.0391
	2.47	0.3230	0.9823	3.04	0.0261
	4.53	0.5400	0.9876	1.83	0.0270
	6.40	0.7010	0.9886	1.41	0.0381
	8.52	0.8748	0.9879	1.13	0.0966
	8.56	0.8701	0.9864	1.13	0.105
	9.65	0.9263	0.9857	1.06	0.194
	9.99	0.9427	0.9839	1.04	0.281
477.6	0.841	0.0448	0.6356	14.2	0.381
	1.72	0.1215	0.7989	6.58	0.229

(cont.)

AUXILIARY INFORMATION

METHOD/APPARATUS/PROCEDURE:	SOURCE AND PURITY OF MATERIALS:
Stirred static cell fitted with glass window. Temperature measured with thermocouple and pressure with Bourdon gauge. After equilibrium established gas and liquid phases sampled and analysed using gas chromatography using a thermal conductivity detector. Details in source and ref. (1).	1. Matheson sample, purity 99.9 mole per cent. Distilled. 2. Obtained from Koch-Light Lab., purity 99 mole per cent or greater. No impurity detected by chromatography.

	ESTIMATED ERROR: $\delta T/K = \pm 0.06$; $\delta P/MPa = \pm 0.007$; (up to 6.9 MPa); ± 0.02 (above 6.9 MPa).
	REFERENCES: 1. Ng, H.-J.; Robinson, D. B. *J. Chem. Eng. Data* <u>1978</u>, *23*, 325.

COMPONENTS:	ORIGINAL MEASUREMENTS:

1. Hydrogen sulfide; H_2S;
   [7783-06-4]
 	
2. Propylcyclohexane; C_9H_{18};
   [1678-92-8]
	Huang, S. S.-S.; Robinson, D. B.
	J. Chem. Eng. Data
	<u>1985</u>, *30*, 154-157.

EXPERIMENTAL VALUES:

T/K	P/MPa	Mole fraction of hydrogen sulfide		K-values	
		in liquid, x_{H_2S}	in vapor, y_{H_2S}	K_{H_2S}	$K_{C_9H_{18}}$
477.6	3.68	0.2805	0.8922	3.18	0.150
	6.32	0.4515	0.9179	2.03	0.150
	8.65	0.5909	0.9200	1.56	0.196
	10.70	0.6850	0.9178	1.34	0.261
	10.83	0.6888	0.9160	1.33	0.270
	11.58	0.7249	0.9142	1.26	0.312
	12.16	0.7463	0.9070	1.22	0.367
	12.31	0.7620	0.8999	1.18	0.421
	12.75	0.7912	0.8912	1.13	0.521
	13.06	0.8270	0.8786	1.06	0.702

COMPONENTS:	ORIGINAL MEASUREMENTS:
1. Hydrogen sulfide; H_2S; [7783-06-4]. 2. (1-Methylethyl)-cyclohexane, (*iso*-propylcyclohexane); C_9H_{18}; [696-29-7]	Eakin, B.E.; DeVaney, W.E. *Am. Inst. Chem. Engnrs. Symp. Ser.* <u>1974</u>, No. 140, *70*, 80-90.

VARIABLES:	PREPARED BY:
Temperature, pressure	C.L. Young.

EXPERIMENTAL VALUES:

T/K	$P/10^5Pa$	Mole fraction of hydrogen sulfide in liquid, x_{H_2S}	in gas, y_{H_2S}
310.9	1.12	0.0394	0.98835
	1.62	0.0533	0.99184
	2.09	0.0710	0.99316
	3.24	0.1058	0.99582
	4.32	0.1378	0.99579
	5.34	0.1747	0.99705
366.5	2.50	0.0389	0.9315
	3.69	0.0568	0.9530
	4.30	0.0652	0.9649
	7.12	0.1113	0.9756
	9.90	0.1508	0.9810
	12.25	0.1834	0.9880
477.6	7.93	0.0398	0.5779
	10.71	0.0655	0.7079
	13.27	0.0842	0.7490
	18.41	0.1188	0.8078
	24.13	0.1570	0.8571
	29.65	0.1937	0.8649

AUXILIARY INFORMATION

METHOD/APPARATUS/PROCEDURE:	SOURCE AND PURITY OF MATERIALS:
Rocking stainless steel equilibrium vessel, fitted with liquid and gas sampling ports. Pressure measured with Bourdon gauge. Components charged into cell under pressure, equilibrated and samples withdrawn. Samples analysed by G.C. using thermal conductivity detector and Porapak P column.	1. Matheson Gas sample purity better than 99 mole per cent. 2. Phillips Petroleum sample, purity better than 99.9 mole per cent.
	ESTIMATED ERROR:
	$\delta T/K = \pm0.06$ (at 310.9K and 366.5K) ±0.12 (at 477.6K) $\delta P/MPa = \pm0.2$. δx_{H_2S}, $\delta y_{H_2S} = \pm0.001$. (estimated by compiler)
	REFERENCES:

COMPONENTS:	ORIGINAL MEASUREMENTS:
1. Hydrogen sulfide; H_2S; [7783-06-4] 2. Decahydronaphthalene, (Decalin); $C_{10}H_{18}$; [91-17-8]	Lenoir, J-Y.; Renault, P.; Renon, H. *J. Chem. Eng. Data*, 1971, *16*, 340-2.
VARIABLES: Temperature	PREPARED BY: C. L. Young

EXPERIMENTAL VALUES:

T/K	Henry's constant H_{H_2S}/atm	Mole fraction at 1 atm* x_{H_2S}
298.2	29.8	0.0336
323.2	37.6	0.0266

* Calculated by compiler assuming a linear function of P_{H_2S} vs x_{H_2S}, i.e., $x_{H_2S}(1 \text{ atm}) = 1/H_{H_2S}$

AUXILIARY INFORMATION

METHOD/APPARATUS/PROCEDURE:	SOURCE AND PURITY OF MATERIALS:
A conventional gas-liquid chromatographic unit fitted with a thermal conductivity detector was used. The carrier gas was helium. The value of Henry's law constant was calculated from the retention time. The value applies to very low partial pressures of gas and there may be a substantial difference from that measured at 1 atm. pressure. There is also considerable uncertainty in the value of Henry's constant since surface adsorption was not allowed for although its possible existence was noted.	(1) L'Air Liquide sample, minimum purity 99.9 mole percent. (2) Touzart and Matignon or Serlabo sample, purity 99 mole per cent.
	ESTIMATED ERROR: $\delta T/K = \pm 0.1$; δH/atm $= \pm 6\%$ (estimated by compiler).
	REFERENCES:

COMPONENTS:	ORIGINAL MEASUREMENTS:
1. Hydrogen sulfide; H_2S; [7783-06-4] 2. 1,1'-Bicyclohexyl; $C_{12}H_{22}$; [92-51-3]	Tremper, K.K.; Prausnitz, J.M. *J. Chem.Engng.Data* <u>1976</u>,*21*,295-9
VARIABLES: Temperature	PREPARED BY: C.L. Young

EXPERIMENTAL VALUES:

T/K	Henry's Constant[a] /atm	Mole fraction[b] of of hydrogen sulfide at 1 atm partial pressure, x_{H_2S}
300	43.0	0.0233
325	55.9	0.0179
350	72.0	0.0139
375	91.6	0.0109
400	112.0	0.00893
425	129.0	0.00775
450	135.0	0.00741
475	129.0	0.00775

a Authors stated measurements were made at several pressures
   and values of solubility used were all within the Henry's-
   Law region.

b Calculated by compiler assuming linear relationship between
   mole fraction and pressure.

AUXILIARY INFORMATION

METHOD/APPARATUS/PROCEDURE	SOURCE AND PURITY OF MATERIALS:
Volumetric apparatus similar to that described by Dymond and Hildebrand (1). Pressure measured with a null detector and precision gauge. Details in ref. (2).	Solvent degassed, no details given.
	ESTIMATED ERROR: $\delta T/K = \pm 0.1$; $\delta x_{H_2S} = \pm 1\%$.
	REFERENCES: 1. Dymond, J,; Hildebrand, J.H. *Ind. Eng. Chem. Fundam.* <u>1967</u>,*6*,130 2. Cukor, P.M.; Prausnitz, J.M. *Ind. Eng.Chem.Fundam.* <u>1971</u>,*10*,638

COMPONENTS:	ORIGINAL MEASUREMENTS:
1. Hydrogen sulfide; H_2S; [7783-06-4] 2. Aromatic compounds	Bell, R.P. *J. Chem. Soc.* <u>1931</u>, 1371-1382.

VARIABLES:	PREPARED BY:
	C.L. Young

EXPERIMENTAL VALUES:

Solvent	T/K	Partition coefficient, s^+	Mole fraction [§] of hydrogen sulfide liquid, x_{NH_3}
Benzene; C_6H_6; [71-43-2]	293.15	15.68	0.0563
Methylbenzene, (Toluene); C_7H_8; [108-88-3]		16.90	0.0672
Bromobenzene; C_6H_5Br; [108-86-1]		12.92	0.0376
Chlorobenzene; C_6H_5Cl; [108-90-7]		13.80	0.0388

s^+ is defined as $s = 22.4 \times \dfrac{293}{273} \times c$ where c is the "solubility in equivalents/litre".

[§] for a partial pressure of 1 atmosphere

AUXILIARY INFORMATION

METHOD/APPARATUS/PROCEDURE:	SOURCE AND PURITY OF MATERIALS:
Volumetric apparatus consisting of bulb (~50cm³ capacity) extended at the top as a graduated tube and joined at bottom to a capillary u-tube. Liquid saturated with gas at atmospheric pressure. Gas withdrawn in a current of air, absorbed in sodium hydroxide and hydrogen peroxide. Excess hydrogen peroxide removed by heating and excess sodium hydroxide titrated.	1. Prepared by reaction of sodium sulfide on hydrochloric acid. 2. Merck and Kahlbaum samples dried over calcium chloride and fractionally distilled.

	ESTIMATED ERROR:
	$\delta T/K = \pm 0.1$; $\delta x_{H_2S} = \pm 1\%$. (estimated by compiler)

	REFERENCES:

COMPONENTS:	ORIGINAL MEASUREMENTS:
1. Hydrogen sulfide; H_2S; [7783-06-4] 2. Benzene; C_6H_6; [71-43-2]	Gerrard, W. *J. Appl. Chem. Biotechnol.* <u>1972</u>, *22*, 623-650.

VARIABLES:	PREPARED BY:
Temperature	P.G.T. Fogg

EXPERIMENTAL VALUES:

T/K	P(total)/mmHg	P(total)/bar	Mole ratio	Mole fraction of H_2S[*] x_{H_2S}
278.15	766	1.021	0.088	0.081
283.15	766	1.021	0.073	0.068
293.15	766	1.021	0.060	0.057

[*] calculated by the compiler for the stated total pressure.

AUXILIARY INFORMATION

METHOD/APPARATUS/PROCEDURE:	SOURCE AND PURITY OF MATERIALS:
Hydrogen sulfide was bubbled into a weighed amount of component 2 in a bubbler tube as described in detail in the source. The amount of gas absorbed at equilibrium for the observed temperature and pressure was found by weighing. Pressure was measured with a mercury manometer.	It was stated that "All materials purified and attested by conventional methods."

ESTIMATED ERROR:
$\delta\, x_{H_2S}$ = ± 4% (author)

REFERENCES:

COMPONENTS:	ORIGINAL MEASUREMENTS:
1. Hydrogen sulfide; H_2S; [7783-06-4] 2. Benzene; C_6H_6; [71-43-2]	Patyi, L.; Furmer, I. E.; Makranczy, J.; Sadilenko, A. S.; Stepanova, Z. G.; Berengarten, M. G. *Zh. Prikl. Khim.* 1978, *51*, 1296-1300.

VARIABLES:	PREPARED BY:
	C. L. Young

EXPERIMENTAL VALUES:

T/K	$\alpha^\dagger$	Mole fraction of hydrogen sulfide at a partial pressure of 101.325 kPa x_{H_2S}
298.15	9.3	0.0358

† volume of gas (measured at 101.325 kPa and 273.15 K) dissolved per volume of solvent.

AUXILIARY INFORMATION

METHOD/APPARATUS/PROCEDURE:	SOURCE AND PURITY OF MATERIALS:
Volumetric method. Pressure measured when known amounts of gas were added, in increments, to a known amount of liquid in a vessel of known dimensions. Corrections were made for the partial pressure of solvent. Details in ref. (1).	Purity better than 99 mole per cent as determined by gas chromatography.

	ESTIMATED ERROR:
	$\delta T/K = \pm 0.1$; $\delta\alpha = \pm 4\%$ or less.

	REFERENCES:
	1. Bodor, E.; Bor, G. J.; Mohai, B.; Sipos, G. *Veszpremi. Vegyip. Egy. Kozl.* 1957, *1*, 55.

COMPONENTS:	ORIGINAL MEASUREMENTS:
1. Hydrogen sulfide; H_2S; [7783-06-4] 2. Methylbenzene, (toluene); C_7H_8; [108-88-3]	Gerrard, W. *J. Appl. Chem. Biotechnol.*, 1972, 22, 623-650

VARIABLES:	PREPARED BY:
Temperature	C.L. Young

EXPERIMENTAL VALUES:

T/K	P*/mmHg	P*/kPa	Mole ratio	Mole fraction[+] of hydrogen sulfide in liquid, x_{H_2S}
265.15	754	100.5	0.150	0.130
267.15	754	100.5	0.138	0.121
273.15	754	100.5	0.115	0.103
283.15	754	100.5	0.082	0.076
293.15	754	100.5	0.071	0.066

[+]calculated by compiler

*total pressure

AUXILIARY INFORMATION

METHOD/APPARATUS/PROCEDURE:	SOURCE AND PURITY OF MATERIALS:
Hydrogen sulfide was bubbled into a weighed amount of component 2. in a bubbler tube as described in detail in the source. The amount of gas absorbed at equilibrium and at the observed temperature and pressure was determined by weighing. Pressure was measured with a mercury manometer. The amount of gas absorbed at successively lower pressures was measured. Eventually the pressure was reduced to the vapor pressure os component 2. The refractive index and infrared spectrum of the liquid showed it to be essentially pure component 2.	1 and 2. Components were purified and attested by conventional methods.
	ESTIMATED ERROR: $\delta T/K = \pm 0.2$; $\delta x_{H_2S} = \pm 4\%$
	REFERENCES:

COMPONENTS:	ORIGINAL MEASUREMENTS:
1. Hydrogen sulfide; H_2S; [7783-06-4] 2. Methylbenzene (toluene); C_7H_8; [108-88-3]	Ng, H.-J.; Kalra, H.; Robinson, D. B.; Kubota, H. *J. Chem. Eng. Data* <u>1980</u>, *25*, 51-55.

VARIABLES:	PREPARED BY:
Temperature, pressure	C. L. Young

EXPERIMENTAL VALUES:

T/K	P/psia	P/MPa	Mole fraction of hydrogen sulfide in liquid, x_{H_2S}	in vapor, y_{H_2S}
310.93	29.5	0.203	0.077	0.971
	79.3	0.547	0.226	0.990
	150	1.03	0.427	0.996
	230	1.59	0.643	0.998
	317	2.19	0.836	0.998
	370	2.55	0.954	0.999
352.59	59.6	0.411	0.071	0.913
	165	1.14	0.238	0.971
	296	2.04	0.442	0.985
	459	3.16	0.651	0.991
	632	4.36	0.817	0.995
	762	5.25	0.914	0.996
	818	5.64	0.953	0.998
394.26	54.5	0.376	0.031	0.677
	154	1.06	0.135	0.891
	350	2.41	0.305	0.944
	589	4.06	0.502	0.963
	853	5.88	0.686	0.975
	1114	7.68	0.816	0.975
	1309	9.03	0.891	0.975
	1413	9.74	0.932	0.978

(cont.)

AUXILIARY INFORMATION

METHOD/APPARATUS/PROCEDURE:	SOURCE AND PURITY OF MATERIALS:
Stirred static cell fitted with glass window. Temperature measured with thermocouple and pressure measured with Bourdon gauge. After equilibrium established gas and liquid phases sampled and analysed by gas chromatography using a thermal conductivity detector. Details in ref. (1).	1. Thio-Pet Chemicals sample, distilled, final purity at least 99.9 mole per cent as determined by GC. 2. Phillips Petroleum research grade sample, purity 99.94 mole per cent.

ESTIMATED ERROR:

$\delta T/K = \pm 0.06$; $\delta P/MPa = \pm 0.02$;

δx_{H_2S}, $\delta y_{H_2S} = \pm 0.003$.

REFERENCES:

1. Ng, H.-J.; Robinson, D. B. *J. Chem. Eng. Data* <u>1978</u>, *23*, 325.

COMPONENTS:	ORIGINAL MEASUREMENTS:
1. Hydrogen sulfide; H_2S; [7782-06-4]	Ng, H.-J.; Kalra, H.; Robinson, D. B.; Kubota, H.
2. Methylbenzene (toluene); C_7H_8; [108-88-3]	*J. Chem. Eng. Data* <u>1980</u>, *25*, 51-55.

EXPERIMENTAL VALUES:

T/K	P/psia	P/MPa	Mole fraction of hydrogen sulfide in liquid, x_{H_2S}	in vapor, y_{H_2S}
477.59	200	1.38	0.037	0.352
	399	2.75	0.142	0.666
	619	4.27	0.247	0.752
	1036	7.14	0.433	0.807
	1468	10.12	0.598	0.826
	1679	11.58	0.689	0.797

COMPONENTS:	ORIGINAL MEASUREMENTS:
1. Hydrogen sulfide; H_2S; [7783-06-4] 2. 1,2-Dimethylbenzene, (o-xylene); C_8H_{10}; [95-47-6]	Gerrard, W. J. Appl. Chem. Biotechnol., 1972, 22, 623-650

VARIABLES:	PREPARED BY:
Temperature	C.L. Young

EXPERIMENTAL VALUES:

T/K	P*/mmHg	P*/kPa	Mole ratio	Mole fraction[+] of hydrogen sulfide in liquid, x_{H_2S}
265.15	754	100.5	0.155	0.134
267.15	754	100.5	0.145	0.127
273.15	754	100.5	0.120	0.107
283.15	754	100.5	0.084	0.077
293.15	754	100.5	0.075	0.070

[+]calculated by compiler

*total pressure

AUXILIARY INFORMATION

METHOD/APPARATUS/PROCEDURE:	SOURCE AND PURITY OF MATERIALS:
Hydrogen sulfide was bubbled into a weighed amount of component 2. in a bubbler tube as described in detail in the source. The amount of gas absorbed at equilibrium and at the observed temperature and pressure was determined by weighing. Pressure was measured with a mercury manometer. The amount of gas absorbed at successively lower pressures was measured. Eventually the pressure was reduced to the vapor pressure of component 2. The refractive index and infrared spectrum of the liquid showed it to be essentially pure component 2.	1 and 2. Components were purified and attested by conventional methods.
	ESTIMATED ERROR:
	$\delta T/K = \pm 0.2$; $\delta x_{H_2S} = \pm 4\%$
	REFERENCES:

COMPONENTS:	ORIGINAL MEASUREMENTS:
1. Hydrogen sulfide; H_2S; [7783-06-4] 2. 1,3-Dimethylbenzene; C_8H_{10}; [108-38-3]	Huang, S. S.-S.; Robinson, D.B. *Fluid Phase Equilibria* 1984, *17*, 373 - 382.
VARIABLES:	PREPARED BY:
Temperature, pressure	P.G.T. Fogg

EXPERIMENTAL VALUES:

T/K	P_{total}/MPa	Mole fraction of H_2S in liquid phase	Mole fraction of H_2S in gas phase
310.9	0.146	0.0403	0.9834
	0.383	0.1334	0.9949
	0.779	0.3139	0.9980
	1.17	0.4651	0.9987
	1.75	0.6873	0.9992
	1.92	0.7514	0.9994
	2.13	0.8394	0.9995
	2.48	0.9501	0.9997
	2.63	0.9931	0.9999
352.6	0.365	0.0690	0.9624
	1.10	0.2525	0.9874
	1.72	0.3899	0.9912
	3.24	0.6637	0.9960
	4.41	0.8233	0.9969
	5.41	0.9319	0.9980
	6.14	0.9888	0.9994
394.3	0.256	0.0206	0.7957
	0.869	0.1106	0.9381
	2.44	0.3242	0.9741
	4.63	0.5637	0.9834
	7.03	0.7656	0.9860
	8.94	0.8783	0.9858
	10.21	0.9490	0.9824
477.6	0.938	0.0310	0.5592
	1.59	0.0826	0.7125
	3.87	0.2582	0.8610
	6.31	0.4240	0.8956
	9.13	0.5859	0.9035
	11.32	0.6908	0.8989
	13.12	0.8450	0.8610

AUXILIARY INFORMATION

METHOD/APPARATUS/PROCEDURE	SOURCE AND PURITY OF MATERIALS
A stirred static cell, fitted with a glass window, was used. Temperature was measured with a thermocouple and pressure with a Bourdon gauge. After equilibrium was established gas and liquid phases were analysed by gas chromatography. Techniques were similar to those described in refs. (1) & (2).	1. from Matheson; purity 99.9 mol%. 2. Matheson *Chromatoquality*; purity > 99 mol%; no impurites detected by chromatography.

REFERENCES

1. Ng, H-J.; Robinson, D.B. *Fluid Phase Equilibria*, 1979, *2*, 283.
2. Ng, H-J., Robinson, D.B. *J. Chem. Eng. Data*, 1980, *25*, 51.

COMPONENTS:	ORIGINAL MEASUREMENTS:
1. Hydrogen sulfide; H_2S; [7783-06-4] 2. 1,3,5-Trimethylbenzene, (mesitylene); C_9H_{12}; [108-67-8]	Eakin, B.E.; DeVaney, W.E. *Am. Inst. Chem. Engnrs.Symp.Ser.* 1974, No. 140, *70*, 80-90.

VARIABLES:	PREPARED BY:
Temperature, pressure	C.L. Young

EXPERIMENTAL VALUES:

T/K	$P/10^5$Pa	Mole fraction of hydrogen sulfide,	
		in liquid, x_{H_2S}	in gas, y_{H_2S}
310.9	1.31	0.0511	0.99457
	2.14	0.0846	0.99663
	3.24	0.1314	0.99701
	4.62	0.1866	0.99787
	5.58	0.2354	0.99816
366.5	2.59	0.0411	0.9369
	5.26	0.0838	0.9590
	7.85	0.1292	0.9758
	10.52	0.1781	0.9871
	13.19	0.2321	0.9888
477.6	8.18	0.0391	0.6359
	14.07	0.0774	0.7517
	19.93	0.1270	0.8408
	26.20	0.1825	0.8715
	32.61	0.2215	0.8945

AUXILIARY INFORMATION

METHOD/APPARATUS/PROCEDURE:	SOURCE AND PURITY OF MATERIALS:
Rocking stainless steel equilibrium vessel, fitted with liquid and gas sampling ports. Pressure measured with Bourdon gauge. Components charged into cell under pressure, equilibrated and samples withdrawn. Samples analysed by G.C. using thermal conductivity detector and Porapak P column.	1. Matheson Gas sample purity better than 99 mole per cent. 2. Eastman Kodak Co. sample, boiling point range 160-163°C.

ESTIMATED ERROR:

$\delta T/K = \pm 0.06$ (at 310.9K and 366.5K) ± 0.12 (at 477.6K) δP/MPa $= \pm 0.2$.
δx_{H_2S}, $\delta y_{H_2S} = \pm 0.001$. (estimated by compiler).

REFERENCES:

COMPONENTS:	ORIGINAL MEASUREMENTS:
1. Hydrogen sulfide; H_2S; [7783-06-4] 2. Alkylbenzenes	Patyi, L.; Furmer, I. E. Makranczy, J.; Sadilenko, A. S.; Stepanova, Z. G.; Berengarten, M. G. *Zh. Prikl. Khim.* <u>1978</u>, *51*, 1296-1300.

VARIABLES:	PREPARED BY:
	C. L. Young

EXPERIMENTAL VALUES:

T/K	$\alpha^{\dagger}$	Mole fraction of hydrogen sulfide at a partial pressure of 101.325 kPa x_{H_2S}
Ethylbenzene; C_8H_{10}; [100-41-4]		
298.15	8.0	0.042
Propylbenzene; C_9H_{12}; [103-65-1]		
298.15	8.8	0.052
(1-methylethyl)-benzene (isopropylbenzene); C_9H_{12}; [98-82-8]		
298.15	8.9	0.053
1,3,5-Trimethylbenzene (mesitylene); C_9H_{12}; [108-67-8]		
298.15	8.22	0.0487

$\dagger$ volume of gas (measured at 101.325 kPa and 273.15 K) dissolved per volume of solvent.

AUXILIARY INFORMATION

METHOD/APPARATUS/PROCEDURE:	SOURCE AND PURITY OF MATERIALS:
Volumetric method. Pressure measured when known amounts of gas were added, in increments, to a known amount of liquid in a vessel of known dimensions. Corrections were made for the partial pressure of solvent. Details in ref. (1).	Purity better than 99 mole per cent as determined by gas chromatography.
	ESTIMATED ERROR: $\delta T/K = \pm 0.1$; $\delta\alpha = \pm 4\%$ or less.
	REFERENCES: 1. Bodor, E.; Bor, G. J.; Mohai, B.; Sipos, G. *Veszpremi. Vegyip. Egy. Kozl.* 1957, *1*, 55.

COMPONENTS:	ORIGINAL MEASUREMENTS:
1. Hydrogen sulfide; H₂S; [7783-06-4] 2. 1,3,5-trimethylbenzene; C₉H₁₂; [108-67-8]	Huang, S. S.-S.; Robinson, D.B. *Fluid Phase Equilibria* 1984, 17, 373 - 382.

VARIABLES:	PREPARED BY:
Temperature, pressure	P.G.T. Fogg

EXPERIMENTAL VALUES:

T/K	P_{total}/MPa	Mole fraction of H₂S in liquid phase	Mole fraction of H₂S in gas phase
310.9	0.172	0.0668	0.9963
	0.424	0.1721	0.9985
	0.827	0.3435	0.9991
	1.36	0.5832	0.9994
	2.02	0.8075	0.9997
	2.48	0.9407	0.9998
	2.63	0.9775	0.9999
352.6	0.186	0.0296	0.9732
	0.620	0.1247	0.9922
	1.17	0.2420	0.9958
	1.97	0.3976	0.9970
	3.45	0.6722	0.9981
	4.86	0.8605	0.9988
	5.58	0.9425	0.9992
	6.22	0.9953	0.9999
394.3	0.339	0.0340	0.9204
	1.08	0.1375	0.9762
	2.16	0.2772	0.9866
	4.12	0.5138	0.9907
	6.95	0.7676	0.9918
	9.00	0.8943	0.9900
	10.41	0.9639	0.9872
477.6	0.674	0.0280	0.6247
	1.85	0.1160	0.8607
	3.53	0.2401	0.9211
	5.72	0.3889	0.9391
	9.45	0.6007	0.9427
	11.78	0.7146	0.9367
	13.75	0.8697	0.9210

AUXILIARY INFORMATION

METHOD/APPARATUS/PROCEDURE	SOURCE AND PURITY OF MATERIALS
A stirred static cell, fitted with a glass window, was used. Temperature was measured with a thermocouple and pressure with a Bourdon gauge. After equilibrium was established gas and liquid phases were analysed by gas chromatography. Techniques were similar to those described in refs. (1) & (2).	1. from Matheson; purity 99.9 mol%. 2. from Aldrich Chemicals; purity > 99 mol%; no impurites detected by chromatography.

	ESTIMATED ERROR
	REFERENCES
	1. Ng, H-J.; Robinson, D.B. *Fluid Phase Equilibria*, 1979, 2, 283. 2. Ng, H-J., Robinson, D.B. *J. Chem. Eng. Data*, 1980, 25, 51.

COMPONENTS:	ORIGINAL MEASUREMENTS:
1. Hydrogen sulphide; H_2S; [7783-06-4] 2. 1,1'-Methylenebisbenzene; (Diphenylmethane); $C_{13}H_{12}$; [101-81-5]	Tremper, K.K.; Prausnitz, J.M. *J. Chem. Engng. Data* 1976, *21*, 295-9
VARIABLES: Temperature	PREPARED BY: C.L. Young

EXPERIMENTAL VALUES:

T/K	Henry's Constant[a] /atm	Mole fraction[b] of hydrogen sulfide at 1 atm partial pressure, x_{H_2S}
300	32.9	0.0304
325	43.9	0.0228
350	56.2	0.0178
375	70.1	0.0143
400	86.1	0.0116
425	104.0	0.00962
450	123.0	0.00813
475	132.0	0.00758

a. Authors stated measurements were made at several pressures and values of solubility used were all within the Henry's Law region.

b. Calculated by compiler assuming linear relationship between mole fraction and pressure.

AUXILIARY INFORMATION

METHOD/APPARATUS/PROCEDURE:	SOURCE AND PURITY OF MATERIALS:
Volumetric apparatus similar to that described by Dymond and Hildebrand (1). Pressure measured with a null detector and precision gauge. Details in ref. (2).	Solvent degassed, no other details given.
	ESTIMATED ERROR: $\delta T/K = \pm 0.1$; $\delta x_{H_2S} = \pm 1\%$
	REFERENCES: 1. Dymond, J.; Hildebrand, J.H. *Ind.Chem.Eng.Fundam.* 1967, *6*, 130. 2. Cukor, P.M.; Prausnitz, J.M. *Ind.Chem.Eng.Fundam.* 1971, *10*, 638.

COMPONENTS:	ORIGINAL MEASUREMENTS:
1. Hydrogen sulfide; H_2S; [7783-06-4] 2. 1-Methylnaphthalene, $C_{11}H_{10}$; [1321-94-4]	Tremper, K.K.; Prausnitz, J.M. *J. Chem. Engng. Data* 1976, *21*, 295-9

VARIABLES:	PREPARED BY:
Temperature	C.L. Young

EXPERIMENTAL VALUES:

T/K	Henry's Constant[a] /atm	Mole fraction[b] of hydrogen sulfide at 1 atm partial pressure, x_{H_2S}
300	31.7	0.0315
325	41.6	0.0240
350	55.8	0.0179
375	72.4	0.0138
400	89.4	0.0112
425	105.0	0.00952
450	117.0	0.00855
475	124.0	0.00806

a. Authors stated measurements were made at several pressures and values of solubility used were all within the Henry's Law region.

b. Calculated by compiler assuming linear relationship between mole fraction and pressure.

AUXILIARY INFORMATION

METHOD/APPARATUS/PROCEDURE:	SOURCE AND PURITY OF MATERIALS:
Volumetric apparatus similar to that described by Dymond and Hildebrand (1). Pressure measured with a null detector and precision gauge. Details in ref. (2).	Solvent degassed, no other details given.

ESTIMATED ERROR:

$$\delta T/K = \pm 0.1; \quad \delta x_{H_2S} = \pm 1\%$$

REFERENCES:
1. Dymond, J.; Hildebrand, J.H. *Ind.Chem.Eng.Fundam.* 1967,*6*,130.
2. Cukor, P.M.; Prausnitz, J.M. *Ind.Chem.Eng.Fundam.* 1971,*10*,638.

COMPONENTS:	ORIGINAL MEASUREMENTS:
1. Hydrogen sulfide; H_2S; [7783-06-4] 2. Hydrogen; H_2; [1333-74-0] 3. 9-Methylanthracene; $C_{15}H_{12}$; [779-02-2]	Kragas, T.K.; Kobayashi, R. *Fluid Phase Equilibria* 1984, *16*, 215-236.

VARIABLES:	PREPARED BY:
Pressure of hydrogen	P.G.T. Fogg

EXPERIMENTAL VALUES:

T/K	P_{H_2}/MPa	Henry's Constant for H_2S at infinite dilution / MPa
373.2	0	4.72 *
	2.785	1.70
	4.288	1.10
	5.833	0.844
398.2	0	6.27 *
	1.344	4.68
	2.255	2.79
	3.530	1.80
	5.061	1.31
423.2	0	8.27 *
	1.344	6.14
	2.303	3.63
	3.516	2.35
	5.254	1.63

* values extrapolated to zero pressure of hydrogen given by the authors.

AUXILIARY INFORMATION

METHOD/APPARATUS/PROCEDURE:	SOURCE AND PURITY OF MATERIALS:
A high pressure chromatographic method described in ref. (1) was used. Hydrogen at various pressures was the carrier gas.	1. from Scientific Gas Products; 99.6% pure. 2. from Linde Division of Union Carbide Corporation. 3. from Aldrich Chem. Co.; purified to > 99.95% by zone refining.

ESTIMATED ERROR:

δ(Henry's law constant) = ± 3.6% (authors)

REFERENCES:

1. Kragas, T.K.; Pollin, J.; Martin, R.J.; Kobayashi, R.

Fluid Phase Equilibria 1984, *16*, 205.

COMPONENTS:	ORIGINAL MEASUREMENTS:
1. Hydrogen sulfide; H$_2$S; [7783-06-4] 2. Methanol; CH$_4$O; [67-56-1]	Bezdel, L.S.; Teodorovich, V.P. *Gazovaya Prom.* <u>1958</u>, *No.8*, 38 - 43.
VARIABLES: Temperature, pressure	PREPARED BY: P.G.T. Fogg

EXPERIMENTAL VALUES:

T/K	P_{H_2S}/mmHg	Volume of H$_2$S, corrected to 273.15 K and 760 mmHg, dissolved by 1 kg methanol / dm^3	Mole fraction of H$_2$S in liquid phase* x_{H_2S}
223.2	15.80	5.67	0.00814
	48.70	15.50	0.02194
243.2	1.8	0.27	0.00039
	3.16	0.37	0.00054
	12.45	1.57	0.00227
	11.20	1.84	0.00266
	16.50	2.60	0.00375
	17.50	2.56	0.00369
	47.50	7.07	0.01013
	49.20	7.17	0.01027
	60.25	9.13	0.01304
303.2	13.90	0.30	0.00043
	40.30	0.79	0.00114
	41.70	0.79	0.00114

* calculated by the compiler.   760 mmHg = 1 atm = 1.013 bar.

AUXILIARY INFORMATION

METHOD/APPARATUS/PROCEDURE:	SOURCE AND PURITY OF MATERIALS:
Pre-cooled mixtures of H$_2$ & H$_2$S were bubbled through CH$_3$OH in a thermostatted absorption vessel of volume 200 cm^3. Temperatures were controlled manually by addition of solid CO$_2$ to butyl alcohol in the thermostat and measured by a pentane thermometer. After 6 - 8 h, 3 - 4 samples of the liquid phase were withdrawn at 30 min intervals. These, together with gas samples, were analysed by iodimetry. The concentration at equilibrium was unaffected by the presence of an initial excess of H$_2$S in the solution.	1. dried with CaCl$_2$.
	ESTIMATED ERROR: $\delta T/K$ = ± 0.5 (authors)
	REFERENCES:

COMPONENTS:	ORIGINAL MEASUREMENTS:
1.Hydrogen sulfide; H_2S; [7783-06-4] 2. Methanol; CH_4O; [67-56-1]	Yorizane, M.; Sadamoto, S.; Masuoka, H.; Eto, Y. *Kogyo Kagaku Zasshi* <u>1969</u>, *72*, 2174-7.

VARIABLES:	PREPARED BY:
Temperature, pressure	C. L. Young

EXPERIMENTAL VALUES:

T/K	P/atm	P/MPa	Mole fraction of hydrogen sulfide in liquid, x_{H_2S}
273.15	2.0	0.20	0.092
	4.0	0.41	0.199
	6.0	0.61	0.329
	7.5	0.76	0.453
	8.0	0.81	0.484
	9.1	0.92	0.608
	9.8	0.99	0.743
	10.0	1.01	0.840
258.15	2.0	0.20	0.165
	3.0	0.30	0.231
	3.4	0.34	0.298
	4.2	0.43	0.367
	4.4	0.45	0.403
	5.0	0.51	0.490
	5.4	0.55	0.585
	5.8	0.59	0.662
248.15	2.0	0.20	0.203
	2.5	0.25	0.290
	3.0	0.30	0.327
	3.4	0.34	0.465
	4.0	0.40	0.582
	4.3	0.43	0.733

AUXILIARY INFORMATION

METHOD/APPARATUS/PROCEDURE:	SOURCE AND PURITY OF MATERIALS:
Vapor-liquid equilibrium cell. Diagram given in source. (Original in Japanese.)	1. Purity 99.0 mole per cent.

ESTIMATED ERROR:

$\delta T/K = \pm 0.1$; $\delta x_{H_2S} = \pm 0.001$

(estimated by compiler).

REFERENCES:

COMPONENTS:	ORIGINAL MEASUREMENTS:
1. Hydrogen sulfide; H_2S; [7783-06-4] 2. Hydrogen; H_2; [1333-74-0] 3. Methanol; CH_4O; [67-56-1]	Yorizane, M.; Sadamoto, S.; Masuoka, H.; Eto, Y. *Kogyo Kagaku Zasshi* 1969, 72, 2174 - 2177.

VARIABLES:	PREPARED BY:
Temperature, pressure, composition:	P.G.T. Fogg

EXPERIMENTAL VALUES:

T/K	Total pressure /atm	Mole fraction of H_2 in liquid	Mole fraction of H_2S in liquid	Mole fraction of H_2 in gas	Mole fraction of H_2S in gas
243.2	10	0.003	0.397	0.813	0.187
		0.010	0.668	0.692	0.308
		0.004	0.721	0.702	0.298
		0.006	0.776	0.655	0.345
	30	0.006	0.397	0.813	0.187
		0.005	0.652	0.896	0.104
		0.006	0.704	0.971	0.030
	50	0.010	0.333	0.977	0.023
		0.013	0.621	0.941	0.059
		0.012	0.813	0.929	0.071
273.2	10	0.003	0.043	0.925	0.075
		0.006	0.107	0.813	0.187
		0.002	0.205	0.702	0.298
		0.003	0.379	0.400	0.600
		0.002	0.617	0.237	0.763
		0.002	0.688	0.163	0.837
	30	0.007	0.043	0.974	0.026
		0.007	0.161	0.928	0.072
		0.007	0.195	0.924	0.076
		0.009	0.197	0.904	0.096
		0.007	0.327	0.781	0.219
		0.009	0.409	0.760	0.240
		0.010	0.685	0.672	0.327
	50	0.012	0.050	0.988	0.012
		0.014	0.149	0.962	0.038
		0.014	0.194	0.945	0.055
		0.010	0.221	0.945	0.055
		0.013	0.353	0.868	0.132
		0.015	0.401	0.854	0.146
		0.018	0.675	0.796	0.204

AUXILIARY INFORMATION

METHOD/APPARATUS/PROCEDURE	SOURCE AND PURITY OF MATERIALS
Vapor-liquid equilibrium cell. Diagram given in source. (Original in Japanese)	1. Purity 99.0 mol%.
	ESTIMATED ERROR
	REFERENCES

COMPONENTS:	ORIGINAL MEASUREMENTS:
(1) Hydrogen sulfide; H_2S; [7783-06-4] (2) Methanol; CH_4O; [67-56-1]	Short, I.; Sahgal, A.; Hayduk, W. *J. Chem. Eng. Data* <u>1983</u>, *28*, 63-66.

VARIABLES:	PREPARED BY:
T/K: 263.15, 298.15 P/kPa: 101.325	W. Hayduk

EXPERIMENTAL VALUES:

T/K	Ostwald Coefficient[1] L/cm^3 cm^{-3}	Bunsen Coefficient[2] α/cm^3 (STP) $cm^{-3}atm^{-1}$	Mole Fraction[1] 10^4 x_1
263.15	40.2	41.73	689
298.15	26.93	24.67	276

[1]Original data

[2]Calculated by compiler

The mole fraction solubility of the original data was used to determine the following equations for $\Delta G°$ and $\ln x_1$ and table of smoothed values:

$$\Delta G°/J\ mol^{-1} = -RT \ln x_1 = 858.98\ T - 168276$$
$$\ln x_1 = 2050.8/T - 10.468$$

T/K	$10^{-4}\Delta G°/J\ mol^{-1}$	$10^4\ x_1$
263.15	5.776	689.0
273.15	6.635	518.0
283.15	7.494	397.3
293.15	8.353	310.4
298.15	8.783	276.0

AUXILIARY INFORMATION

METHOD/APPARATUS/PROCEDURE:	SOURCE AND PURITY OF MATERIALS:
A volumetric method using a glass apparatus was employed. Degassed solvent contacted the gas while flowing as a thin film, at a constant rate, through an absorption spiral into a solution buret. A constant solvent flow was obtained by means of a calibrated syringe pump. The solution at the end of the spiral was considered saturated. Dry gas was maintained at atmospheric pressure in a gas buret by mechanically raising the mercury level in the buret at an adjustable rate. The solubility was calculated from the constant slope of volume of gas dissolved and volume of solvent injected. Degassing was accomplished using a two stage vacuum process described by Clever et al. (1).	1. Liquid Carbonic. Specified minimum purity 99.5 per cent. 2. Canlab. Specified as absolute methanol of minimum purity 99.5 per cent.

ESTIMATED ERROR:

$\delta T/K$ = 0.1

$\delta x_1/x_1$ = 0.01

REFERENCES:

1. Clever, H.L.; Battino, R.; Saylor, J.H.; Gross, P.M. *J. Phys. Chem.* <u>1957</u>, *61*, 1078.

COMPONENTS:	ORIGINAL MEASUREMENTS:
1. Hydrogen sulfide; H_2S; [7783-06-4] 2. Ethanol; C_2H_6O; [64-17-5]	Gerrard, W. *J. Appl. Chem. Biotechnol.*, <u>1972</u>, *22*, 623-650
VARIABLES: Temperature	PREPARED BY: C.L. Young

EXPERIMENTAL VALUES:

T/K	P*/mmHg	P*/kPa	Mole ratio	Mole fraction[+] of hydrogen sulfide in liquid, x_{H_2S}
265.15	769	102.5	0.078	0.072
267.15	769	102.5	0.0725	0.0676
273.15	769	102.5	0.054	0.051
283.15	769	102.5	0.033	0.032
293.15	769	102.5	0.018	0.018

[+]calculated by compiler

*total pressure

AUXILIARY INFORMATION

METHOD/APPARATUS/PROCEDURE:	SOURCE AND PURITY OF MATERIALS:
Hydrogen sulfide was bubbled into a weighed amount of component 2. in a bubbler tube as described in detail in the source. The amount of gas absorbed at equilibrium and at the observed temperature and pressure was determined by weighing. Pressure was measured with a mercury manometer. The amount of gas absorbed at successively lower pressures was measured. Eventually the pressure was reduced to the vapor pressure of component 2. The refractive index and infrared spectrum of the liquid showed it to be essentially pure component 2.	1 and 2. Components were purified and attested by conventional methods.

	ESTIMATED ERROR:
	$\delta T/K = \pm 0.2$; $\delta x_{H_2S} = \pm 4\%$

	REFERENCES:

COMPONENTS:	ORIGINAL MEASUREMENTS:
(1) Hydrogen sulfide; H_2S; [7783-06-4] (2) 1-Butanol; $C_4H_{10}O$; [71-36-3]	Short, I.; Sahgal, A.; Hayduk, W. *J. Chem. Eng. Data* 1983, 28, 63-66.
VARIABLES: T/K: 263.15-333.15 P/kPa: 101.325	PREPARED BY: W. Hayduk

EXPERIMENTAL VALUES:

T/K	Ostwald Coefficient[1] L/cm^3 cm^{-3}	Bunsen Coefficient[2] α/cm^3 (STP) cm^{-3}atm^{-1}	Mole Fraction[1] x_1
263.15	22.8	23.67	0.0873 (0.0875)[3]
298.15	8.58	7.86	0.0315 (0.0314)
333.15	3.65	2.99	0.0126 (0.0126)

[1]Original data

[2]Calculated by compiler

[3]The mole fraction solubility of the original data was used to determine the following equations for $\Delta G°$ and $\ln x_1$ and table of smoothed values:

$$\Delta G°/\text{J mol}^{-1} = -RT \ln x_1 - 673.32\ T \ln T - 3552.3\ T$$

$$\ln x_1 = 43.291 - 8.2056 \ln T$$

Std. deviation for $\Delta G° = 63$ J mol^{-1}

T/K	$10^{-4}\Delta G°$/J mol^{-1}	x_1	T/K	$10^{-4}\Delta G°$/J mol^{-1}	x_1
263.15	5.261	0.0875	303.15	8.949	0.0274
273.15	6.147	0.0644	313.15	9.929	0.0210
283.15	7.058	0.0480	323.15	10.93	0.0162
293.15	7.992	0.0361	333.15	11.95	0.0126
298.15	8.468	0.0314			

AUXILIARY INFORMATION

METHOD/APPARATUS/PROCEDURE:

A volumetric method using a glass apparatus was employed. Degassed solvent contacted the gas while flowing as a thin film, at a constant rate, through an absorption spiral into a solution buret. A constant solvent flow was obtained by means of a calibrated syringe pump. The solution at the end of the spiral was considered saturated. Dry gas was maintained at atmospheric pressure in a gas buret by mechanically raising the mercury level in the buret at an adjustable rate. The solubility was calculated from the constant slope of volume of gas dissolved and volume of solvent injected.

Degassing was accomplished using a two stage vacuum process described by Clever et al. (1).

SOURCE AND PURITY OF MATERIALS:

1. Liquid Carbonic. Specified minimum purity 99.5 per cent.

2. Canlab. Specified minimum purity 99.0 per cent.

ESTIMATED ERROR:

δT/K $= 0.1$

$\delta x_1/x_1 = 0.01$

REFERENCES:

1. Clever, H.L.; Battino, R.; Saylor, J.H.; Gross, P.M.
J. Phys. Chem. 1957, 61, 1078.

COMPONENTS:	ORIGINAL MEASUREMENTS:
1. Hydrogen sulfide; H_2S; [7783-06-4] 2. 1-Octanol; $C_8H_{18}O$; [111-87-5]	Gerrard, W. *J. Appl. Chem. Biotechnol.*, 1972, 22, 623-650

VARIABLES:	PREPARED BY:
Temperature	C.L. Young

EXPERIMENTAL VALUES:

T/K	$P*$/mmHg	$P*$/kPa	Mole ratio	Mole fraction[+] of hydrogen sulfide in liquid, x_{H_2S}
265.15	755	100.7	0.105	0.095
267.15	755	100.7	0.097	0.091
273.15	755	100.7	0.080	0.074
283.15	755	100.7	0.064	0.060
293.15	755	100.7	0.054	0.051

[+]calculated by compiler

*total pressure

AUXILIARY INFORMATION

METHOD/APPARATUS/PROCEDURE:	SOURCE AND PURITY OF MATERIALS:
Hydrogen sulfide was bubbled into a weighed amount of component 2. in a bubbler tube as described in detail in the source. The amount of gas absorbed at equilibrium and at the observed temperature and pressure was determined by weighing. Pressure was measured with a mercury manometer. The amount of gas absorbed at successively lower pressures was measured. Eventually the pressure was reduced to the vapor pressure of component 2. The refractive index and infrared spectrum of the liquid showed it to be essentially pure component 2.	1 and 2. Components were purified and attested by conventional methods.
	ESTIMATED ERROR: $\delta T/K = \pm 0.2$; $\delta x_{H_2S} = \pm 4\%$
	REFERENCES:

COMPONENTS:	ORIGINAL MEASUREMENTS:
1. Hydrogen sulfide; H_2S; [7783-06-4] 2. Benzenemethanol (Benzyl alcohol); C_7H_8O; [100-51-6]	Lenoir, J-Y.; Renault, P.; Renon, H. *J. Chem. Eng. Data*, <u>1971</u>, *16*, 340-2.

VARIABLES:	PREPARED BY:
	C. L. Young

EXPERIMENTAL VALUES:

T/K	Henry's constant H_{H_2S}/atm	Mole fraction at 1 atm* x_{H_2S}
298.2	24.0	0.0417

* Calculated by compiler assuming a linear function of P_{H_2S} vs x_{H_2S}, i.e., x_{H_2S}(1 atm) $= 1/H_{H_2S}$

AUXILIARY INFORMATION

METHOD/APPARATUS/PROCEDURE:	SOURCE AND PURITY OF MATERIALS:
A conventional gas-liquid chromato- graphic unit fitted with a thermal conductivity detector was used. The carrier gas was helium. The value of Henry's law constant was calculated from the retention time. The value applies to very low partial pressures of gas and there may be a substantial difference from that measured at 1 atm. pressure. There is also considerable uncertainty in the value of Henry's constant since surface adsorption was not allowed for although its possible existence was noted.	(1) L'Air Liquide sample, minimum purity 99.9 mole per cent. (2) Touzart and Matignon or Serlabo sample, purity 99 mole per cent.
	ESTIMATED ERROR: $\delta T/K = \pm 0.1$; $\delta H/atm = \pm 6\%$ (estimated by compiler).
	REFERENCES:

COMPONENTS:	ORIGINAL MEASUREMENTS:
1. Hydrogen sulfide; H_2S; [7783-06-4] 2. 1,2-Ethanediol, (Ethylene glycol); $C_2H_6O_2$; [107-21-1]	Lenoir, J-Y.; Renault, P.; Renon, H. *J. Chem. Eng. Data*, <u>1971</u>, *16*, 340-2
VARIABLES:	PREPARED BY: C. L. Young

EXPERIMENTAL VALUES:

T/K	Henry's constant H_{H_2S}/atm	Mole fraction at 1 atm* x_{H_2S}
298.2	52.9	0.0189

* Calculated by compiler assuming a linear function of P_{H_2S} vs x_{H_2S}, i.e.,
$x_{H_2S}(1\ atm) = 1/H_{H_2S}$

AUXILIARY INFORMATION

METHOD/APPARATUS/PROCEDURE:	SOURCE AND PURITY OF MATERIALS:
A conventional gas-liquid chromatographic unit fitted with a thermal conductivity detector was used. The carrier gas was helium. The value of Henry's law constant was calculated from the retention time. The value applies to very low partial pressures of gas and there may be a substantial difference from that measured at 1 atm. pressure. There is also considerable uncertainty in the value of Henry's constant since surface adsorption was not allowed for although its possible existence was noted.	(1) L'Air Liquuide sample, minimum purity 99.9 mole per cent. (2) Touzart and Matignon or Serlabo sample, purity 99 mole per cent.
	ESTIMATED ERROR: $\delta T/K = \pm 0.1$; $\delta H/atm = \pm 6\%$ (estimated by compiler).
	REFERENCES:

COMPONENTS:	ORIGINAL MEASUREMENTS:
1. Hydrogen sulfide; H_2S; [7783-06-4] 2. 1,2-Ethanediol; $C_2H_6O_2$; [107-21-1]	Gerrard, W. J. Appl. Chem. Biotechnol., <u>1972</u>, 22, 623-650.
VARIABLES:	PREPARED BY:
Temperature	C.L. Young

EXPERIMENTAL VALUES:

T/K	$P*$/mmHg	$P*$/kPa	Mole ratio	Mole fraction[+] of hydrogen sulfide in liquid, x_{H_2S}
265.15	760	101.3	0.0365	0.0352
267.15	760	101.3	0.033	0.032
273.15	760	101.3	0.024	0.023
283.15	760	101.3	0.017	0.017
293.15	760	101.3	0.013	0.013

[+]calculated by compiler

*total pressure

AUXILIARY INFORMATION

METHOD/APPARATUS/PROCEDURE:	SOURCE AND PURITY OF MATERIALS:
Hydrogen sulfide was bubbled into a weighed amount of component 2. in a bubbler tube as described in detail in the source. The amount of gas absorbed at equilibrium and at the observed temperature and pressure was determined by weighing. Pressure was measured with a mercury manometer. The amount of gas absorbed at successively lower pressures was measured. Eventually the pressure was reduced to the vapor pressure of component 2. The refractive index and infrared spectrum of the liquid showed it to be essentially pure component 2.	1 and 2. Components were purified and attested by conventional methods.
	ESTIMATED ERROR: $\delta T/K = \pm 0.2$; $\delta x_{H_2S} = \pm 4\%$
	REFERENCES:

COMPONENTS:	ORIGINAL MEASUREMENTS:
(1) Hydrogen sulfide; H_2S; [7783-06-4] (2) 1,2-Ethanediol (ethylene glycol); $C_2H_6O_2$; [107-21-1]	Short, I.; Sahgal, A.; Hayduk, W. *J. Chem. Eng. Data* <u>1983</u>, *28*, 63-66.

| VARIABLES:
 T/K: 263.15-333.15
 P/kPa: 101.325 | PREPARED BY:

 W. Hayduk |

EXPERIMENTAL VALUES:

T/K	Ostwald Coefficient[1] $L/cm^3\ cm^{-3}$	Bunsen Coefficient[2] $\alpha/cm^3\ (STP)cm^{-3}atm^{-1}$	Mole Fraction[1] $10^4 x_1$
263.15	13.34	13.85	332 (332.0)[3]
298.15	5.36	4.91	122 (122.0)
333.15	3.44	2.82	71.8 (71.8)

[1]Original data

[2]Calculated by compiler

[3]The mole fraction solubility of the original data was used to determine the following equations for $\Delta G°$ and $\ln x_1$ and table of smoothed values:

$$\Delta G°/J\ mol^{-1} = -RT \ln x_1 = 12462\ T - 1730.3\ T \ln T - 668490$$
$$\ln x_1 = 8146.8/T + 21.087 \ln T - 151.874$$

Std. deviation for $\Delta G° = 28\ J\ mol^{-1}$

T/K	$10^{-4}\Delta G°/J\ mol^{-1}$	$10^4 x_1$	T/K	$10^{-4}\Delta G°/J\ mol^{-1}$	$10^4 x_1$
263.15	7.353	332.0	303.15	11.21	110.4
273.15	8.410	234.7	313.15	12.03	92.8
283.15	9.404	174.7	323.15	12.79	80.50
293.15	10.34	136.1	333.15	13.49	71.80
298.15	10.78	122.0			

AUXILIARY INFORMATION

METHOD/APPARATUS/PROCEDURE:

A volumetric method using a glass apparatus was employed. Degassed solvent contacted the gas while flowing as a thin film, at a constant rate, through an absorption spiral into a solution buret. A constant solvent flow was obtained by means of a calibrated syringe pump. The solution at the end of the spiral was considered saturated. Dry gas was maintained at atmospheric pressure in a gas buret by mechanically raising the mercury level in the buret at an adjustable rate. The solubility was calculated from the constant slope of volume of gas dissolved and volume of solvent injected.

Degassing was accomplished using a two stage vacuum process described by Clever et al. (1).

SOURCE AND PURITY OF MATERIALS:

1. Liquid Carbonic. Specified minimum purity 99.5 per cent.

2. Canlab. Baker Analyzed grade of minimum specified purity 99.8 per cent.

ESTIMATED ERROR:
$\delta T/K = 0.1$
$\delta x_1/x_1 = 0.01$

REFERENCES:

1. Clever, H.L.; Battino, R.; Saylor, J.H.; Gross, P.M. *J. Phys. Chem* <u>1957</u>, *61*, 1078.

COMPONENTS:	ORIGINAL MEASUREMENTS:
1. Hydrogen sulfide; H₂S; [7783-06-4] 2. Ethylene glycols	Byeseda, J.J.; Deetz, J.A.; Manning, W.P. *Proc.Laurance Reid Gas Cond.Conf.* 1985.
VARIABLES:	PREPARED BY: P.G.T. Fogg

EXPERIMENTAL VALUES:

Solvent	Ostwald coeff. L	Mole fraction in liquid* x_{H_2S}
1,2-Ethanediol (*ethylene glycol*); $C_2H_6O_2$; [107-21-1]	6.3	0.0160
2,2'[1,2-ethanediylbis(oxy)]bis- ethanol, (*triethylene glycol*); $C_6H_{14}O_4$; [112-27-6]	11.9	0.0618

$$T/K = 297.1 \qquad P_{H_2S}/psia = 14.73 \qquad P_{H_2S}/bar = 1.016$$

     * calculated by the compiler.

                        AUXILIARY INFORMATION

METHOD/APPARATUS/PROCEDURE:	SOURCE AND PURITY OF MATERIALS:
The H₂S was contained in a thermostatted metal cylinder connected to a pressure gage, vacuum pump and supply of gas. A tight fitting internal piston sealed with an O-ring fitted into the cylinder so that the volume of gas could be changed by controlled movement of the piston. A measured volume of solvent was injected into the cylinder by a syringe. The absorption of gas was found from the movement of the piston which was necessary to maintain constant pressure.	No information
	ESTIMATED ERROR:
	REFERENCES:

COMPONENTS:	ORIGINAL MEASUREMENTS:
1. Hydrogen sulfide; H_2S; [7783-06-4] 2. Oxybispropanol; (Dipropylene glycol); $C_6H_{14}O_3$; [25265-71-8]	Lenoir, J-Y.; Renault, P.; Renon, H.; *J. Chem. Eng. Data*, <u>1971</u>, *16*, 340-2.

VARIABLES:	PREPARED BY:
Temperature	C. L. Young

EXPERIMENTAL VALUES:

T/K	Henry's constant H_{H_2S}/atm	Mole fraction at 1 atm* x_{H_2S}
298.2	22.9	0.0437
323.2	30.7	0.0326
343.2	74.3	0.0135

*   Calculated by compiler assuming a linear function of P_{H_2S} vs x_{H_2S}, i.e., $x_{H_2S}(1 \text{ atm}) = 1/H_{H_2S}$

AUXILIARY INFORMATION

METHOD/APPARATUS/PROCEDURE:	SOURCE AND PURITY OF MATERIALS:
A conventional gas-liquid chromato-graphic unit fitted with a thermal conductivity detector was used. The carrier gas was helium. The value of Henry's law constant was calculated from the retention time. The value applies to very low partial pressures of gas and there may be a substantial difference from that measured at 1 atm. pressure. There is also considerable uncertainty in the value of Henry's constant since surface adsorption was not allowed for although its possible existence was noted.	(1) L'Air Liquide sample, minimum purity 99.9 mole per cent. (2) Touzart and Matignon or Serlabo sample, purity 99 mole per cent.
	ESTIMATED ERROR: $\delta T/K = \pm 0.1$; δH/atm $= \pm 6\%$ (estimated by compiler).
	REFERENCES:

COMPONENTS:	ORIGINAL MEASURMENTS:
1. Hydrogen sulfide; H₂S; [7783-06-4] 2. 2,2'[1,2-ethanediylbis(oxy)]bis-ethanol, (*triethylene glycol*); $C_6H_{14}O_4$; [112-27-6]	Blake, R.J. *Oil & Gas J.* 1967, *65(2)*, 105-108.

VARIABLES:	PREPARED BY:
Temperature, pressure.	P.G.T. Fogg

EXPERIMENTAL VALUES:

The author presented smooth curves showing the variation of the absorption coefficient, α, (volume of gas reduced to 273.15 K and 1.013 bar dissolved by one volume of the solvent at the experimental temperature) with change of T/°F. These curves were for partial pressures of 0, 25, 50, 10 & 150 psig. The compiler has used the curves to prepare the tables below.

Values of α for various temperatures and pressures

T/K	Partial pressure of H₂S/bar				
	1.013	2.737	4.460	7.908	11.355
273.15	21.5	41.3			
283.15	16.4	31.6	50.5		
293.15	12.8	25.9	41.4		
303.15	10.2	21.6	35.0	64.2	
313.15	8.1	18.5	30.2	55.2	
323.15	6.8	16.0	26.2	47.9	75.8
333.15	5.6	14.2	23.1	42.0	63.7
343.15	5.0	12.5	20.1	36.7	54.9
353.15	4.4	11.1	17.7	32.2	47.2
363.15	4.2	10.2	15.4	27.8	40.7
373.15	4.0	9.6	13.5	23.9	34.7

T/K	P_{H_2S}/bar	Mole fraction in liquid x_{H_2S}
273.15	1.013	0.114
	2.737	0.199
283.15	1.013	0.090
	2.737	0.160
	4.460	0.233
293.15	1.013	0.071
	2.737	0.135
	4.460	0.199
	7.908	0.309
303.15	1.013	0.058
	2.737	0.115
	4.460	0.174
	7.908	0.278
313.15	1.013	0.047
	2.737	0.100
	4.460	0.154
	7.908	0.249

AUXILIARY INFORMATION

METHOD/APPARATUS/PROCEDURE:	SOURCE AND PURITY OF MATERIALS:
No information.	No information.
	ESTIMATED ERROR:
	REFERENCES:

COMPONENTS:	ORIGINAL MEASUREMENTS:
1. Hydrogen sulfide; H_2S; [7783-06-4] 2. Polyethylene glycols (α-hydro-ω-hydroxy-poly(oxy-1,2-ethanediyl); $(C_2H_4O)_nH_2O$; [25322-68-3]	Gestrich, W.; Reinke, L. *Chem.-Ing.-Tech.* 1983, *55*, 629-631

VARIABLES:	PREPARED BY:
Temperature, pressure.	P.G.T. Fogg

EXPERIMENTAL VALUES:

Solubilities were measured at temperatures from 70 °C to 150 °C and pressures from 196 to 695 Torr. Over this pressure range the weight of gas absorbed per gram of solvent was found to be proportional to the pressure. Values of this proportionality constant were graphically displayed by the authors in small scale diagrams and also presented as equations of the form:

$$\ln (S/mg) = - A + B/(T/K)$$

In these equations S is the weight of gas absorbed by 1 g of solvent for a pressure of gas of 1 bar.

	A	B	r^2
Polyethylene glycol *P200*	3.7573	1915.0	0.9943
Polyethylene glycols *P300* & *P400*	3.6376	1904.0	0.9894
Polyethylene glycol *P1000*	4.6163	2263.6	0.9995

P200, *P300* etc. correspond to the nomenclature of the suppliers, Chemischen Werke Hüls.

r = correlation coefficient.

AUXILIARY INFORMATION

METHOD/APPARATUS/PROCEDURE:	SOURCE AND PURITY OF MATERIALS:
No information	
	ESTIMATED ERROR:
	REFERENCES:

COMPONENTS:	ORIGINAL MEASUREMENTS:
1. Hydrogen sulfide; H_2S; [7783-06-4] 2. Polyethylene glycols (α-Hydro-ω-hydroxy-poly(oxy-1,2-ethanediyl); $(C_2H_4O)_nH_2O$; [25322-68-3]	Härtel, G.H. *J. Chem. Eng. Data* 1985, *30*, 57-61.

VARIABLES:	PREPARED BY:
Temperature	P.G.T. Fogg

EXPERIMENTAL VALUES:

	T/K	Henry's constant[†] H_{H_2S}/bar	Mole fraction solubility x_{H_2S} (1.013 bar)*
Polyethylene glycol 200	333.15	21.989	0.046
	353.15	33.121	0.031
	373.15	44.812	0.023
Polyethylene glycol 400	333.15	9.498	0.107
	353.15	15.023	0.067
	373.15	20.598	0.049

Henry's constant is given by :

$$H_{H_2S} = \frac{pressure}{mole\ fraction\ solubility}$$

Solubilities were measured at concentrations not greater than x_{H_2S} = 0.16. The variation of mole fraction solubility with pressure was found to be almost linear with a correlation coefficient of better than 0.99.

* Calculated by the compiler assuming that the variation of mole fraction solubility with variation of pressure was linear to 1.013 bar.

[†] There is some ambiguity in the manner in which data are tabulated in the original paper. The compiler considers that these values correspond to experimental measurements by the author.

AUXILIARY INFORMATION

METHOD/APPARATUS/PROCEDURE:	SOURCE AND PURITY OF MATERIALS:
Absorption was found from the decrease in pressure when a known volume of gas came into contact with a known mass of degassed solvent. Pressure changes were found by use of a transducer with a mercury manometer to provide a reference pressure.	1. From Matheson, Heusenstamm, FRG; minimum purity 99.5%.
	ESTIMATED ERROR: δP = ± 0.25 mbar (author)
	REFERENCES:

COMPONENTS:	ORIGINAL MEASUREMENTS:
1. Hydrogen sulfide; H_2S; [7783-06-4] 2. 1,1'-Oxybisethane, (diethyl ether); $C_4H_{10}O$; [60-29-7]	Parsons, L.B. *J. Amer. Chem. Soc.* <u>1925</u>, 47, 1820-1830.

VARIABLES:	PREPARED BY:
	P.G.T. Fogg

EXPERIMENTAL VALUES:

P(total /mmHg	/bar	T/K	Concentrations of H_2S / Normality	Average concentration* /mol dm^{-3}	x_{H_2S}
740	0.987	299.2	0.506 0.520 0.505 0.515	0.256	0.026

* estimated by the compiler on the assumption that dissolution of gas
   caused negligible change of volume.

AUXILIARY INFORMATION

METHOD/APPARATUS/PROCEDURE:	SOURCE AND PURITY OF MATERIALS:
The ether appears to have been saturated with hydrogen sulfide by bubbling the gas through the solvent at room temperature and a total pressure equal to barometric pressure. The gas dissolved by a measured volume of solution was then allowed to react with a 3% aqueous solution of cadmium sulfate so as to produce a precipitate of cadmium sulfide and a solution of sulfuric acid. There was good agreement between the amount of hydrogen sulfide calculated from the weight of cadmium sulfide and the amount calculated from the quantity of sulfuric acid.	1. Prepared from 25% solution of sodium sulfide by treatment with concentrated acid. Washed with distilled water and dried with $CaCl_2$ and P_2O_5. 2. Distilled from fused $CaCl_2$ and then treated with sodium wire; redistilled.
	ESTIMATED ERROR:
	REFERENCES:

COMPONENTS:	ORIGINAL MEASUREMENTS:
1. Hydrogen sulfide; H$_2$S; [7783-06-4] 2. 1,1'-Oxybisoctane; C$_{14}$H$_{34}$O$_3$; [629-82-3]	Gerrard, W. *J. Appl. Chem. Biotechnol.* 1972, 22, 623-650.

VARIABLES:	PREPARED BY:
Temperature	P.G.T. Fogg

EXPERIMENTAL VALUES:

T/K	P(total)/mmHg	P(total)/bar	Mole ratio	Mole fraction of H$_2$S[*] x_{H_2S}
265.15	761	1.015	0.237	0.192
267.15	761	1.015	0.223	0.182
273.15	761	1.015	0.184	0.155
283.15	761	1.015	0.138	0.121
293.15	761	1.015	0.110	0.099

[*] calculated by the compiler for the stated total pressure.

AUXILIARY INFORMATION

METHOD/APPARATUS/PROCEDURE:	SOURCE AND PURITY OF MATERIALS:
Hydrogen sulfide was bubbled into a weighed amount of component 2 in a bubbler tube as described in detail in the source. The amount of gas absorbed at equilibrium for the observed temperature and pressure was found by weighing. Pressure was measured with a mercury manometer.	It was stated that "All materials purified and attested by conventional methods."
	ESTIMATED ERROR: $\delta\, x_{H_2S} = \pm\ 4\%$ (author)
	REFERENCES:

COMPONENTS:	ORIGINAL MEASUREMENTS:
1. Hydrogen sulfide; H$_2$S; [7783-06-4] 2. Ethoxybenzene, (ethylphenyl ether); C$_8$H$_{10}$O; [103-73-1]	Gerrard, W. *J. Appl. Chem. Biotechnol.*, <u>1972</u>, *22*, 623-650

VARIABLES:	PREPARED BY:
Temperature	C.L. Young

EXPERIMENTAL VALUES:

T/K	P*/mmHg	P*/kPa	Mole ratio	Mole fraction[+] of hydrogen sulfide in liquid, x_{H_2S}
265.15	748	99.7	0.175	0.149
267.15	748	99.7	0.164	0.141
273.15	748	99.7	0.135	0.119
283.15	748	99.7	0.100	0.091
293.15	748	99.7	0.078	0.072

[+]calculated by compiler

*total pressure

AUXILIARY INFORMATION

METHOD/APPARATUS/PROCEDURE:	SOURCE AND PURITY OF MATERIALS:
Hydrogen sulfide was bubbled into a weighed amount of component 2. in a bubbler tube as described in detail in the source. The amount of gas absorbed at equilibrium and at the observed temperature and pressure was determined by weighing. Pressure was measured with a mercury manometer. The amount of gas absorbed at successively lower pressures was measured. Eventually the pressure was reduced to the vapor pressure of component 2. The refractive index and infrared spectrum of the liquid showed it to be essentially pure component 2.	1 and 2. Components were purified and attested by conventional methods.

	ESTIMATED ERROR:
	$\delta T/K = \pm 0.2$; $\delta x_{H_2S} = \pm 4\%$

	REFERENCES:

COMPONENTS:	ORIGINAL MEASUREMENTS:
1. Hydrogen sulfide; H_2S; [7783-06-4] 2. Polyglycol ethers	Sciamanna, S.F. Ph.D. Dissertation, Department of Chemical Engineering, University of California, Berkeley, U.S.A. 1986. (Directed by Lynn, S.)

VARIABLES:	PREPARED BY:
Temperature	P.G.T. Fogg

EXPERIMENTAL VALUES:

Solvent	A/K	B	H'/MPa 298.15 K	373.15 K
1,1'-Oxybis(2-methoxyethane), (*diethylene glycol dimethyl ether*; *diglyme*); $C_6H_{14}O_3$; [111-96-6]	−2095.9 ± 1.8%	13.483 ± 0.8%	2.50	10.27
2,5,8,11-Tetraoxadodecane, (*triethylene glycol dimethyl ether*; *triglyme*); $C_8H_{18}O_4$; [112-49-2]	−2147.2 ± 3.4%	13.443 ± 1.5%	2.69	11.42
2,5,8,11,14-Pentaoxapentadecane, (*tetraethylene glycol dimethyl ether*; *tetraglyme*); $C_{10}H_{22}O_5$; [143-24-8]	−2257.1 ± 0.4%	13.616 ± 0.2%	2.75	12.61
2-(2-methoxyethoxy)ethanol, (*diethylene glycol monomethyl ether*; *methyl carbitol*); $C_5H_{12}O_3$; [111-77-3]	−1943.2 ± 1.0%	13.255 ± 0.4%	2.97	11.02
3,6,9,12-Tetraoxahexadecan-1-ol, (*tetraethylene glycol monobutyl ether*); $C_{12}H_{26}O_5$; [1559-34-8]	−1817.2 ± 1.7%	12.500 ± 0.7%	4.10	13.97

Values of H' are Henry's law constants, extrapolated to zero pressure, as defined by the equation:

H' = partial pressure of gas / weight fraction solubility

If Henry's law is taken to be of the form:

H = partial pressure of gas / mole fraction solubility.

then the Henry's laws constant, H, extrapolated to zero partial pressure is given by:

$$H / kPa = \exp [A/T + B]$$

The values of A and B are based upon experimental measurements of solubilities carried out in the temperature range 288.2 K to 373.2 K at pressures which did not exceed 133 kPa.

AUXILIARY INFORMATION

METHOD/APPARATUS/PROCEDURE	SOURCE AND PURITY OF MATERIALS
Dried solvent was added to a flask of known weight. The solvent was then heated, degassed, and the flask reweighed. Subsequent steps were automated and data stored in a computer. The solvent vapor pressure was recorded after each increment of temperature of 5 K from 288.2 to 373.2 K. The flask was then cooled, a predetermined mass of gas added and total pressures recorded at intervals of 5 K. The process was repeated with further additions of gas.	2-(2-methoxyethoxy)ethanol was referred to by the trade name ®*Dowanol DM* ; 3,6,9,12-tetraoxahexadecan-1-ol by the trade name ®*Dowanol TBH*.
	ESTIMATED ERROR: As indicated above.
	REFERENCES:

COMPONENTS:	ORIGINAL MEASUREMENTS:
1. Hydrogen sulfide; H_2S; [7783-06-4] 2. 2,5,8,11,14-Pentaoxapentadecane; (*tetraethylene glycol dimethyl ether*); $C_{10}H_{22}O_5$; [143-24-8]	Härtel, G.H. *J.Chem.Eng.Data* <u>1985</u>, *30*, 57-61.
VARIABLES: Temperature	PREPARED BY: P.G.T. Fogg

EXPERIMENTAL VALUES:

T/K	Henry's constant H_{H_2S}/bar	Mole fraction solubility x_{H_2S} (1.013 bar)[*]
293.15	3.390	0.299
313.15	6.179	0.164
333.15	8.310	0.122
353.15	11.321	0.089
373.15	16.901	0.060

Henry's constant is defined as :

$$H_{H_2S} = \frac{pressure}{mole\ fraction\ solubility}$$

Solubilities were measured at concentrations not greater than x_{H_2S} = 0.16. The variation of mole fraction solubility with pressure was found to be almost linear with a correlation coefficient of better than 0.99.

[*] Calculated by the compiler assuming that the variation of mole fraction solubility with variation of pressure was linear to 1.013 bar. (The value for 293.15 K lies outside the range for which linearity was experimentally demonstrated by the author)

AUXILIARY INFORMATION

METHOD/APPARATUS/PROCEDURE	SOURCE AND PURITY OF MATERIALS
Absorption was found from the decrease in pressure when a known volume of gas came into contact with a known mass of degassed solvent. Pressure changes were found by use of a transducer with a mercury manometer to provide a reference pressure.	1. From Matheson, Heusenstamm, FRG; minimum purity 99.5%.
	ESTIMATED ERROR δP = ± 0.25 mbar (author)
	REFERENCES

COMPONENTS:	ORIGINAL MEASUREMENTS:
1. Hydrogen sulfide; H₂S; [7783-06-4] 2. 2,5,8,11,14-Pentaoxapentadecane (*tetraethylene glycol dimethyl ether*); C₁₀H₂₂O₅; [143-24-8] 3. Tricyclodecanedimethanol C₁₂H₁₈O₂; [26896-48-0]	Härtel, G.H. *J.Chem.Eng.Data* <u>1985</u>, *30*, 57-61.

VARIABLES:	PREPARED BY:
Temperature	P.G.T. Fogg

EXPERIMENTAL VALUES:

The solvent consisted of a mixture of tetraethylene glycol dimethyl ether (85 wt%) and tricyclodecanedimethanol (15 wt%)

T/K	Henry's constant H_{H_2S}/bar	Mole fraction solubility x_{H_2S} (1.013 bar)[*]
293.15	5.169	0.196
313.15	8.487	0.119
333.15	12.097	0.084
353.15	15.464	0.066
373.15	21.551	0.047

Henry's constant is defined as :

$$H_{H_2S} = \frac{\text{pressure}}{\text{mole fraction solubility}}$$

Solubilities were measured at concentrations not greater than x_{H_2S} = 0.16. The variation of mole fraction solubility with pressure was found to be almost linear with a correlation coefficient of better than 0.99.

[*] Calculated by the compiler assuming that the variation of mole fraction solubility with variation of pressure was linear to 1.013 bar. (The value for 293.15 K lies outside the range for which linearity was experimentally demonstrated by the author)

AUXILIARY INFORMATION

METHOD/APPARATUS/PROCEDURE	SOURCE AND PURITY OF MATERIALS
Absorption was found from the decrease in pressure when a known volume of gas came into contact with a known mass of degassed solvent. Pressure changes were found by use of a transducer with a mercury manometer to provide a reference pressure.	1. From Matheson, Heusenstamm, FRG; minimum purity 99.5%. ESTIMATED ERROR δP = ± 0.25 mbar (author) REFERENCES

COMPONENTS:	ORIGINAL MEASUREMENTS:
1. Hydrogen sulfide; H_2S; [7783-06-4] 2. Polar solvents	Sweeney, C.W. *Chromatographia* 1984, *18*, 663-7.

VARIABLES:	PREPARED BY:
Temperature	P.G.T. Fogg

EXPERIMENTAL VALUES:

Solvent	Henry's constant/bar	
	298.15 K	323.15 K
Phosphoric acid, tributyl ester (*tributyl phosphate*); $C_{12}H_{27}PO_4$; [126-73-8]	4.14	8.27
4-Methyl-1,3-dioxolan-2-one (*propylene carbonate*); $C_4H_6O_3$; [108-32-7]	22.6	37.4
1-Methyl-2-pyrrolidinone (*N-methylpyrrolidone*); C_5H_9NO; [872-50-4]	5.81	12.3
2,5,8,11,14-Pentaoxapentadecane (*tetraethylene glycol dimethyl ether*); $C_{10}H_{22}O_5$; [143-24-8]	3.91	7.46

The Henry's constant, H, was defined as :

$$H = \frac{f_{H_2S}}{x_{H_2S}} \quad (x_{H_2S} \to 0)$$

where f_{H_2S} is the fugacity of H_2S in the gas phase and x_{H_2S} the mole fraction of H_2S in the liquid phase.

AUXILIARY INFORMATION

METHOD/APPARATUS/PROCEDURE	SOURCE AND PURITY OF MATERIALS
Henry's constants were calculated from retention volumes measured with a modified 204 Series Pye-Unicam gas chromatograph. Helium was used as carrier gas and the support material was PTFE. Further details are given in refs. 1 - 3. The author noted that the Henry's constant which he reported for dissolution in N-methylpyrolidone at 298 K was 16-20% lower than literature values and suggested that the discrepancy could be due to adsorption at the gas-liquid interface in these chromatographic measurements.	1. from Cambrian Gases, London; 97.5 - 99.9% pure. 2. from Aldrich Chemicals, Gillingham, U.K.; re-distilled.

	ESTIMATED ERROR
	$\delta T/K = \pm 0.05$; $\delta H/H = \pm 0.05$ (author)

REFERENCES:

1. Conder, J.R.; Young C.L., *"Physicochemical Measurements by Gas Chromatography"*, Wiley, Chichester, U.K.

2. Ng, S.; Harris, H.G.; Prausnitz, J.M. *J. Chem. Eng. Data* 1969, *14*, 482.

3. Lin, P.J.; Parcher, J.F. *J. Chromatogr. Sci.* 1982, *20*, 33.

COMPONENTS:	ORIGINAL MEASUREMENTS:
1. Hydrogen sulfide; H_2S; [7783-06-4] 2. 18-Methyl-2,5,8,11,14,17-hexaoxanonadecane (*pentaethylene glycol methyl isopropyl ether*) $C_{14}H_{30}O_6$; [63095-29-4]	Härtel, G.H. *J. Chem. Eng. Data* <u>1985</u>, *30*, 57-61.

VARIABLES:	PREPARED BY:
Temperature	P.G.T. Fogg

EXPERIMENTAL VALUES:

T/K	Henry's constant H_{H_2S}/bar	Mole fraction solubility x_{H_2S} (1.013 bar)*
293.15	2.804	0.361
313.15	4.859	0.208
333.15	6.667	0.152
353.15	11.739	0.086
373.15	17.921	0.057

Henry's constant is given by :

$$H_{H_2S} = \frac{pressure}{mole\ fraction\ solubility}$$

Solubilities were measured at concentrations not greater than $x_{H_2S} = 0.16$. The variation of mole fraction solubility with pressure was found to be almost linear with a correlation coefficient of better than 0.99.

* Calculated by the compiler assuming that the variation of mole fraction solubility with variation of pressure was linear to 1.013 bar. (Values for 293.15 K and 313.15 K lie outside the range for which linearity has been experimentally demonstrated).

AUXILIARY INFORMATION

METHOD/APPARATUS/PROCEDURE:	SOURCE AND PURITY OF MATERIALS:
Absorption was found from the decrease in pressure when a known volume of gas came into contact with a known mass of degassed solvent. Pressure changes were found by use of a transducer with a mercury manometer to provide a reference pressure.	1. From Matheson, Heusenstamm, FRG; minimum purity 99.5%.
	ESTIMATED ERROR: δP = ± 0.25 mbar (author)
	REFERENCES:

COMPONENTS:	ORIGINAL MEASUREMENTS:
1. Hydrogen sulfide; H$_2$S; [7783-06-4] 2. *Sepasolv MPE*[†]	Wolfer, W.; Schwarz, E.; Vodrazka, W.; Volkamer, K. *Oil Gas J.* <u>1980</u>, *78(3)*, 66-70.

VARIABLES:	PREPARED BY:
Temperature	P.G.T. Fogg

EXPERIMENTAL VALUES:

The authors stated that BASF had developed *Sepasolv MPE*, a special mixture of oligoethylene glycol methyl isopropyl ethers with a mean relative molecular mass of 316. Solubilities of several gases were presented on a small scale graph with solubility, α, in units of m^3/m^3 bar, plotted against temperature. No experimental points were shown. The solubility of H$_2$S was plotted over the temperature range -5 °C to 145 °C. The compiler found that the line plotted for this gas fits the equation:

$$\log_{10} (\alpha \text{ bar}) = -2.24 + 1090 \text{ K/T}$$

The compiler considers that α/pressure is the volume of gas absorbed, reduced to 273.15 K and 1 bar (or alternatively 1 atm), absorbed by one volume of solvent at the temperature of measurement.

The pressure and temperatures at which measurements were made was not stated although the graphical information was intended to show behaviour of the solvent from low pressures to pressures greater than 1 bar. Use of the data implies an assumption that the reduced volume of gas absorbed is proportional to partial pressure of gas. The higher the pressure the greater the errors introduced by this assumption.

[†]*Sepasolv MPE* is a registered trademark of BASF.

AUXILIARY INFORMATION

METHOD/APPARATUS/PROCEDURE:	SOURCE AND PURITY OF MATERIALS:
No information.	No information.
	ESTIMATED ERROR: δα/α = ±10% (estimated by the compiler)
	REFERENCES:

COMPONENTS:	ORIGINAL MEASUREMENTS:
(1) Hydrogen sulfide; H_2S; [7783-06-4] (2) Furan, tetrahydro-; C_4H_8O; [109-99-9]	Short, I.; Sahgal, A.; Hayduk, W. *J. Chem. Eng. Data* <u>1983</u>,

VARIABLES:	PREPARED BY:
T/K: 263.15-298.15 P/kPa: 101.325	W. Hayduk

EXPERIMENTAL VALUES:

T/K	Ostwald Coefficient[1] $L/cm^3\ cm^{-3}$	Bunsen Coefficient[2] $\alpha/cm^3\ (STP)cm^{-3}atm^{-1}$	Mole Fraction[1] x_1
263.15	76.9	79.82	0.222
298.15	33.4	30.60	0.1014

[1]Original data

[2]Calculated by compiler

The mole fraction solubility of the original data was used to determine the following equations for $\Delta G°$ and $\ln x_1$ and table of smoothed values:

$$\Delta G°/J\ mol^{-1} = -RT \ln x_1 = 671.24\ T - 144.137$$
$$\ln x_1 = 1756.6/T - 8.1802$$

T/K	$10^{-4}\Delta G°/J\ mol^{-1}$	x_1
263.15	3.250	0.2220
273.15	3.921	0.1739
283.15	4.592	0.1385
293.15	5.264	0.1121
298.15	5.600	0.1014

AUXILIARY INFORMATION

METHOD/APPARATUS/PROCEDURE:	SOURCE AND PURITY OF MATERIALS:
A volumetric method using a glass apparatus was employed. Degassed solvent contacted the gas while flowing as a thin film, at a constant rate, through an absorption spiral into a solution buret. A constant solvent flow was obtained by means of a calibrated syringe pump. The solution at the end of the spiral was considered saturated. Dry gas was maintained at atmospheric pressure in a gas buret by mechanically raising the mercury level in the buret at an adjustable rate. The solubility was calculated from the constant slope of volume of gas dissolved and volume of solvent injected.	1. Liquid Carbonic. Specified minimum purity 99.5 per cent. 2. Canlab. Baker analyzed grade of minimum specified purity 99.0 per cent.

	ESTIMATED ERROR:
Degassing was accomplished using a two stage vacuum process described by Clever et al. (1).	$\delta T/K = 0.1$ $\delta x_1/x_1 = 0.01$

REFERENCES:

1. Clever, H.L.; Battino, R.; Saylor, J.H.; Gross, P.M. *J. Phys. Chem.* <u>1957</u>, *61*, 1078.

COMPONENTS:	ORIGINAL MEASUREMENTS:
1. Hydrogen sulfide; H_2S; [7783-06-4] 2. 1,4-Dioxane; $C_4H_8O_2$; [123-91-1]	Gerrard, W. *J. Appl. Chem. Biotechnol.* <u>1972</u>, *22*, 623-650.

VARIABLES:	PREPARED BY:
Temperature	P.G.T. Fogg

EXPERIMENTAL VALUES:

T/K	P(total)/mmHg	P(total)/bar	Mole ratio	Mole fraction of H_2S[*] x_{H_2S}
283.15	752	1.003	0.147	0.128
293.15	752	1.003	0.100	0.091

[*] calculated by the compiler for the stated total pressure.

AUXILIARY INFORMATION

METHOD/APPARATUS/PROCEDURE:	SOURCE AND PURITY OF MATERIALS:
Hydrogen sulfide was bubbled into a weighed amount of component 2 in a bubbler tube as described in detail in the source. The amount of gas absorbed at equilibrium for the observed temperature and pressure was found by weighing. Pressure was measured with a mercury manometer.	It was stated that "All materials purified and attested by conventional methods."
	ESTIMATED ERROR: δx_{H_2S} = ± 4% (author)
	REFERENCES:

COMPONENTS:	ORIGINAL MEASUREMENTS:
1. Hydrogen sulfide; H_2S; [7783-06-4] 2. Crown Ethers	Linford, R.G.; Thornhill, D.G.T. *J. Chem. Thermodynamics* <u>1985</u>, *17*, 701-702.

VARIABLES:	PREPARED BY:
	P.G.T. Fogg

EXPERIMENTAL VALUES:

Crown ether	Mole fraction of H_2S in liquid, x_{H_2S}
1,4,7,10-Tetraoxacyclododecane; (*12-crown-4*); $C_8H_{16}O_4$; [294-93-9]	0.31
1,4,7,10,13-Pentaoxacyclopentadecane; (*15-crown-5*); $C_{10}H_{20}O_5$; [33100-27-5]	0.33

$$T/K \ = \ 295.2$$

$$P/bar \ = \ 1.01325$$

AUXILIARY INFORMATION

METHOD/APPARATUS/PROCEDURE:	SOURCE AND PURITY OF MATERIALS:
Solubilities were determined by a method based upon that described by Gerrard (1) who measured increases in weight when gases were bubbled through solvents under test. In the case of these systems equilibrium was reached in 0.2 h.	1. supplied by Cambrian Chemicals; purity 99.6 mole %. 2. from Borregaard, Sarpsburg, Norway and supplied by Trafford Chemicals, Altrincham, U.K.
	ESTIMATED ERROR:
	REFERENCES: 1. Gerrard, W. *J.Appl.Chem.Biotechnol.* <u>1972</u>, *22*, 623.

COMPONENTS:	ORIGINAL MEASUREMENTS:

1. Hydrogen sulfide; H₂S; [7783-06-4]
 	
2. 4-Methyl-1,3-dioxolan-2-one,
   (propylene carbonate); C₄H₆O₃;
   [108-32-7]
	Shakhova, S.F.; Bondareva, T.I.;
	Zubchenko, Yu. P.
	
	Khim. Prom. 1966, (10), 753-4.

VARIABLES:	PREPARED BY:
Temperature, pressure	C.L. Young.

EXPERIMENTAL VALUES:

T/K	P/mmHg	P/kPa	α^+ vol/vol	Mole fraction of hydrogen sulfide in liquid, x_{H_2S}
273.15	200	26.7	5.46	0.0206
	400	53.3	11.2	0.0408
	600	80.0	16.7	0.0596
	760	101	20.9	0.0730
288.15	200	26.7	3.59	0.0138
	400	53.3	7.00	0.0266
	600	80.0	10.4	0.0386
	760	101	12.8	0.0480
298.15	200	26.7	2.90	0.0112
	400	53.3	5.60	0.0218
	600	80.0	8.30	0.0314
	760	101	10.4	0.0380
313.15	200	26.7	2.00	0.0080
	400	53.3	3.90	0.0150
	600	80.0	5.65	0.0214
	760	101	6.90	0.0264

+ appears to be volume of gas reduced to T/K = 273.15
   and P = 1 atmosphere absorbed by unit volume of liquid
   (measured at room temperature).

AUXILIARY INFORMATION

METHOD/APPARATUS/PROCEDURE:	SOURCE AND PURITY OF MATERIALS:
Glass equilibrium cell fitted with magnetic stirrer. Samples of liquid analysed by stripping out hydrogen sulfide. Details in source and ref. (1).	No details given.

ESTIMATED ERROR:

$\delta T/K = \pm 0.1$; $\delta P/kPa = \pm 1.0$; $\delta x_{H_2S} = \pm 5\%$ (estimated by compiler).

REFERENCES:

1. Shenderei, E.R.; Zel'venskii, Ya.D.
   Ivanovskii, F.P.

   Khim. Prom. 1960, 370.

COMPONENTS:	ORIGINAL MEASUREMENTS:

1. Hydrogen sulfide; H_2S; [7783-06-4]
 	
2. 4-Methyl-1,3-dioxolan-2-one
     (Propylene Carbonate); $C_4H_6O_3$
     [108-32-7]
	Lenoir, J-Y.; Renault, P.; Renon, H.
	
	J. Chem. Eng. Data; <u>1971</u>, *16*, 340-2

VARIABLES:	PREPARED BY:
Temperature	C.L. Young

EXPERIMENTAL VALUES:

T/K	Henry's Constant H_{H_2S} / atm	Mole fraction at 1 atm* x_{H_2S}
298.15	20.9	0.0478
323.15	33.8	0.0296
343.15	40.1	0.0249

* Calculated by compiler assuming a linear function of P_{H_2S} vs x_{H_2S}, ie. x_{H_2S} (1 atm) = $1/H_{H_2S}$

AUXILIARY INFORMATION

METHOD/APPARATUS/PROCEDURE:	SOURCE AND PURITY OF MATERIALS:
A conventional gas-liquid chromatographic unit fitted with a thermal conductivity detector was used. The carrier gas was helium. The value of Henry's law constant was calculated from the retention time. The value applies to very low partial pressures of gas and there may be a substantial difference from that measured at 1 atm. pressure. There is also considerable uncertainty in the value of Henry's constant since surface adsorption was not allowed for although its possible existence was noted.	(1) L'Air Liquide sample, minimum purity 99.9 mole per cent. (2) Touzart and Matignon or Serlabo sample, purity 99 mole per cent.
	ESTIMATED ERROR: $\delta T/K = \pm 0.1$; $\delta H/atm = \pm 6\%$ (estimated by compiler).
	REFERENCES:

COMPONENTS:	ORIGINAL MEASUREMENTS:
1. Hydrogen sulfide; H_2S; [7783-06-4] 2. 4-Methyl-1,3-dioxolan-2-one, (*propylene carbonate*); $C_4H_6O_3$; [108-32-7]	Isaacs, E.E.; Otto, F.D.; Mather, A.E. *Can.J.Chem.Eng.* <u>1977</u>, 55, 751 - 752.

VARIABLES:	PREPARED BY:
Temperature	P.G.T. Fogg

EXPERIMENTAL VALUES:

T/K	P_{H_2S}/kPa	Mole ratio in liquid H_2S/propylene carbonate	Mole fraction* x_{H_2S}
313.2	2378.4	2.631	0.725
	2140.9	1.943	0.660
	1430.7	0.943	0.485
	616.2	0.231	0.188
	244.7	0.062	0.058
373.2	4960.0	1.202	0.546
	3594.4	0.622	0.383
	2189.1	0.308	0.235
	639.1	0.062	0.058
	184.7	0.0066	0.0066

* calculated by the compiler.

AUXILIARY INFORMATION

METHOD/APPARATUS/PROCEDURE:	SOURCE AND PURITY OF MATERIALS:
The equilibrium cell consisted of a Jerguson gauge with a 250 cm^3 gas reservoir. Temperatures were measured by thermocouples and controlled to ± 0.5 K by an air-bath. Pressures were measured by a Heise Bourdon tube gauge. The cell was charged with solvent. H_2S was added to give an appropriate pressure. Nitrogen was added, when necessary, to ensure that total pressure > 350 kPa. Gases were circulated by a magnetic pump for at least 8 h. The gas phase was analysed by gas chromatography. Samples of the liquid phase were withdrawn, heated at barometric pressure and the quantity of gas evolved found from P-V-T values.	1. no information 2. from Eastman Kodak Co.; minimum purity 98%.
	ESTIMATED ERROR: $\delta T/K = \pm 0.5$ δ(mole ratio) = ± 0.02 or 4%, whichever is the larger (authors).
	REFERENCES:

COMPONENTS:	ORIGINAL MEASUREMENTS:
1. Hydrogen sulfide; H_2S; [7783-06-4] 2. 4-Methyl-1,3-dioxolan-2-one, (Propylene carbonate); $C_4H_8O_3$; [108-32-7]	Rivas, O.R.; Prausnitz, J.M. *Am. Inst. Chem. Engnrs. J.* 1979, *25*, 975-984.

VARIABLES:	PREPARED BY:
Temperature	C.L. Young

EXPERIMENTAL VALUES:

T/K	Henry's constant, H /MPa	Mole fraction of + hydrogen sulfide in liquid, x_{H_2S}
263.15	1.17	0.0866
298.15	2.56	0.0396
323.15	3.91	0.0259
348.15	5.55	0.0183
373.15	7.48	0.0136

+ at a partial pressure of 101.3 kPa calculated by compiler assuming Henry's law applies at that pressure.

AUXILIARY INFORMATION

METHOD/APPARATUS/PROCEDURE:	SOURCE AND PURITY OF MATERIALS:
Volumetric apparatus with a fused quartz precision bourdon pressure gauge. Solubility apparatus carefully thermostatted. Solvent degassed *in situ*. Apparatus described in ref. (1) and modifications given in source.	1. and 2. Purity at least 99 mole per cent.

ESTIMATED ERROR:

$\delta T/K = \pm 0.05$; $\delta x_{H_2S} = \pm 1\%$.

REFERENCES:

1. Cukor, P.M.; Prausnitz, J.M.

   Ind. Eng. Chem. Fundam. 1971, *10*, 638.

COMPONENTS:	ORIGINAL MEASUREMENTS:

COMPONENTS:

- (1) Hydrogen sulfide; H_2S;
       [7783-06-4]

- (2) Acetic acid; $C_2H_4O_2$;
       [64-19-7]

ORIGINAL MEASUREMENTS:

Short, I.; Sahgal, A.; Hayduk, W.

J. Chem. Eng. Data 1983, *28*, 63-66.

VARIABLES:

T/K: 298.15, 333.15

P/kPa: 101.325

PREPARED BY:

W. Hayduk

EXPERIMENTAL VALUES:

T/K	Ostwald Coefficient[1] $L/cm^3\ cm^{-3}$	Bunsen Coefficient[2] $\alpha/cm^3\ (STP)cm^{-3}atm^{-1}$	Mole Fraction[1] $10^4\ x_1$
298.15	12.47	11.42	287
333.15	6.97	5.72	150.7

[1] Original data

[2] Calculated by compiler

The mole fraction solubility of the original data was used to determine the following equations for $\Delta G°$ and $\ln x_1$ and table of smoothed values:

$$\Delta G°/J\ mol^{-1} = -RT\ \ln x_1 = 794.52\ T - 150015$$
$$\ln x_1 = 1828.2/T - 9.6826$$

T/K	$10^{-4}\Delta G°/J\ mol^{-1}$	$10^4\ x_1$
298.15	8.687	287.0
303.15	9.084	259.4
313.15	9.879	214.0
323.15	10.67	178.6
333.15	11.47	150.7

AUXILIARY INFORMATION

METHOD/APPARATUS/PROCEDURE:

A volumetric method using a glass apparatus was employed. Degassed solvent contacted the gas while flowing as a thin film, at a constant rate, through an absorption spiral into a solution buret. A constant solvent flow was obtained by means of a calibrated syringe pump. The solution at the end of the spiral was considered saturated. Dry gas was maintained at atmospheric pressure in a gas buret by mechanically raising the mercury level in the buret at an adjustable rate. The solubility was calculated from the constant slope of volume of gas dissolved and volume of solvent injected.

Degassing was accomplished using a two stage vacuum process described by Clever et al. (1).

SOURCE AND PURITY OF MATERIALS:

1. Liquid Carbonic. Specified minimum purity 99.5 per cent.

2. Canlab. Specified minimum purity 99.7 per cent.

ESTIMATED ERROR:

$\delta T/K$ = 0.1

$\delta x_1/x_1$ = 0.01

REFERENCES:

1. Clever, H.L.; Battino, R.; Saylor, J.H.; Gross, P.M.
    J. Phys. Chem. 1957, *61*, 1078.

COMPONENTS:	ORIGINAL MEASUREMENTS:
1. Hydrogen sulfide; H_2S; [7783-06-4] 2. Acetic acid anhydride; (Acetic anhydride); $C_4H_6O_3$; [108-24-7]	Gerrard, W. *J. Appl. Chem. Biotechnol.*, <u>1972</u>, 22, 623-650

VARIABLES:	PREPARED BY:
Temperature	C.L. Young

EXPERIMENTAL VALUES:

T/K	$P*$/mmHg	$P*$/kPa	Mole ratio	Mole fraction[+] of hydrogen sulfide in liquid, x_{H_2S}
265.15	758	101.1	0.151	0.131
267.15	758	101.1	0.144	0.126
273.15	758	101.1	0.120	0.107
283.15	758	101.1	0.085	0.078
293.15	758	101.1	0.066	0.062

[+]calculated by compiler

*total pressure

AUXILIARY INFORMATION

METHOD/APPARATUS/PROCEDURE:	SOURCE AND PURITY OF MATERIALS:
Hydrogen sulfide was bubbled into a weighed amount of component 2. in a bubbler tube as described in detail in the source. The amount of gas absorbed at equilibrium and at the observed temperature and pressure was determined by weighing. Pressure was measured with a mercury manometer. The amount of gas absorbed at successively lower pressures was measured. Eventually the pressure was reduced to the vapor pressure of component 2. The refractive index and infrared spectrum of the liquid showed it to be essentially pure component 2.	1 and 2. Components were purified and attested by conventional methods.

ESTIMATED ERROR:
$\delta T/K = \pm 0.2$; $\delta x_{H_2S} = \pm 4\%$

REFERENCES:

COMPONENTS:	ORIGINAL MEASUREMENTS:
1. Hydrogen sulfide; H_2S; [7783-06-4] 2. Hexanoic acid; $C_6H_{12}O_2$; [142-62-1]	Gerrard, W. *J. Appl. Chem. Biotechnol.*, 1972, 22, 623-650
VARIABLES: Temperature	PREPARED BY: C.L. Young

EXPERIMENTAL VALUES:

T/K	$P*$/mmHg	$P*$/kPa	Mole ratio	Mole fraction[+] of hydrogen sulfide in liquid, x_{H_2S}
265.15	766	102.1	0.116	0.104
267.15	766	102.1	0.109	0.098
273.15	766	102.1	0.089	0.081
283.15	766	102.1	0.069	0.065
293.15	766	102.1	0.057	0.054

[+]calculated by compiler

*total pressure

AUXILIARY INFORMATION

METHOD/APPARATUS/PROCEDURE:	SOURCE AND PURITY OF MATERIALS:
Hydrogen sulfide was bubbled into a weighed amount of component 2. in a bubbler tube as described in detail in the source. The amount of gas absorbed at equilibrium and at the observed temperature and pressure was determined by weighing. Pressure was measured with a mercury manometer.	1 and 2. Components were purified and attested by conventional methods.
	ESTIMATED ERROR: $\delta T/K = \pm 0.2$; $\delta x_{H_2S} = \pm 4\%$
	REFERENCES:

COMPONENTS:	ORIGINAL MEASUREMENTS:
1. Hydrogen sulfide; H_2S; [7783-06-4] 2. 2-Hydroxybenzoic acid, methyl ester (methyl salicylate); $C_8H_8O_3$; [119-36-8]	Gerrard, W. J. Appl. Chem. Biotechnol. <u>1972</u>, 22, 623-650.

VARIABLES:	PREPARED BY:
Temperature	P.G.T. Fogg

EXPERIMENTAL VALUES:

T/K	P(total)/mmHg	P(total)/bar	Mole ratio	Mole fraction of H_2S[*] x_{H_2S}
265.15	753	1.004	0.147	0.128
267.15	753	1.004	0.136	0.120
273.15	753	1.004	0.111	0.100
283.15	753	1.004	0.082	0.076
293.15	753	1.004	0.064	0.060

[*] calculated by the compiler for the stated total pressure.

AUXILIARY INFORMATION

METHOD/APPARATUS/PROCEDURE:	SOURCE AND PURITY OF MATERIALS:
Hydrogen sulfide was bubbled into a weighed amount of component 2 in a bubbler tube as described in detail in the source. The amount of gas absorbed at equilibrium for the observed temperature and pressure was found by weighing. Pressure was measured with a mercury manometer.	It was stated that "All materials purified and attested by conventional methods."
	ESTIMATED ERROR: δx_{H_2S} = ± 4% (author)
	REFERENCES:

COMPONENTS:	ORIGINAL MEASUREMENTS:
(1) Hydrogen sulfide; H_2S [7783-06-4] (2) 2-Propanone (acetone); C_3H_6O; [67-64-1]	Short, I.; Sahgal, A.; Hayduk, W. J. Chem. Eng. Data 1983, 28, 63-66.

VARIABLES:	PREPARED BY:
T/K: 263.15, 298.15 P/kPa: 101.325	W. Hayduk

EXPERIMENTAL VALUES:

T/K	Ostwald Coefficient[1] L/cm^3 cm^{-3}	Bunsen Coefficient[2] α/cm^3 (STP) $cm^{-3}atm^{-1}$	Mole Fraction[1] x_1
263.15	57.7	59.89	0.1607
298.15	24.6	22.54	0.0698

[1]Original data

[2]Calculated by compiler

The mole fraction solubility of the original data was used to determine the following equations for $\Delta G°$ and $\ln x_1$ and table of smoothed values:

$$\Delta G°/J\ mol^{-1} = -RT \ln x_1 = 732.91\ T - 153390$$
$$\ln x_1 = 1869.3/T - 8.9319$$

T/K	$10^{-4} \Delta G°/J\ mol^{-1}$	x_1
263.15	3.948	0.1607
273.15	4.681	0.1239
283.15	5.414	0.0973
293.15	6.146	0.0777
298.15	6.513	0.0698

AUXILIARY INFORMATION

METHOD/APPARATUS/PROCEDURE:	SOURCE AND PURITY OF MATERIALS:
A volumetric method using a glass apparatus was employed. Degassed solvent contacted the gas while flowing as a thin film, at a constant rate, through an absorption spiral into a solution buret. A constant solvent flow was obtained by means of a calibrated syringe pump. The solution at the end of the spiral was considered saturated. Dry gas was maintained at atmospheric pressure in a gas buret by mechanically raising the mercury level in the buret at an adjustable rate. The solubility was calculated from the constant slope of volume of gas dissolved and volume of solvent injected. Degassing was accomplished using a two stage vacuum process described by Clever et al. (1).	1. Liquid Carbonic. Specified minimum purity 99.5 per cent. 2. Canlab. Specified minimum purity 99.5 per cent.

	ESTIMATED ERROR:
	$\delta T/K$ = 0.1 $\delta x_1/x_1$ = 0.01

	REFERENCES:
	1. Clever, H.L.; Battino, R.; Saylor, J.H.; Gross, P.M. J. Phys. Chem. 1957, 61, 1078.

COMPONENTS:	ORIGINAL MEASUREMENTS:
1. Hydrogen sulfide; H_2S; [7783-06-4] 2. Phenol; C_6H_6O; [108-95-2]	Lenoir, J-Y.; Renault, P.; Renon, H. *J. Chem. Eng. Data*, <u>1971</u>, *16*, 340-3

VARIABLES:	PREPARED BY:
	C. L. Young

EXPERIMENTAL VALUES:

T/K	Henry's constant H_{H_2S}/atm	Mole fraction at 1 atm* x_{H_2S}
323.2	50.0	0.0200

* Calculated by compiler assuming a linear function of P_{H_2S} vs x_{H_2S}, i.e., $x_{H_2S}(1\ atm) = 1/H_{H_2S}$

AUXILIARY INFORMATION

METHOD/APPARATUS/PROCEDURE:	SOURCE AND PURITY OF MATERIALS:
A conventional gas-liquid chromato-graphic unit fitted with a thermal conductivity detector was used. The carrier gas was helium. The value of Henry's law constant was calculated from the retention time. The value applies to very low partial pressures of gas and there may be a substantial difference from that measured at 1 atm. pressure. There is also considerable uncertainty in the value of Henry's constant since surface adsorption was not allowed for although its possible existence was noted.	(1) L'Air Liquide sample, minimum purity 99.9 mole per cent. (2) Touzart and Matignon or Serlabo sample, purity 99 mole per cent.
	ESTIMATED ERROR: $\delta T/K = \pm 0.1$; $\delta H/atm = \pm 6\%$ (estimated by compiler).
	REFERENCES:

COMPONENTS:	ORIGINAL MEASUREMENTS:
1. Hydrogen sulfide; H_2S; [7783-06-4] 2. Hydrogen; H_2; [1333-74-0] 3. Dibenzofuran; $C_{12}H_8O$; [132-64-9]	Kragas, T.K.; Kobayashi, R. *Fluid Phase Equilibria* 1984, *16*, 215-236.

VARIABLES:	PREPARED BY:
Pressure of hydrogen	P.G.T. Fogg

EXPERIMENTAL VALUES:

T/K	P_{H_2}/MPa	Henry's Constant for H_2S at infinite dilution / MPa
373.2	0.731	6.68
	1.524	3.34
	2.551	2.03
	3.434	1.51
	4.440	1.17
	5.102	0.986
	5.723	0.907
398.2	1.703	3.62
	2.165	2.91
	2.792	2.32
	3.833	1.74
	4.681	1.47
	4.902	1.38
	5.881	1.17

AUXILIARY INFORMATION

METHOD/APPARATUS/PROCEDURE:	SOURCE AND PURITY OF MATERIALS:
A high pressure chromatographic method described in ref. (1) was used. Hydrogen at various pressures was the carrier gas.	1. from Scientific Gas Products; 99.6% pure. 2. from Linde Division of Union Carbide Corporation. 3. from Aldrich Chem. Co.; purified to > 99.95% by zone refining.
	ESTIMATED ERROR: δ(Henry's law constant) = ± 3.6%
	REFERENCES: 1. Kragas, T.K.; Pollin, J.; Martin, R.J.; Kobayashi, R. *Fluid Phase Equilibria* 1984, *16*, 205.

COMPONENTS:	ORIGINAL MEASUREMENTS:
1. Hydrogen sulfide; H_2S; [7783-06-4] 2. Various liquids	Devyatykh, G.G.; Ezheleva, A.E.; Zorin, A.D.; Zueva, M. V. *Zh. Neorgan. Khim.* <u>1963</u>, *8(6)*, 1307-1313. *Russ. J. Inorg. Chem.* <u>1963</u>, *8,* 678-682.

VARIABLES:	PREPARED BY:
Temperature, pressure	P.G.T. Fogg

EXPERIMENTAL VALUES:

Solvent	Distribution constant vol.H_2S / vol.solvent	Heat of solution / kcal mol^{-1}
2-Ethoxyethanol; $C_4H_{10}O_2$; [110-80-5]	12.5	− 0.9
Nitrobenzene; $C_6H_5NO_2$; [98-95-3]	12.4	− 2.29
2-Furancarboxaldehyde; $C_5H_4O_2$; [98-01-1]	17.3	− 3.5
1,1'-Oxybis[2-chloroethane]; $C_4H_8Cl_2O$; [111-44-4]	10.5	− 3.3
1,2-Benzenedicarboxylic acid, didecyl ester; $C_{28}H_{46}O_4$; [84-77-5]	34.3	− 3.6
Triethoxysilane; $C_6H_{16}O_3Si$; [998-30-1]	3.52	− 4.7
Silicic acid, tetraethyl ester; $C_8H_{28}O_4Si$; [78-10-4]	5.21	− 4.5
Liquid paraffin	18.32	− 2.6
Silicone 702-DF	37.0	− 3.0
Silicone PFMS-4F	8.4	− 2.8
Silicone VKZh-94B	16.5	− 5.6

Temperature = 293.2 K

AUXILIARY INFORMATION

METHOD/APPARATUS/PROCEDURE:	SOURCE AND PURITY OF MATERIALS:
A chromatographic method was used. The support phase consisted of Nichrome spirals. The carrier gas was either nitrogen or hydrogen. the volume, V_1, of the liquid phase was calculated from the weight of the column before and after filling with liquid and allowing to drain. The free volume, V_g, was equated with the retention volume for hydrogen gas. the distribution constant, K, was calculated from the James and Martin equation i.e.	1. from FeS and HCl; purified by vacuum distillation. H_2 & N_2 : passed through activated carbon and through molecular sieve.

METHOD/APPARATUS/PROCEDURE (continued):

$$V_R = V_g + KV_1$$

where V_R is the retention volume for hydrogen sulfide.

ESTIMATED ERROR:

$\delta T/K = \pm 0.5$ (authors)

REFERENCES:

COMPONENTS:	ORIGINAL MEASUREMENTS:
1. Hydrogen sulfide; H₂S; [7783-06-4] 2. Various liquids	Devyatykh, G.G.; Ezheleva, A.E.; Zorin, A.D.; Zueva, M. V. *Zh. Neorgan. Khim.* <u>1963</u>, *8(6)*, 1307-1313. *Russ. J. Inorg. Chem.* <u>1963</u>, *8*, 678-682.

EXPERIMENTAL VALUES:

Distribution constants were measured between 278.2 K and 323.2 K with a total pressure of hydrogen sulfide and carrier gas of about 760 mmHg (1.013 bar). In the case of measurements on 2-ethoxyethanol the partial pressure of hydrogen sulfide at the inlet to the column varied from 1.4 to 21.9 mmHg. The partial pressure at the exit to the column varied from 0.2 to 20.7 mmHg depending upon the inlet pressure and the length of the column. The corresponding partial pressures during experiments on other solvents were not stated but are likely to have been similar.

The authors stated that, at a fixed temperature, the distribution constants did not depart from the mean values by more than ± 3%. These mean values were reported at one temperature only, but heats of solution, said to have been calculated from the variation of distribution constants with temperature, were given. The detail of these calculations were not given by the authors but the compiler considers that they used an equation of the form:

$$\ln K = (- \Delta H/RT) + A$$

where K is the distribution constant for a temperature T, ΔH is the heat of solution of hydrogen sulfide in the solvent and A is a constant for the solvent. On the basis of this equation distribution constants in the range 278.2 K to 323.2 K may be estimated from a value at 293.2 K and the corresponding heat of solution.

The equation for K may be written in the form:

$$K = \exp [A + (B/T)]$$

The following values of A and B have been calculated by the compiler:

Solvent	A	B/K
2-Ethoxyethanol	0.980	453
Nitrobenzene	- 1.415	1153
2-Furancarboxaldehyde	- 3.160	1762
1,1'-Oxybis[2-chloroethane]	- 3.316	1662
1,2-Benzenedicarboxylic acid, didecyl ester	- 2.647	1813
Triethoxysilane	- 6.813	2367
Silicic acid, tetraethyl ester	- 6.077	2266
Liquid paraffin	- 1.557	1309
Silicone 702-DF	- 1.541	1511
Silicone PFMS-4F	- 2.680	1410
Silicone VKZh-94B	- 6.814	2820

In the opinion of the compiler, these distribution constants can not be equated with Ostwald coefficients unless the assumption is made that equilibrium was established between gas and liquid phases under the conditions of the experiment. In addition, possible adsorption of the hydrogen sulfide at the stationary phase / carrier gas interface may have lowered the accuracy of the results.

COMPONENTS:	ORIGINAL MEASUREMENTS:
1. Hydrogen sulfide; H_2S; [7783-06-4] 2. Substituted methanes	Bell, R.P. *J. Chem. Soc.* <u>1931</u>, 1371-1382.

VARIABLES:	PREPARED BY:
	C.L. Young.

EXPERIMENTAL VALUES:

Solvent	T/K	Partition coefficient, s^{+}	Mole fraction[§] of hydrogen sulfide in liquid, x_{H_2S}
Tetrachloromethane, (Carbon tetrachloride); CCl_4; [56-23-5]	293.15	10.79	0.0419
Trichloromethane; (Chloroform); $CHCl_3$ [67-66-3]		32.8	0.103
Tribromomethane, (Bromoform); $CHBr_3$; [75-25-2]		16.76	0.0581

[+] s defined as $s = 22.4 \times \dfrac{293}{273} \times c$ where c is the "solubility in equivalents/litre".

[§] for a partial pressure of 101.325 kPa.

AUXILIARY INFORMATION

METHOD/APPARATUS/PROCEDURE:	SOURCE AND PURITY OF MATERIALS:
Volumetric apparatus consisting of bulb ($\sim$50cm^3 capacity) extended at the top as a graduated tube and joined at bottom to a capillary u-tube. Liquid saturated with gas at atmospheric pressure. Gas withdrawn in a current of air, absorbed in sodium hydroxide and hydrogen peroxide. Excess hydrogen peroxide removed by heating and excess sodium hydroxide titrated.	1. Prepared by reaction of sodium sulfide on hydrochloric acid. 2. Merck and Kahlbaum samples dried over calcium chloride and fractionally distilled.

	ESTIMATED ERROR:
	$\delta T/K = \pm 0.1$; $\delta x_{H_2S} = \pm 1\%$. (estimated by compiler)

	REFERENCES:

COMPONENTS:	ORIGINAL MEASUREMENTS:
1. Hydrogen sulfide; H_2S; [7783-06-4] 2. Halogenated ethanes	Bell, R.P. *J. Chem. Soc.*, <u>1931</u>, 1371-1382

VARIABLES:	PREPARED BY:
	C. L. Young

EXPERIMENTAL VALUES:

Solvent	T/K	Partition Coefficient, S^+	Mole fraction[§] of hydrogen sulfide in liquid, x_{H_2S}
Pentachloroethane; (Pentachloroethane); C_2HCl_5; [76-01-7]	293.15	10.63	0.0514
Bromoethane; (Ethyl bromide); C_2H_5Br; [74-96-4]	293.15	43.3	0.126
1,1,2,2-Tetrachloroethane; (s-Tetrachloroethane); $C_2H_2Cl_4$; [79-34-5]	293.15	16.66	0.0702
1,1,2,2-Tetrabromoethane; (s-Tetrabromoethane); $C_2H_2Br_4$; [79-27-6]	293.15	9.49	0.0446

S^+ defined as $= 22.4 + \dfrac{293}{273} \times C$ where C is the "solubility in equivalents/litre"

§ for a partial pressure of 101.325 kPa

AUXILIARY INFORMATION

METHOD/APPARATUS/PROCEDURE:	SOURCE AND PURITY OF MATERIALS:
Volumetric apparatus consisting of bulb (~50 cm³ capacity) extended at the top as a graduated tube and joined at bottom to a capillary U-tube. Liquid saturated with gas at atmospheric pressure. Gas withdrawn in a current of air, absorbed in sodium hydroxide and hydrogen peroxide. Excess hydrogen peroxide removed by heating and excess sodium hydroxide titrated.	1. Prepared by reaction of sodium sulfide on hydrochloric acid. 2. Merck and Kahlbaum samples dried over calcium chloride and fractionally distilled, except tetrachloroethane which was fractionally distilled from commercial sample.
	ESTIMATED ERROR: $\delta T/K = \pm 0.1$; $\delta x_{H_2S} = \pm 1\%$ (estimated by compiler)
	REFERENCES:

COMPONENTS:	ORIGINAL MEASUREMENTS:
1. Hydrogen sulfide; H_2S; [7783-06-4] 2. Ethene derivatives	Bell, R.P. *J. Chem. Soc.*, 1931, 1371-1382

VARIABLES:	PREPARED BY:
	C. L. Young

EXPERIMENTAL VALUES:

Solvent	T/K	Partition Coefficient, S^+	Mole fraction[§] of hydrogen sulfide in liquid, x_{H_2S}
Tetrachloroethene, (Tetrachloroethylene); C_2Cl_4; [127-18-4]	293.15	8.90	0.0372
Trichloroethene, (Trichloroethylene); C_2HCl_3; [79-01-6]		13.16	0.0482
1,2-Dibromoethane; (Ethylene bromide); $C_2H_4Br_2$; [106-93-4]		17.80	0.0608
1,2-Dichloroethane; (Ethylene chloride); $C_2H_4Cl_2$; [107-06-2]		23.0	0.0719

S^+ defined as $S = 22.4 \times \dfrac{293}{273} \times C$ where C is the "solubility in equivalents/litre"

[§] for partial pressure of 101.325 kPa

AUXILIARY INFORMATION

METHOD/APPARATUS/PROCEDURE:	SOURCE AND PURITY OF MATERIALS:
Volumetric apparatus consisting of bulb (~50 cm^3 capacity) extended at the top as a graduated tube and joined at bottom to a capillary U-tube. Liquid saturated with gas at atmospheric pressure. Gas withdrawn in a current of air, absorbed in sodium hydroxide and hydrogen peroxide. Excess hydrogen peroxide removed by heating and excess sodium hydroxide titrated.	1. Prepared by reaction of sodium sulfide on hydrochloric acid. 2. Merck and Kahlbaum samples dried over calcium chloride and fractionally distilled.
	ESTIMATED ERROR: $\delta T/K = \pm 0.1$; $\delta x_{H_2S} = \pm 1\%$ (estimated by compiler)
	REFERENCES:

COMPONENTS:	ORIGINAL MEASUREMENTS:
1. Hydrogen sulfide; H_2S; [7783-06-4] 2. Compounds containing chlorine	Gerrard, W. *J. Appl. Chem. Biotechnol.* 1972, 22, 623-650.

VARIABLES:	PREPARED BY:
Temperature	P.G.T. Fogg

EXPERIMENTAL VALUES:

T/K	P(total)/mmHg	P(total)/bar	Mole ratio	Mole fraction of H_2S [*] x_{H_2S}
		Trichloromethane (*chloroform*); $CHCl_3$; [67-66-3]		
265.15	765	1.020	0.141	0.124
		2,2,2-Trichloroethanol; $C_2H_3Cl_3O$; [115-20-8]		
273.15	758	1.011	0.063	0.059
		2,2-Dichloroacetic acid; $C_2H_2Cl_2O_2$; [79-43-6]		
283.15	758	1.011	0.056	0.053

[*] calculated by the compiler for the stated total pressure.

AUXILIARY INFORMATION

METHOD/APPARATUS/PROCEDURE:	SOURCE AND PURITY OF MATERIALS:
Hydrogen sulfide was bubbled into a weighed amount of component 2 in a bubbler tube as described in detail in the source. The amount of gas absorbed at equilibrium for the observed temperature and pressure was found by weighing. Pressure was measured with a mercury manometer.	It was stated that "All materials purified and attested by conventional methods."
	ESTIMATED ERROR: δx_{H_2S} = ± 4% (author)
	REFERENCES:

COMPONENTS:	ORIGINAL MEASUREMENTS:
1. Hydrogen sulfide; H_2S; [7783-06-4] 2. Brominated hydrocarbons	Gerrard, W. *J. Appl. Chem. Biotechnol.* <u>1972</u>, 22, 623-650.
VARIABLES: Temperature	PREPARED BY: P.G.T. Fogg

EXPERIMENTAL VALUES:

T/K	P(total)/mmHg	P(total)/bar	Mole ratio	Mole fraction of H_2S[*] x_{H_2S}
		1-Bromobutane; C_4H_9Br; [109-65-9]		
265.15	760	1.013	0.156	0.135
		1,2-Dibromoethane; $C_2H_4Br_2$; [106-93-4]		
284.15	760	1.013	0.083	0.077
293.15	760	1.013	0.062	0.058
		Tribromomethane (*bromoform*); $CHBr_3$; [75-25-2]		
283.15	765	1.020	0.082	0.076
293.15	765	1.020	0.0625	0.059

[*] calculated by the compiler for the stated total pressure.

AUXILIARY INFORMATION

METHOD/APPARATUS/PROCEDURE:	SOURCE AND PURITY OF MATERIALS:
Hydrogen sulfide was bubbled into a weighed amount of component 2 in a bubbler tube as described in detail in the source. The amount of gas absorbed at equilibrium for the observed temperature and pressure was found by weighing. Pressure was measured with a mercury manometer.	It was stated that "All materials purified and attested by conventional methods."
	ESTIMATED ERROR: δx_{H_2S} = ± 4% (author)
	REFERENCES:

COMPONENTS:	ORIGINAL MEASUREMENTS:
1. Hydrogen sulfide; H_2S; [7783-06-4] 2. 1-Bromooctane; $C_8H_{17}Br$; [111-83-1]	Gerrard, W. J. Appl. Chem. Biotechnol. <u>1972</u>, 22, 623-650.

VARIABLES:	PREPARED BY:
Temperature	P.G.T. Fogg

EXPERIMENTAL VALUES:

T/K	P(total)/mmHg	P(total)/bar	Mole ratio	Mole fraction of H_2S [*] x_{H_2S}
265.15	755	1.007	0.167	0.143
267.15	755	1.007	0.156	0.135
273.15	755	1.007	0.128	0.113
283.15	755	1.007	0.097	0.088
293.15	755	1.007	0.080	0.074

[*] calculated by the compiler for the stated total pressure.

AUXILIARY INFORMATION

METHOD/APPARATUS/PROCEDURE:	SOURCE AND PURITY OF MATERIALS:
Hydrogen sulfide was bubbled into a weighed amount of component 2 in a bubbler tube as described in detail in the source. The amount of gas absorbed at equilibrium for the observed temperature and pressure was found by weighing. Pressure was measured with a mercury manometer.	It was stated that "All materials purified and attested by conventional methods."

	ESTIMATED ERROR:
	δx_{H_2S} = ± 4% (author)

	REFERENCES:

COMPONENTS:	ORIGINAL MEASUREMENTS:
1. Hydrogen sulfide; H₂S; [7783-06-4] 2. Chlorobenzene; C₆H₅Cl; [108-90-7]	Gerrard, W. *J. Appl. Chem. Biotechnol.*, 1972, 22, 623-650
VARIABLES: Temperature	PREPARED BY: C.L. Young

EXPERIMENTAL VALUES:

T/K	P^*/mmHg	P^*/kPa	Mole ratio	Mole fraction[+] of hydrogen sulfide in liquid, x_{H_2S}
265.15	752	100.3	0.124	0.110
267.15	752	100.3	0.115	0.103
273.15	752	100.3	0.094	0.086
283.15	752	100.3	0.075	0.070
293.15	752	100.3	0.057	0.054

[+]calculated by compiler

*total pressure

AUXILIARY INFORMATION

METHOD/APPARATUS/PROCEDURE:	SOURCE AND PURITY OF MATERIALS:
Hydrogen sulfide was bubbled into a weighed amount of component 2. in a bubbler tube as described in detail in the source. The amount of gas absorbed at equilibrium and at the observed temperature and pressure was determined by weighing. Pressure was measured with a mercury manometer. The amount of gas absorbed at successively lower pressures was measured. Eventually the pressure was reduced to the vapor pressure of component 2. The refractive index and infrared spectrum of the liquid showed it to be essentially pure component 2.	1 and 2. Components were purified and attested by conventional methods.
	ESTIMATED ERROR: $\delta T/K = \pm 0.2$; $\delta x_{H_2S} = \pm 4\%$
	REFERENCES:

COMPONENTS:	ORIGINAL MEASUREMENTS:
1. Hydrogen sulfide; H_2S; [7783-06-4] 2. Chlorobenzene; C_6H_5Cl; [108-90-7] or Bromobenzene; C_6H_5Br; [108-86-1]	Patyi, L.; Furmer, I. E.; Makranczy, J.; Sadilenko, A. S.; Stepanova, Z. G.; Berengarten, M. G. Zh. Prikl. Khim. 1978, 51, 1296-1300.

VARIABLES:	PREPARED BY:
	C. L. Young

EXPERIMENTAL VALUES:

T/K	$\alpha^{\dagger}$	Mole fraction of hydrogen sulfide at a partial pressure of 101.325 kPa x_{H_2S}
Chlorobenzene; C_6H_5Cl; [108-90-7]		
298.15	9.0	0.039
Bromobenzene; C_6H_5Br; [108-86-1]		
298.15	8.3	0.038

$\dagger$ volume of gas (measured at 101.325 kPa and 273.15 K) dissolved per volume of solvent.

AUXILIARY INFORMATION

METHOD/APPARATUS/PROCEDURE:	SOURCE AND PURITY OF MATERIALS:
Volumetric method. Pressure measured when known amounts of gas were added, in increments, to a known amount of liquid in a vessel of known dimensions. Corrections were made for the partial pressure of solvent. Details in ref. (1).	Purity better than 99 mole per cent as determined by gas chromatography.
	ESTIMATED ERROR: $\delta T/K = \pm 0.1$; $\delta\alpha = \pm 4\%$ or less.
	REFERENCES: 1. Bodor, E.; Bor, G. J.; Mohai, B.; Sipos, G. Veszpremi. Vegyip. Egy. Kozl. 1957, 1, 55.

COMPONENTS:	ORIGINAL MEASUREMENTS:
(1) Hydrogen sulfide; H_2S; [7783-06-4] (2) Chlorobenzene; C_6H_5Cl; [108-90-7]	Short, I.; Sahgal, A.; Hayduk, W. J. Chem. Eng. Data 1983, 28, 63-66.

VARIABLES:	PREPARED BY:
T/K: 263.15-333.15 P/kPa: 101.325	W. Hayduk

EXPERIMENTAL VALUES:

T/K	Ostwald Coefficient[1] L/cm^3 cm^{-3}	Bunsen Coefficient[2] α/cm^3 (STP)$cm^{-3}atm^{-1}$	Mole Fraction[1] x_1
263.15	24.9	25.85	0.1039 (0.1039)[3]
298.15	12.62	11.56	0.0505 (0.0505)
333.15	7.78	6.38	0.0294 (0.0294)

[1]Original data

[2]Calculated by compiler

[3]The mole fraction solubility of the original data was used to determine the following equations for $\Delta G°$ and $\ln x_1$ and table of smoothed values:

$$\Delta G°/J\ mol^{-1} = -RT \ln x_1 = 1963.0T - 191.80\ T \ln T - 186400$$

$$\ln x_1 = 2271.6/T + 2.3374 \ln T - 23.923$$

T/K	$10^{-4}\Delta G°/J\ mol^{-1}$	x_1	T/K	$10^{-4}\Delta G°/J\ mol^{-1}$	x_1
263.15	4.890	0.1039	303.15	7.643	0.0463
273.15	5.588	0.0826	313.15	8.315	0.0393
283.15	6.280	0.0670	323.15	8.891	0.0338
293.15	6.965	0.0553			
298.15	7.305	0.0505			

AUXILIARY INFORMATION

METHOD/APPARATUS/PROCEDURE:	SOURCE AND PURITY OF MATERIALS:
A volumetric method using a glass apparatus was employed. Degassed solvent contacted the gas while flowing as a thin film, at a constant rate, through an absorption spiral into a solution buret. A constant solvent flow was obtained by means of a calibrated syringe pump. The solution at the end of the spiral was considered saturated. Dry gas was maintained at atmospheric pressure in a gas buret by mechanically raising the mercury level in the buret at an adjustable rate. The solubility was calculated from the constant slope of volume of gas dissolved and volume of solvent injected. Degassing was accomplished using a two stage vacuum process described by Clever et al. (1).	1. Liquid Carbonic. Specified minimum purity 99.5 per cent. 2. Canlab. Baker Analyzed grade of minimum specified purity 99.0 per cent.
	ESTIMATED ERROR: $\delta T/K$ = 0.1 $\delta x_1/x_1$ = 0.01
	REFERENCES: 1. Clever, H.L.; Battino, R.; Saylor, J.H.; Gross, P.M. J. Phys. Chem. 1957, 61, 1078.

COMPONENTS:	ORIGINAL MEASUREMENTS:
1. Hydrogen sulfide; H_2S; [7783-06-4] 2. Bromobenzene; C_6H_5Br; [108-86-1]	Gerrard, W. *J. Appl. Chem. Biotechnol.*, <u>1972</u>, *22*, 623-650

VARIABLES:	PREPARED BY:
Temperature	C.L. Young

EXPERIMENTAL VALUES:

T/K	$P*$/mmHg	$P*$/kPa	Mole ratio	Mole fraction[+] of hydrogen sulfide in liquid, x_{H_2S}
266.15	752	100.3	0.124	0.110
273.15	752	100.3	0.103	0.093
283.15	752	100.3	0.076	0.071
293.15	752	100.3	0.059	0.056

[+]calculated by compiler

*total pressure

AUXILIARY INFORMATION

METHOD/APPARATUS/PROCEDURE:	SOURCE AND PURITY OF MATERIALS:
Hydrogen sulfide was bubbled into a weighed amount of component 2. in a bubbler tube as described in detail in the source. The amount of gas absorbed at equilibrium and at the observed temperature and pressure was determined by weighing. Pressure was measured with a mercury manometer. The amount of gas absorbed at successively lower pressures was measured. Eventually the pressure was reduced to the vapor pressure of component 2. The refractive index and infrared spectrum of the liquid showed it to be essentially pure component 2.	1 and 2. Components were purified and attested by conventional methods.
	ESTIMATED ERROR: $\delta T/K = \pm 0.2$; $\delta x_{H_2S} = \pm 4\%$
	REFERENCES:

COMPONENTS:	ORIGINAL MEASUREMENTS:
1. Hydrogen sulfide; H_2S; [7783-06-4] 2. Iodobenzene; C_6H_5I; [591-50-4]	Gerrard, W. *J. Appl. Chem. Biotechnol.*, 1972, 22, 623-650

VARIABLES:	PREPARED BY:
Temperature	C.L. Young

EXPERIMENTAL VALUES:

T/K	$P*$/mmHg	$P*$/kPa	Mole ratio	Mole fraction[+] of hydrogen sulfide in liquid, x_{H_2S}
265.15	752	100.3	0.139	0.122
273.15	752	100.3	0.106	0.096
283.15	752	100.3	0.080	0.074
293.15	752	100.3	0.060	0.057

[+]calculated by compiler

*total pressure

AUXILIARY INFORMATION

METHOD/APPARATUS/PROCEDURE:	SOURCE AND PURITY OF MATERIALS:
Hydrogen sulfide was bubbled into a weighed amount of component 2. in a bubbler tube as described in detail in the source. The amount of gas absorbed at equilibrium and at the observed temperature and pressure was determined by weighing. Pressure was measured with a mercury manometer. The amount of gas absorbed at successively lower pressures was measured. Eventually the pressure was reduced to the vapor pressure of component 2. The refractive index and infrared spectrum of the liquid showed it to be essentially pure component 2.	1 and 2. Components were purified and attested by conventional methods.
	ESTIMATED ERROR: $\delta T/K = \pm 0.2$; $\delta x_{H_2S} = \pm 4\%$
	REFERENCES:

COMPONENTS:	ORIGINAL MEASUREMENTS:
1. Hydrogen sulfide; H_2S; [7783-06-4] 2. Benzenamine (*aniline*); C_6H_7N; [62-53-3]	Bancroft, W.D.; Belden, B.C. *J. Phys. Chem.* <u>1930</u>, *34*, 2123-4.

VARIABLES:
 	
   Pressure
	PREPARED BY:
	
	     P.G.T. Fogg

EXPERIMENTAL VALUES:

P_{H_2S}/mmHg	P_{H_2S}/bar[*]	Wt. of H_2S per unit volume of benzenamine /mg cm^{-3}	Mole fraction of H_2S in liquid[*]
a 102	0.136	2.74	0.0073
199	0.265	5.32	0.0141
296	0.395	8.13	0.0213
390	0.520	10.6	0.0276
484	0.645	13.2	0.0342
579	0.772	15.8	0.0406
676	0.901	18.6	0.0475
874	1.165	24.0	0.0604
1160	1.547	31.6	0.0781
b 750	1.000	20.8	0.0528
591	0.788	16.4	0.0421
513	0.684	14.3	0.0369
406	0.541	11.3	0.0294
310	0.413	8.45	0.0221
178	0.237	4.64	0.0123

[*] calculated by the compiler.   T/K = 295.2

Set *a* coresponds to successive increases in pressure; set *b* to decreases.

AUXILIARY INFORMATION

METHOD/APPARATUS/PROCEDURE:	SOURCE AND PURITY OF MATERIALS:
The apparatus and procedure described in (1) were used. A glass vessel containing a sample of benzenamine was connected via a vacuum line to a manometer and gas burette containing H_2S. Experiments appear to have been carried out at room temperature as there is no mention of the use of a thermostat.	1. Prepared by heating Sb_2S_3 with conc. HCl; washed with water and dried over P_2O_5. 2. Freshly distilled before use.
	ESTIMATED ERROR: $\delta T/K = \pm 1$
	REFERENCES: 1. Bancroft, W.D.; Barnett, C.E. *J. Phys. Chem.* <u>1930</u>, *34*, 449.

COMPONENTS:	ORIGINAL MEASUREMENTS:
1. Hydrogen sulfide; H_2S; [7783-06-4] 2. Benzenamine (Aniline); C_6H_7N; [62-53-3]	Lenoir, J-Y.; Renault, P.; Renon, H. J. Chem. Eng. Data, 1971, 16, 340-2.

VARIABLES:	PREPARED BY:
	C. L. Young

EXPERIMENTAL VALUES:

T/K	Henry's constant H_{H_2S}/atm	Mole fraction at 1 atm* x_{H_2S}
298.2	15.1	0.0662

* Calculated by compiler assuming a linear function of P_{H_2S} vs x_{H_2S}, i.e., x_{H_2S}(1 atm) = $1/H_{H_2S}$

AUXILIARY INFORMATION

METHOD/APPARATUS/PROCEDURE:	SOURCE AND PURITY OF MATERIALS:
A conventional gas-liquid chromatographic unit fitted with a thermal conductivity detector was used. The carrier gas was helium. The value of Henry's law constant was calculated from the retention time. The value applies to very low partial pressures of gas and there may be substantial difference from that measured at 1 atm. pressure. There is also considerable uncertainty in the value of Henry's constant since surface adsorption was not allowed for although its possible existence was noted.	(1) L'Air Liquide sample, minimum purity 99.9 mole per cent. (2) Touzart and Matignon or Serlabo sample, purity 99 mole per cent.
	ESTIMATED ERROR: $\delta T/K = \pm 0.1$; δH/atm = $\pm 6\%$ (estimated by compiler).
	REFERENCES:

COMPONENTS:	ORIGINAL MEASUREMENTS:
1. Hydrogen sulfide; H$_2$S; [7783-06-4] 2. Benzenamine (Aniline); C$_6$H$_7$N; [62-53-3]	Gerrard, W. *J. Appl. Chem. Biotechnol.*, <u>1972</u>, *22*, 623-650

VARIABLES:	PREPARED BY:
Temperature	C.L. Young

EXPERIMENTAL VALUES:

T/K	P*/mmHg	P*/kPa	mole ratio	Mole fraction[+] of hydrogen sulfide in liquid, x_{H_2S}
265.15	755	100.7	0.143	0.125
267.15	755	100.7	0.136	0.120
273.15	755	100.7	0.110	0.099
283.15	755	100.7	0.081	0.075
293.15	755	100.7	0.065	0.061

[+]calculated by compiler

*total pressure

AUXILIARY INFORMATION

METHOD/APPARATUS/PROCEDURE:	SOURCE AND PURITY OF MATERIALS:
Hydrogen sulfide was bubbled into a weighed amount of component 2. in a bubbler tube as described in detail in the source. The amount of gas absorbed at equilibrium and at the observed temperature and pressure was determined by weighing. Pressure was measured with a mercury manometer.	1 and 2. Components were purified and attested by conventional methods.
	ESTIMATED ERROR: $\delta T/K = \pm 0.2$; $\delta x_{H_2S} = \pm 4\%$
	REFERENCES:

COMPONENTS:	ORIGINAL MEASUREMENTS:
1. Hydrogen sulfide; H_2S; [7783-06-4] 2. Benzenamine and Methylbenzenamines	Patyi, L.; Furmer, I. E.; Makranczy, J.; Sadilenko, A. S.; Stepanova, Z. G.; Berengarten, M. G. *Zh. Prikl. Khim.* 1978, *51*, 1296-1300.

VARIABLES:	PREPARED BY:
	C. L. Young

EXPERIMENTAL VALUES:

T/K	$\alpha^{\dagger}$	Mole fraction of hydrogen sulfide at a partial pressure of 101.325 kPa x_{H_2S}
Benzenamine (Aniline); C_6H_7N; [62-53-3]		
298.15	11.0	0.043
N-Methylbenzenamine (Methylaniline); C_7H_9N; [100-61-8]		
298.15	9.8	0.045
N,N-Dimethylbenzenamine (Dimethylaniline); $C_8H_{11}N$; [121-69-7]		
298.15	10.5	0.056

†volume of gas (measured at 101.325 kPa and 273.15 K) dissolved per volume of solvent.

AUXILIARY INFORMATION

METHOD/APPARATUS/PROCEDURE:	SOURCE AND PURITY OF MATERIALS:
Volumetric method. Pressure measured when known amounts of gas were added, in increments, to a known amount of liquid in a vessel of known dimensions. Corrections were made for the partial pressure of solvent. Details in ref. (1).	Purity better than 99 mole per cent as determined by gas chromatography.

ESTIMATED ERROR:
$\delta T/K = \pm 0.1$; $\delta\alpha = \pm 4\%$ or less.

REFERENCES:
1. Bodor, E.; Bor, G. J.; Mohai, B.; Sipos, G. *Veszpremi. Vegyip. Egy. Kozl.* 1957, *1*, 55.

COMPONENTS:	ORIGINAL MEASUREMENTS:
1. Hydrogen sulfide; H_2S; [7783-06-4] 2. N,N-Dimethylbenzenamine; (Dimethylaniline); $C_8H_{11}N$; [121-69-7]	Gerrard, W. *J. Appl. Chem. Biotechnol.*, <u>1972</u>, *22*, 623-650
VARIABLES: Temperature	PREPARED BY: C.L. Young

EXPERIMENTAL VALUES:

T/K	$P*$/mmHg	$P*$/kPa	Mole ratio	Mole fraction[+] of hydrogen sulfide in liquid, x_{H_2S}
278.15	755	100.7	0.132	0.117
283.15	755	100.7	0.114	0.102
293.15	755	100.7	0.091	0.083

[+]calculated by compiler

*total pressure

AUXILIARY INFORMATION

METHOD/APPARATUS/PROCEDURE:	SOURCE AND PURITY OF MATERIALS:
Hydrogen sulfide was bubbled into a weighed amount of component 2. in a bubbler tube as described in detail in the source. The amount of gas absorbed at equilibrium and at the observed temperature and pressure was determined by weighing. Pressure was measured with a mercury manometer.	1 and 2. Components were purified and attested by conventional methods.
	ESTIMATED ERROR: $\delta T/K = \pm 0.2$; $\delta x_{H_2S} = \pm 4\%$
	REFERENCES:

COMPONENTS:	ORIGINAL MEASUREMENTS:
1. Hydrogen sulfide; H_2S; [7783-06-4] 2. Ethylbenzenamines	Patyi, L.; Furmer, I. E.; Makranczy, J.; Sadilenko, A. S.; Stepanova, Z. G.; Berengarten, M. G. *Zh. Prikl. Khim.* 1978, *51*, 1296-1300.
VARIABLES:	PREPARED BY: C. L. Young

EXPERIMENTAL VALUES:

T/K	α [†]	Mole fraction of hydrogen sulfide at a partial pressure of 101.325 kPa x_{H_2S}
N-Ethylbenzenamine (Ethylaniline); $C_8H_{11}N$; [103-69-5]		
298.15	10.3	0.055
N,N-Diethylbenzenamine (Diethylaniline); $C_{10}H_{15}N$; [91-66-7]		
298.15	8.9	0.060

[†] volume of gas (measured at 101.325 kPa and 273.15 K)
dissolved per volume of solvent.

AUXILIARY INFORMATION

METHOD/APPARATUS/PROCEDURE:	SOURCE AND PURITY OF MATERIALS:
Volumetric method. Pressure measured when known amounts of gas were added, in increments, to a known amount of liquid in a vessel of known dimensions. Corrections were made for the partial pressure of solvent. Details in ref. (1).	Purity better than 99 mole per cent as determined by gas chromatography.
	ESTIMATED ERROR: $\delta T/K = \pm 0.1$; $\delta\alpha = \pm 4\%$ or less.

	REFERENCES:
	
	1. Bodor, E.; Bor, G. J.; Mohai,
	   B.; Sipos, G.
	   Veszpremi. Vegyip. Egy. Kozl.
	   1957, *1*, 55.

COMPONENTS:	ORIGINAL MEASUREMENTS:
1. Hydrogen sulfide; H_2S; [7783-06-4] 2. Pyridine; C_5H_5N; [110-86-1]	Gerrard, W. *J. Appl. Chem. Biotechnol*, <u>1972</u>, *22*, 623-650

VARIABLES:	PREPARED BY:
Temperature	C.L. Young

EXPERIMENTAL VALUES:

T/K	$P*/mmHg$	$P*/kPa$	Mole ratio	Mole fraction[+] of hydrogen sulfide in liquid, x_{H_2S}
265.15	752	100.3	0.266	0.210
267.15	752	100.3	0.243	0.195
273.15	752	100.3	0.187	0.158
283.15	752	100.3	0.134	0.118
293.15	752	100.3	0.103	0.093

[+]calculated by compiler

*total pressure

AUXILIARY INFORMATION

METHOD/APPARATUS/PROCEDURE:	SOURCE AND PURITY OF MATERIALS:
Hydrogen sulfide was bubbled into a weighed amount of component 2. in a bubbler tube as described in detail in the source. The amount of gas absorbed at equilibrium and at the observed temperature and pressure was determined by weighing. Pressure was measured with a mercury manometer.	1 and 2. Components were purified and attested by conventional methods.
	ESTIMATED ERROR: $\delta T/K = \pm 0.2$; $\delta x_{H_2S} = \pm 4\%$
	REFERENCES:

COMPONENTS:	ORIGINAL MEASUREMENTS:
1. Hydrogen sulfide; H_2S; [7783-06-4] 2. Quinoline; C_9H_7N; [91-22-5]	Gerrard, W. *J. Appl. Chem. Biotechnol.*, <u>1972</u>, *22*, 623-650

VARIABLES:	PREPARED BY:
Temperature	C.L. Young

EXPERIMENTAL VALUES:

T/K	P*/mmHg	P*/kPa	Mole ratio	Mole fraction[+] of hydrogen sulfide in liquid, x_{H_2S}
265.15	748	99.7	0.246	0.197
267.15	748	99.7	0.224	0.183
273.15	748	99.7	0.175	0.149
283.15	748	99.7	0.129	0.114
293.15	748	99.7	0.098	0.089

[+]calculated by compiler

*total pressure

AUXILIARY INFORMATION

METHOD/APPARATUS/PROCEDURE:	SOURCE AND PURITY OF MATERIALS:
Hydrogen sulfide was bubbled into a weighed amount of component 2. in a bubbler tube as described in detail in the source. The amount of gas absorbed at equilibrium and at the observed temperature and pressure was determined by weighing. Pressure was measured with a mercury manometer.	1 and 2. Components were purified and attested by conventional methods.

ESTIMATED ERROR:

$\delta T/K = \pm 0.2$; $\delta x_{H_2S} = \pm 4\%$

REFERENCES:

COMPONENTS:	ORIGINAL MEASUREMENTS:
1. Hydrogen sulfide; H_2S; [7783-06-4] 2. Acetonitrile, C_2H_5N; [75-05-8] 3. Water; H_2O, [7732-18-5] Perchloric acid, lithium salt; $LiClO_4$; [7791-03-9]	Evans, J.F.; Blount, H.N. *Anal. Lett.* <u>1974</u>, *7(6)*, 445-451.

VARIABLES:	PREPARED BY:
Concentrations of water and of lithium perchlorate	P.G.T. Fogg

EXPERIMENTAL VALUES:

Concn. of H_2O /mol dm^{-3}	Concn. of $LiClO_4$ /mol dm^{-3}	Concn. of H_2S /mol dm^{-3}	Method of analysis	Number of analyses	Mole fraction x_{H_2S}
0	0	0.532 ± 0.004	I	7	0.027[*]
0	0	0.525 ± 0.006	G	10	
0.111	0	0.536 ± 0.003	I	4	
0.278	0	0.544 ± 0.003	I	4	
0.556	0	0.519 ± 0.001	I	4	
0	0.0999	0.537 ± 0.002	I	4	
0	0.201	0.547 ± 0.005	I	4	

Total pressure = 759 torr = 1.012 bar
Temperature = 298.2 K

I - iodimetry;   G - gravimetry.

[*] calculated by the compiler from the density of acetonitrile in ref. (1) and an average value of 0.528 mol dm^{-3} for the concentration of H_2S.

AUXILIARY INFORMATION

METHOD/APPARATUS/PROCEDURE:	SOURCE AND PURITY OF MATERIALS:
Hydrogen sulfide was bubbled through anhydrous acetonitrile and then through the liquid under test in a thermostatted absorption vessel. Samples of solution were subsequently withdrawn, diluted and analysed. Some samples were analysed by titration with acidic aqueous standard iodine solution. Others were analysed by addition of silver nitrate solution followed by the weighing of the silver sulfide formed. The absence of a spectral band at 266.5 nm was taken to indicate that formation of thioacetamide did not occur under the experimental conditions.	1. from Matheson Gas Products; dried with activated alumina. 2. u.v. grade from Burdick & Jackson Labs., Muskegon, Michigan; fractionally distilled from CaH_2 and passed through an activated alumina column to remove further traces of water; final water content 2.2×10^{-3} mol dm^{-3}. 3. $LiClO_4$ - *Alfa*: recrystallised from distilled water; air dried; crushed; vacuum dried at 160 °C.
	ESTIMATED ERROR: $\delta P/\text{bar} = \pm 7 \times 10^{-4}$; $\delta T/K = \pm 0.5$
	REFERENCES: 1. Riddick, J.A.; Bunger, W.B. *Techniques of Chemistry, Vol.II,* Wiley-Interscience, New York, <u>1970</u>.

COMPONENTS:	ORIGINAL MEASUREMENTS:

1. Hydrogen sulfide; H_2S; [7783-06-4]
 	
2. Acetonitrile; C_2H_3N; [75-05-8]
   N,N-Dimethylformamide; C_3H_7NO; [68-12-2]
   N,N-Dimethylacetamide; C_4H_9NO; [127-19-5]
	Hayduk, W.; Pahlevanzadeh, H.
	
	Can. J. Chem. Eng. 1987, 65, 299-307.

VARIABLES:	PREPARED BY:
Temperature	P.G.T. Fogg

EXPERIMENTAL VALUES:

T/K	Total press. /kPa	Mole fraction solubility at total press.	Mole fraction solubility at partial press. of 101.325 kPa.	Ostwald* coeff. L
Acetonitrile				
268.15	100.0	0.100	0.103	49.3
298.15	100.7	0.0421	0.0476	22.9
333.15	101.7	0.0110	0.0211	10.6
N,N-Dimethylformamide				
268.15	100.7	0.252	0.253	97.2
298.15	102.0	0.116	0.116	41.1
333.15	100.8	0.0520	0.0541	19.3
N,N-Dimethylacetamide				
268.15	100.7	0.303	0.304	104.1
298.15	100.9	0.145	0.146	44.7
333.15	100.8	0.0646	0.0661	19.9

* The Oswald coefficient was calculated as the volume of gas at a gas partial pressure of 1 atm. which will completely dissolve in one volume of solvent.

AUXILIARY INFORMATION

METHOD/APPARATUS/PROCEDURE:	SOURCE AND PURITY OF MATERIALS:
Solubilities were measured using a constant solvent flow apparatus described in (1). Solvent vapor pressures were calculated from Antoine constants given in (2). Densities of solvents were as reported in (3). The gas molar volumes were calculated using the second virial coefficient given in (4).	1. from Matheson of Canada. 2. from Aldrich Chemicals; HPLC grade; minimum purity 99.9%

	ESTIMATED ERROR:

REFERENCES:
1. Asatani, H.; Hayduk, W. *Can.J.Chem.Eng.* 1983, 61, 227.
2. Reid, R.C.; Prausnitz, J.M.; Sherwood, T.K. *The Properties of Gases and Liquids*, 1977, McGraw-Hill, New York.
3. Zhang, G.; Hayduk, W. *Can.J.Chem. Eng.* 1984, 62, 713.
4. Dymond, J.H.; Smith, E.B. *The Virial Coefficients of Pure Gases and Mixtures*, 1980, Oxford University Press, New York.

COMPONENTS:	ORIGINAL MEASUREMENTS:
1. Hydrogen sulfide; H_2S; [7783-06-4]	Gerrard, W.
2. Benzonitrile; C_7H_5N; [100-47-0]	*J. Appl. Chem. Biotechnol.*, 1972, 22, 623-650

VARIABLES:	PREPARED BY:
Temperature	C.L. Young

EXPERIMENTAL VALUES:

T/K	$P*$/mmHg	$P*$/kPa	Mole ratio	Mole fraction[+] of hydrogen sulfide in liquid, x_{H_2S}
265.15	749	99.9	0.186	0.157
267.15	749	99.9	0.166	0.142
273.15	749	99.9	0.133	0.117
283.15	749	99.9	0.101	0.092
293.15	749	99.9	0.088	0.081

[+]calculated by compiler

*total pressure

AUXILIARY INFORMATION

METHOD/APPARATUS/PROCEDURE:	SOURCE AND PURITY OF MATERIALS:
Hydrogen sulfide was bubbled into a weighed amount of component 2. in a bubbler tube as described in detail in the source. The amount of gas absorbed at equilibrium and at the observed temperature and pressure was determined by weighing. Pressure was measured with a mercury manometer.	1 and 2. Components were purified and attested by conventional methods.
	ESTIMATED ERROR: $\delta T/K = \pm0.2$; $\delta x_{H_2S} = \pm4\%$
	REFERENCES:

COMPONENTS:	ORIGINAL MEASUREMENTS:
1. Hydrogen sulfide; H_2S; [7783-06-4] 2. 3-Methyl-1H-pyrazole; $C_4H_6N_2$; [1453-58-3] 1,3-Dimethyl-1H-pyrazole; $C_5H_8N_2$; [694-48-4]	Egorova, V.I.; Grishko, N.I. Neokladnova, L.N.; Furmanov, A.S. Podvigailova, I.G. *Deposited Document* __1975__, *VINITI 2907-76*
VARIABLES: Temperature	PREPARED BY: P.G.T. Fogg

EXPERIMENTAL VALUES:

Solvent	T/K	Absorption coeff.[†] α	Henry's law constant[*] H/mmHg	Mole fraction[**] P_{H_2S} = 1 atm x_{H_2S}
3-Methyl-1H-pyrazole	283.2		7910	0.096
	298.2	9.1	11220	0.068
	313.2		13200	0.058
1,3-Dimethyl-1H-pyrazole	283.2		5490	0.138
	298.2	21.5	8260	0.092
	313.2		10420	0.073

$$760 \text{ mmHg} = 1.013 \text{ bar} = 1.013 \times 10^5 \text{ Pascal}$$

[*] defined as:
$$H = P_{H_2S} / x_{H_2S}$$

[**] calculated by the compiler.

[†] volume of gas, reduced to 273.15 K and 1.013 bar, absorbed by one volume of liquid when the partial pressure of gas is 1.013 bar.

AUXILIARY INFORMATION

METHOD/APPARATUS/PROCEDURE:	SOURCE AND PURITY OF MATERIALS:
The volume of gas absorbed by the solvent held in a thermostatted cell was measured at several pressures to a total pressure of about 760 mmHg. The temperature was controlled to ± 0.1 °C.	2. $C_4H_6N_2$: b.pt. 204 °C $C_5H_8N_2$: b.pt. 131-145 °C
	ESTIMATED ERROR:
	REFERENCES:

COMPONENTS:	ORIGINAL MEASUREMENTS:
1. Hydrogen sulfide; H_2S; [7783-06-4] 2. N,N-Dimethylformamide; C_3H_7NO; [68-12-2]	DuPont de Nemours and Co. (Inc.) *Chem. Eng. News* <u>1955</u>, *33*, 2366.

VARIABLES:	PREPARED BY:
	P.G.T. Fogg

EXPERIMENTAL VALUES:

T/K	P_{H_2S}	Vol. of H_2S absorbed[†] per volume of DMF	Mole fraction[*] x_{H_2S} (1 atm)
298.2	1 atm (1.013 bar)	35	0.109

[*] calculated by the compiler using the density of DMF at 298.2 K from (1).

[†] the volume of the gas has been reduced to 273.15 K and 1.013 bar.

AUXILIARY INFORMATION

METHOD/APPARATUS/PROCEDURE:	SOURCE AND PURITY OF MATERIALS:
Not stated.	Not stated.
	ESTIMATED ERROR:
	REFERENCES: 1. *Lange's Handbook of Chemistry 12th edition*, McGraw-Hill, New York. <u>1979</u>.

COMPONENTS:	ORIGINAL MEASUREMENTS:
1. Hydrogen sulfide; H_2S; [7783-06-4] 2. N,N-Dimethylformamide; C_3H_7NO; [68-12-2]	Gerrard, W. J. Appl. Chem. Biotechnol. 1972, 22, 623-650.

VARIABLES:	PREPARED BY:
Temperature	P.G.T. Fogg

EXPERIMENTAL VALUES:

T/K	P(total)/mmHg	P(total)/bar	Mole ratio	Mole fraction of H_2S[*] x_{H_2S}
265.15	750	1.000	0.387	0.279
267.15	750	1.000	0.360	0.265
273.15	750	1.000	0.286	0.222
283.15	750	1.000	0.205	0.170
293.15	750	1.000	0.151	0.131

[*] calculated by the compiler for the stated total pressure.

AUXILIARY INFORMATION

METHOD/APPARATUS/PROCEDURE:	SOURCE AND PURITY OF MATERIALS:
Hydrogen sulfide was bubbled into a weighed amount of component 2 in a bubbler tube as described in detail in the source. The amount of gas absorbed at equilibrium for the observed temperature and pressure was found by weighing. Pressure was measured with a mercury manometer.	It was stated that "All materials purified and attested by conventional methods."
	ESTIMATED ERROR: δx_{H_2S} = ± 4% (author)
	REFERENCES:

COMPONENTS:	ORIGINAL MEASUREMENTS:
1. Hydrogen sulfide; H_2S; [7783-06-4] 2. N,N-Dimethylformamide; C_3H_7NO; [68-12-2]	Byeseda, J.J.; Deetz, J.A.; Manning, W.P. *Proc.Laurance Reid Gas Cond.Conf.* 1985.
VARIABLES:	PREPARED BY: P.G.T. Fogg

EXPERIMENTAL VALUES:

T/K	P_{H_2S}/psia	P_{H_2S}/bar*	Ostwald coeff. L	Mole fraction in liquid* x_{H_2S}
297.1	14.73	1.016	33.6	0.097

* calculated by compiler

AUXILIARY INFORMATION

METHOD/APPARATUS/PROCEDURE:	SOURCE AND PURITY OF MATERIALS:
The H_2S was contained in a thermostatted metal cylinder connected to a pressure gage, vacuum pump and supply of gas. A tight fitting internal piston sealed with an O-ring fitted into the cylinder so that the volume of gas could be changed by controlled movement of the piston. A measured volume of solvent was injected into the cylinder by a syringe. The absorption of gas was found from the movement of the piston which was necessary to maintain constant pressure.	
	ESTIMATED ERROR:
	REFERENCES:

COMPONENTS:	ORIGINAL MEASUREMENTS:
1. Hydrogen sulfide; H$_2$S; [7783-06-4] 2. 1-Methyl-2-pyrrolidinone; C$_5$H$_9$NO; [872-50-4]	Lenoir, J-Y.; Renault, P.; Renon, H. *J. Chem. Eng. Data* <u>1971</u>, *16*, 340-2

VARIABLES:	PREPARED BY:
	C.L. Young

EXPERIMENTAL VALUES:

T/K	Henry's constant H_{H_2S}/atm	Mole fraction at 1 atm* x_{H_2S}
298.15	5.56	0.180

* Calculated by compiler assuming a linear function of P_{H_2S} vs x_{H_2S}, i.e. x_{H_2S} (1 atm) $= 1/H_{H_2S}$

AUXILIARY INFORMATION

METHOD/APPARATUS/PROCEDURE:	SOURCE AND PURITY OF MATERIALS:
A conventional gas-liquid chromatographic unit fitted with a thermal conductivity detector was used. The carrier gas was helium. The value of Henry's law constant was calculated from the retention time. The value applies to very low partial pressures of gas and there may be a substantial difference from that measured at 1 atm. pressure. There is also considerable uncertainty in the value of Henry's constant since surface adsorption was not allowed for although its possible existence was noted.	(1) L'Air Liquide sample, minimum purity 99.9 mole per cent. (2) Touzart and Matignon or Serlabo sample, purity 99 mole per cent.
	ESTIMATED ERROR: $\delta T/K = \pm 0.1$; δH/atm $= \pm 6\%$ (estimated by compiler).
	REFERENCES:

COMPONENTS:	ORIGINAL MEASUREMENTS:
1. Hydrogen sulfide; H_2S; [7783-06-4] 2. 1-Methyl-2-pyrrolidinone; C_5H_9NO; [872-50-4]	Rivas, O.R.; Prausnitz, J.M. *Am. Inst. Chem. Engnrs. J.* 1979, *25*, 975-984.

VARIABLES:
 	
   Temperature
	PREPARED BY:
	
	   C.L. Young

EXPERIMENTAL VALUES:

T/K	Henry's constant, H /MPa	Mole fraction of hydrogen[+] sulfide in liquid, x_{H_2S}
263.15	0.31	0.327
298.15	0.76	0.133
323.15	1.35	0.0751
348.15	2.17	0.0467
373.15	3.21	0.0316

+ at a partial pressure of 101.3 kPa calculated by compiler assuming Henry's law applies at that pressure.

AUXILIARY INFORMATION

METHOD/APPARATUS/PROCEDURE:	SOURCE AND PURITY OF MATERIALS:
Volumetric apparatus with a fused quartz precision bourdon pressure gauge. Solubility apparatus carefully thermostatted. Solvent degassed *in situ*. Apparatus described in ref. (1) and modifications given in source.	1. and 2. Purity at least 99 mole per cent.

ESTIMATED ERROR:

$\delta T/K = \pm 0.05$; $\delta x_{H_2S} = \pm 1\%$.

REFERENCES:

1. Cukor, P.M.; Prausnitz, J.M.

   Ind. Eng. Chem. Fundam. 1971, *10*, 638.

COMPONENTS:	ORIGINAL MEASUREMENTS:
1. Hydrogen sulfide; H_2S; [7783-06-4] 2. 1-Methyl-2-pyrrolidinone; C_5H_9NO; [872-50-4]	Rivas, O.R.; Prausnitz, J.M. *Ind. Eng. Chem. Fundam.* <u>1979</u>, *18*, 289-292.

VARIABLES:	PREPARED BY:
Temperature	C.L. Young

EXPERIMENTAL VALUES:

T/K	Henry's constant / atm.	Mole fraction at 1atm partial pressure
296.65	7.2	0.139
308.15	9.5	0.105

* Calculated by compiler assuming mole fraction solubility linear with pressure.

AUXILIARY INFORMATION

METHOD/APPARATUS/PROCEDURE:	SOURCE AND PURITY OF MATERIALS:
Volumetric apparatus with a fused quartz precision bourdon pressure gauge. Solubility apparatus carefully thermostatted. Solvent degassed *in situ*. Apparatus described in ref (1) and modifications given in source.	No details given.

ESTIMATED ERROR:

$\delta T/K = \pm 0.05$; $\delta x_{H_2S} = \pm 1\%$.

REFERENCES:

1. Cukor, P.M.; Prausnitz, J.M. *Ind. Eng. Chem. Fundam.* <u>1971</u>, *10*, 638.

COMPONENTS:	ORIGINAL MEASUREMENTS:
1. Hydrogen sulfide; H_2S; [7783-06-4] 2. 1-Methyl-2-pyrrolidinone; C_5H_9NO; [872-50-4] 3. 2-(2-Aminoethoxy)-ethanol,(Diglycolamine); $C_4H_{11}NO_2$; [929-06-6]	Rivas, O.R.; Prausnitz, J.M.; *Ind. Eng. Chem. Fundam.* <u>1979</u>, *18,* 289-292.

VARIABLES:	PREPARED BY:
Temperature, liquid composition	C.L. Young.

EXPERIMENTAL VALUES:

T/K	P/mmHg	P/kPa	Component	Mole fraction in liquid, x	in gas, y
348.7	239.8	31.97	1 2 3	0.0207 0.8585 0.1208	0.9714 0.0276 0.0010
	355.4	47.38	1 2 3	0.0297 0.8506 0.1197	0.9808 0.0186 0.0006
382.65	196.4	26.19	1 2 3	0.0101 0.8678 0.1221	0.8812 0.1140 0.0048
	368.3	49.10	1 2 3	0.0179 0.8609 0.1212	0.9367 0.0607 0.0026

AUXILIARY INFORMATION

METHOD/APPARATUS/PROCEDURE:	SOURCE AND PURITY OF MATERIALS:
Volumetric apparatus with a fused quartz precision bourdon pressure gauge. Solubility apparatus carefully thermostatted. Solvent degassed *in situ*. Apparatus described in ref (1) and modifications given in source.	No details given.

ESTIMATED ERROR:

$\delta T/K = \pm 0.05$; $\delta x_{H_2S} = \pm 1\%$.

REFERENCES:

1. Cukor, P.M.; Prausnitz, J.M.
   Ind. Eng. Chem. Fundam. <u>1971</u>,
   10, 638.

COMPONENTS:	ORIGINAL MEASUREMENTS:
1. Hydrogen sulfide; H_2S; [7783-06-4] 2. 1-Methyl-2-pyrrolidinone, (N-Methylpyrrolidone); C_5H_9NO; [872-50-4]	Yarym-Agaev, N. L.; Matvienko, V. G.; Povalyaeva, N. V. *Zh. Prikl. Khim.* <u>1980</u>, *53*, 2456-2461. *J. Appl. Chem. (Russ).* <u>1980</u>, *53*, <u>1810-1814</u>.
VARIABLES:	PREPARED BY: C. L. Young

EXPERIMENTAL VALUES:

T/K	P_{H_2S}/atm	$P^a_{H_2S}$/MPa	Solubility[b]	Mole fraction of[a] hydrogen sulfide, x_{H_2S}
273.2	1.00	0.101	90	0.285
	2.00	0.203	180	0.443
	3.00	0.304	270	0.544
	4.00	0.405	360	0.614
	5.00	0.507	462	0.672
	6.00	0.608	632	0.737
	7.00	0.709	880	0.796
	8.00	0.811	1290	0.851
	9.00	0.912	2220	0.908
278.2	1.00	0.101	76	0.252
	2.00	0.203	150	0.399
	3.00	0.304	225	0.499
	4.00	0.405	300	0.570
	5.00	0.507	378	0.626
	6.00	0.608	484	0.682
	7.00	0.709	640	0.739
	8.00	0.811	830	0.786
	9.00	0.912	1170	0.838
	10.0	1.013	1820	0.890
283.2	1.00	0.101	66	0.226
	2.00	0.203	130	0.365
	3.00	0.304	193	0.461
	4.00	0.405	257	0.532
	5.00	0.507	322	0.588

(cont.)

AUXILIARY INFORMATION

METHOD/APPARATUS/PROCEDURE:	SOURCE AND PURITY OF MATERIALS:
Apparatus for high pressure work consisted of a thermostatted steel absorption vessel containing a weighed quantity of solvent and connected through valves to a thermostatted cylinder of liquid H_2S. High pressures were generated by heating the cylinder in a second thermostat and estimated from data given in ref. (1) and (2). The gas which dissolved in the liquid was estimated from the increase in weight of the vessel with allowance for the H_2S present in the gas phase. A flow technique was used at partial pressures of H_2S of equal to atmospheric pressure or less. The composition of the N_2 + H_2S flow stream was adjusted to obtain the desired partial pressure of H_2S	1. Prepared by heating sulfur with paraffin and purified by freezing in liquid nitrogen. 2. Technical grade, distilled at 15-20 mmHg; contained 0.05% water; RI 1.4705 (20°C)

	ESTIMATED ERROR: $\delta T/K = \pm 0.1$; $\delta P/atm = \pm 0.05$; $\delta x_{H_2S} = \pm 4\%$ (estimated by compiler)

REFERENCES:

1. *Technical Encyclopedia. Handbook of Physical Chemical and Technological Data, Vol.5*, Izd. Sov. Entsiklop. Moscow 1930.

2. *Chemist's Handbook Vol.1*, Leningrad-Moscow 1962.

COMPONENTS:	ORIGINAL MEASUREMENTS:

1. Hydrogen sulfide; H_2S; [7783-06-4]
 	
2. 1-Methyl-2-pyrrolidinone,
   (N-Methylpyrrolidone); C_5H_9NO;
   [872-50-4]
	Yarym-Agaev, N. L.; Matvienko, V. G.;
	Povalyaeva, N. V.
	Zh. Prikl. Khim., <u>1980</u>, *53*,2456-2461
	J. Appl. Chem. (Russ), <u>1980</u>, *53*,
	1810-1814.

EXPERIMENTAL VALUES:

T/K	P_{H_2S}/atm	$P^a_{H_2S}$/MPa	Solubility[b]	Mole fraction of[a] hydrogen sulfide, x_{H_2S}
283.2	6.00	0.608	390	0.633
	7.00	0.709	482	0.681
	8.00	0.811	616	0.732
	9.00	0.912	800	0.780
	10.0	1.013	1060	0.824
	11.0	1.115	1440	0.864
	12.0	1.216	2200	0.907
288.2	1.00	0.101	57	0.201
	2.00	0.203	112	0.331
	3.00	0.304	167	0.425
	4.00	0.405	222	0.496
	5.00	0.507	278	0.552
	6.00	0.608	334	0.596
	7.00	0.709	402	0.640
	8.00	0.811	478	0.679
	9.00	0.912	574	0.718
	10.0	1.013	700	0.756
	11.0	1.115	890	0.798
	12.0	1.216	1160	0.837
	13.0	1.317	1590	0.876
293.2	1.00	0.101	49	0.178
	2.00	0.203	96	0.298
	3.00	0.304	141	0.384
	4.00	0.405	188	0.454
	5.00	0.507	235	0.510
	6.00	0.608	282	0.555
	7.00	0.709	334	0.596
	8.00	0.811	394	0.636
	9.00	0.912	460	0.671
	10.0	1.013	550	0.709
	11.0	1.115	674	0.749
	12.0	1.216	834	0.787
	13.0	1.317	1066	0.825
	14.0	1.419	1400	0.861
	15.0	1.520	1860	0.892
298.2	1.00	0.101	42	0.156
	2.00	0.203	84	0.271
	3.00	0.304	121	0.349
	4.00	0.405	167	0.425
	5.00	0.507	210	0.482
	6.00	0.608	252	0.527
	7.00	0.709	296	0.567
	8.00	0.811	340	0.601
	9.00	0.912	390	0.633
	10.0	1.013	448	0.665
	11.0	1.115	526	0.699
	12.0	1.216	625	0.734
	13.0	1.317	750	0.768
	14.0	1.419	910	0.801
	15.0	1.520	1120	0.832
	16.0	1.621	1380	0.859
	17.0	1.723	1740	0.885
	18.0	1.824	2280	0.910
303.2	1.00	0.101	38	0.144
	2.00	0.203	75	0.249
	3.00	0.304	113	0.333
	4.00	0.405	150	0.399
	5.00	0.507	187	0.453

(cont.)

COMPONENTS:	ORIGINAL MEASUREMENTS:

1. Hydrogen sulfide; H_2S; [7783-06-4]
 	
2. 1-Methyl-2-pyrrolidinone,
   (N-Methylpyrrolidone); C_5H_9NO;
   [872-50-4]
	Yarym-Agaev, N. L.; Matvienko, V. G.;
	Povalyaeva, N. V.
	Zh. Prikl. Khim., <u>1980</u>, *53*, 2456-2461.
	J. Appl. Chem. (Russ), <u>1980</u>, *53*,
	1810-1814.

EXPERIMENTAL VALUES:

T/K	P_{H_2S}/atm	$P^a_{H_2S}$/MPa	Solubility[b]	Mole fraction of[a] hydrogen sulfide, x_{H_2S}
303.2	6.00	0.608	225	0.499
	7.00	0.709	264	0.539
	8.00	0.811	302	0.572
	9.00	0.912	340	0.601
	10.0	1.013	384	0.630
	11.0	1.115	436	0.659
	12.0	1.216	506	0.691
	13.0	1.317	594	0.724
	14.0	1.419	694	0.754
	15.0	1.520	810	0.781
	16.0	1.621	940	0.806
	17.0	1.723	1100	0.830
	18.0	1.824	1340	0.856
	19.0	1.925	1660	0.880
	20.0	2.027	2120	0.904
313.2	1.00	0.101	29	0.114
	2.00	0.203	57	0.201
	3.00	0.304	85	0.273
	4.00	0.405	113	0.333
	5.00	0.507	141	0.384
	6.00	0.608	169	0.428
	7.00	0.709	197	0.466
	8.00	0.811	225	0.499
	9.00	0.912	256	0.531
	10.0	1.013	286	0.559
	11.0	1.115	320	0.586
	12.0	1.216	364	0.617
	13.0	1.317	412	0.646
	14.0	1.419	466	0.673
	15.0	1.520	526	0.699
	16.0	1.621	590	0.723
	17.0	1.723	660	0.745
	18.0	1.824	740	0.766
	19.0	1.925	840	0.788
	20.0	2.027	970	0.811
273.2	0.075	0.0076	7.5	0.032
	0.278	0.0282	26.7	0.106
	0.488	0.0494	45.0	0.166
	0.779	0.0789	73.1	0.244
	0.976	0.0989	88.7	0.282
260.1	1.00	0.101	147	0.394
273.2	1.00	0.101	90.9	0.287
283.8	1.00	0.101	64.6	0.222
289.2	1.00	0.101	55.8	0.198
300.8	1.00	0.101	40.5	0.152
303.2	1.00	0.101	37.1	0.141
313.2	1.00	0.101	29.0	0.114
314.4	1.00	0.101	27.8	0.110
323.2	1.00	0.101	21.5	0.087
333.2	1.00	0.101	17.8	0.073
342.9	1.00	0.101	14.7	0.061
355.4	1.00	0.101	11.0	0.046
364.7	1.00	0.101	8.99	0.0382
399.2	1.00	0.101	5.10	0.0221

[a] Calculated by compiler

[b] Volume of hydrogen sulfide measured at 0 °C and
1 atmosphere dissolved in 1 g of N-methylpyrrolidone.

COMPONENTS:	ORIGINAL MEASUREMENTS:
1. Hydrogen sulfide; H_2S; [7783-06-4] 2. 1-Methyl-2-pyrrolidinone, (*N-methylpyrrolidone*); C_5H_9NO; [872-50-4]	Murrieta-Guevara, F; Rodriguez, A.T. *J.Chem.Eng.Data* <u>1984</u>, *29*, 456 - 460.

VARIABLES:	PREPARED BY:
Concentration	P.G.T. Fogg

EXPERIMENTAL VALUES:

T/K	Mole fraction in liquid phase x_{H_2S}	P_{H_2S}/atm	P_{H_2S}/bar[*]
298.15	0.034	0.207	0.210
	0.057	0.395	0.400
	0.087	0.638	0.646
	0.119	0.911	0.923
	0.150	1.191	1.207
	0.192	1.619	1.640

[*] calculated by compiler.

AUXILIARY INFORMATION

METHOD/APPARATUS/PROCEDURE:	SOURCE AND PURITY OF MATERIALS:
The apparatus consisted of a glass equilibrium cell fitted with inlet tube ending in a porous glass disc. The cell was attached to a gas storage cylinder. Pressure in the cylinder was measured by a Bourdon tube gage and in the cell by a pressure transducer. Measured amounts of solvent and of H_2S were passed into the cell and vapor was then continuously circulated by a magnetic pump with liquid stirred by a magnetic stirrer. Temperatures were controlled by an air bath. Equilibrium pressures were measured after about 30 min.	1. from Matheson; purity 99.5% 2. purity > 99.5%

	ESTIMATED ERROR: $\delta T/K = \pm 0.1$; $\delta P = \pm 0.1\%$; $\delta x_{H_2S} = \pm 5\%$ (authors)
	REFERENCES:

COMPONENTS:	ORIGINAL MEASUREMENTS:
1. Hydrogen sulfide; H₂S; [7783-06-4] 2. 1-Methyl-2-pyrrolidinone, (*N-methylpyrrolidone*); C₅H₉NO; [872-50-4] 3. 2-Aminoethanol, (*monoethanolamine*); C₂H₇NO; [141-43-5]	Murrieta-Guevara, F; Rodriguez, A.T. *J.Chem.Eng.Data* <u>1984</u>, *29*, 456 - 460.

VARIABLES:	PREPARED BY:
Concentration	P.G.T. Fogg

EXPERIMENTAL VALUES:

Composition of solvent before addition of H₂S	Mole fraction in liquid phase x_{H_2S}	P_{H_2S}/atm	P_{H_2S}/bar*
NMP + 5.1 wt% MEA	0.089	0.199	0.202
	0.129	0.377	0.382
	0.167	0.661	0.670
	0.208	1.052	1.066
	0.237	1.387	1.405
	0.264	1.735	1.758
NMP + 14.3 wt% MEA	0.080	0.134	0.136
	0.147	0.308	0.312
	0.207	0.619	0.627
	0.252	1.108	1.123
	0.274	1.439	1.458
	0.285	1.586	1.607

T/K = 273.2   *calculated by compiler.

NMP = 1-methyl-2-pyrrolidinone; (*N-methylpyrrolidone*)

MEA = 2-aminoethanol (*monoethanolamine*)

AUXILIARY INFORMATION

METHOD/APPARATUS/PROCEDURE:	SOURCE AND PURITY OF MATERIALS:
The apparatus consisted of a glass equilibrium cell fitted with inlet tube ending in a porous glass disc. The cell was attached to a gas storage cylinder. Pressure in the cylinder was measured by a Bourdon tube gage and in the cell by a pressure transducer. Measured amounts of solvent and of H₂S were passed into the cell and vapor was then continuously circulated by a magnetic pump with liquid stirred by a magnetic stirrer. Temperatures were controlled by an air bath. Equilibrium pressures were measured after about 30 min.	1. from Matheson; purity 99.5% 2. & 3. purity > 99.5%
	ESTIMATED ERROR: $\delta T/K = \pm 0.1$; $\delta P = \pm 0.1\%$; $\delta x_{H_2S} = \pm 5\%$ (authors)
	REFERENCES:

COMPONENTS:	ORIGINAL MEASUREMENTS:
1. Hydrogen sulfide; H_2S; [7783-06-4] 2. Hexahydro-1-methyl-2H-azepin-2-one (*N-methyl-ε-caprolactam*); $C_7H_{13}NO$; [2556-73-2]	Wehner, K.; Burk, W. Kisan, W. *Chem. Tech. (Leipzig)* <u>1977</u>, *29(8)*, 445-448.

VARIABLES:	PREPARED BY:
	P.G.T. Fogg

EXPERIMENTAL VALUES:

T/K	Kuenen coefficient, S, [*] $/cm^3 g^{-1} atm^{-1}$	Mole fraction (1.013 bar)[**] x_{H_2S}
293.2	37.8	0.178

The authors also gave a small scale graph showing tha variation of Kuenen coefficient with temperature from about 280 K to about 400 K.

$$1 \text{ atm } = 1.013 \text{ bar } = 1.013 \times 10^5 \text{ Pascal}$$

[*] This is the volume of gas, reduced to 273.5 K and 1 atm, which is dissolved by 1 g of solvent, divided by the partial pressure of the gas in atmospheres.

[**] Calculated by the compiler on the assumption that the value of S is applicable to a partial pressure of 1 atm.

AUXILIARY INFORMATION

METHOD/APPARATUS/PROCEDURE:	SOURCE AND PURITY OF MATERIALS:
No details given except that measurements were made by staff at VEB Leuna-Werke 'Walter Ulbricht'.	No information given.
	ESTIMATED ERROR:
	REFERENCES:

COMPONENTS:	ORIGINAL MEASUREMENTS:
1. Hydrogen sulfide; H_2S; [7783-06-4] 2. Nitrobenzene; $C_6H_5NO_2$; [98-95-3]	Lenoir, J-Y.; Renault, P.; Renon, H. *J. Chem. Eng. Data*, <u>1971</u>, *16*, 340-2

VARIABLES:	PREPARED BY:
	C. L. Young

EXPERIMENTAL VALUES:

T/K	Henry's constant H_{H_2S}/atm	Mole fraction at 1 atm* x_{H_2S}
298.2	18.7	0.0535

*   Calculated by compiler assuming a linear function of P_{H_2S} vs x_{H_2S}, i.e.,
    x_{H_2S}(1 atm) = $1/H_{H_2S}$

AUXILIARY INFORMATION

METHOD/APPARATUS/PROCEDURE:	SOURCE AND PURITY OF MATERIALS:
A conventional gas-liquid chromatographic unit fitted with a thermal conductivity detector was used. The carrier gas was helium. The value of Henry's law constant was calculated from the retention time. The value applies to very low partial pressures of gas and there may be a substantial difference from that measured at 1 atm. pressure. There is also considerable uncertainty in the value of Henry's constant since surface adsorption was not allowed for although its possible existence was noted.	(1) L'Air Liquide sample, minimum purity 99.9 mole per cent. (2) Touzart and Matignon or Serlabo sample, purity 99 mole per cent.
	ESTIMATED ERROR: $\delta T/K = \pm0.1$; δH/atm = $\pm6\%$ (estimated by compiler).
	REFERENCES:

COMPONENTS:	ORIGINAL MEASUREMENTS:
1. Hydrogen sulfide; H_2S; [7783-06-4] 2. Nitrobenzene; $C_6H_5NO_2$; [98-95-3]	Gerrard, W. *J. Appl. Chem. Biotechnol.*, 1972, 22, 623-650

VARIABLES:	PREPARED BY:
Temperature	C.L. Young

EXPERIMENTAL VALUES:

T/K	P*/mmHg	P*/kPa	Mole ratio	Mole fraction[+] of hydrogen sulfide in liquid, x_{H_2S}
265.15	748	99.7	0.146	0.127
267.15	748	99.7	0.127	0.113
273.15	748	99.7	0.098	0.089
283.15	748	99.7	0.074	0.069
293.15	748	99.7	0.056	0.053

[+]calculated by compiler

*total pressure

AUXILIARY INFORMATION

METHOD/APPARATUS/PROCEDURE:	SOURCE AND PURITY OF MATERIALS:
Hydrogen sulfide was bubbled into a weighed amount of component 2. in a bubbler tube as described in detail in the source. The amount of gas absorbed at equilibrium and at the observed temperature and pressure was determined by weighing. Pressure was measured with a mercury manometer.	1 and 2. Components were purified and attested by conventional methods.
	ESTIMATED ERROR: $\delta T/K = \pm 0.2$; $\delta x_{H_2S} = \pm 4\%$
	REFERENCES:

COMPONENTS:	ORIGINAL MEASUREMENTS:
1. Hydrogen sulfide; H_2S; [7783-06-4] 2. Sulfur; S; [7704-34-9]	Fanelli, R. *Ind. Eng. Chem.* 1949, *41*, 2031-2033.
VARIABLES: Temperature	PREPARED BY: P.G.T. Fogg

EXPERIMENTAL VALUES:

T/K	Wt. of H_2S × 100 / wt of sulfur	T/K	Wt. of H_2S × 100 / wt of sulfur
399.2	0.057	632.2	0.186
414.2	0.067	642.2	0.188
423.2	0.088	644.2	0.189
431.2	0.117	646.2	0.186
440.2	0.133	661.2	0.182
448.2	0.139	671.2	0.173
475.2	0.160	673.2	0.171
506.2	0.177	700.2	0.141
546.2	0.179	710.2	0.116
573.2	0.190	713.2	0.098
595.2	0.189	717.2	0.065
611.2	0.187	717.8	0.049
623.2	0.189		

The total pressure was equal to barometric pressure (unspecified).

AUXILIARY INFORMATION

METHOD/APPARATUS/PROCEDURE:	SOURCE AND PURITY OF MATERIALS:
Hydrogen sulfide from a cylinder was bubbled through 50-60 g of molten sulfur in a glass vessel of capacity 40 cm^3 for a period of 16 to 90 hr. Temperatures were measured by a thermocouple in a well inside the vessel. An electric furnace controlled by Fenwall thermoswitches was used for heating. The amount of absorbed hydrogen sulfide was found by connecting the vessel to a Schiff gas burette and collecting and measuring the gas evolved as the sulfur cooled. The gas remaining in the solid sulfur was found from the increase in weight of vessel and contents.	1. from a cylinder; filtered by glass wool.
	ESTIMATED ERROR:
	REFERENCES:

COMPONENTS:	ORIGINAL MEASUREMENTS:
1. Hydrogen sulfide; H_2S; [7783-06-4] 2. Carbon disulfide; CS_2; [75-15-0]	Gattow, G.; Krebs, B. Z. anorg. allgem. Chem. 1963, 325, 15-25

VARIABLES:	PREPARED BY:
Temperature, concentration	P.G.T. Fogg

EXPERIMENTAL VALUES:

The total pressure was given by:

$$\log_{10}(P/\text{mmHg}) = -\frac{A}{T/K} + B$$

The vapor pressure of pure CS_2 is very small at the temperatures of measurement. e.g. at 195 K it is 0.63 mmHg. The total pressure was therefore close to the partial pressure of H_2S.

Mol % H_2S in liquid	Temperature range /K	A	B
5	153-195	540	3.56
10	153-195	729	5.605
20	158-195	850	6.545
30	163-195	942	7.11
40	170-195	944	7.16
50	171-195	974	7.34
60	173-195	975	7.37
70	178-195	976	7.38
80	180-195	979	7.41
90	183-195	980	7.44
100	188-213	1017	7.66

AUXILIARY INFORMATION

METHOD/APPARATUS/PROCEDURE:	SOURCE AND PURITY OF MATERIALS:
Vapor pressures were measured by a mercury manometer in an apparatus connected to a high-vacuum system. Temperatures were controlled by a petroleum ether bath cooled by a cold-finger containing liquid air. Quantities of liquid H_2S were measured out by volume at - 78 °C. Required volumes had been calculated using the density given in (1). Measured quantities of CS_2 were introduced into the apparatus and frozen in liquid air. H_2S was vacuum distilled into the vessel containing CS_2. Mixtures were brought to the temperature of measurement, magnetically stirred and vapor pressures recorded.	1. purified by vacuum sublimation. 2. dried over P_2O_5.

ESTIMATED ERROR:
$\delta T/K = \pm 0.1$ (authors)

REFERENCES:

1. Steele, B.D.; McIntosh, D.;
   Archibald, E.H.
   Z. physik. Chem. 1906, 55, 129.

COMPONENTS:	ORIGINAL MEASUREMENTS:
1. Hydrogen sulfide; H_2S; [7783-06-4] 2. Sulfinylbismethane, (Dimethylsulfoxide); C_2H_6SO; [67-68-5]	Lenoir, J-Y.; Renault, P.; Renon, H. *J. Chem. Eng. Data*, <u>1971</u>, *16*, 340-2

VARIABLES:	PREPARED BY:
	C. L. Young

EXPERIMENTAL VALUES:

T/K	Henry's constant H_{H_2S}/atm	Mole fraction at 1 atm* x_{H_2S}
298.2	10.9	0.0917

* Calculated by compiler assuming a linear function of P_{H_2S} vs x_{H_2S}, i.e., x_{H_2S}(1 atm) = $1/H_{H_2S}$

AUXILIARY INFORMATION

METHOD/APPARATUS/PROCEDURE:	SOURCE AND PURITY OF MATERIALS:
A conventional gas-liquid chromatographic unit fitted with a thermal conductivity detector was used. The carrier gas was helium. The value of Henry's law constant was calculated from the retention time. The value applies to very low partial pressures of gas and there may be a substantial difference from that measured at 1 atm. pressure. There is also considerable uncertainty in the value of Henry's constant since surface adsorption was not allowed for although its possible existence was noted.	(1) L'Air Liquide sample, minimum purity 99.9 mole per cent. (2) Touzart and Matignon or Serlabo sample, purity 99 mole per cent.
	ESTIMATED ERROR: $\delta T/K = \pm 0.1$; δH/atm = $\pm 6\%$ (estimated by compiler).
	REFERENCES:

COMPONENTS:	ORIGINAL MEASUREMENTS:
1. Hydrogen sulfide; H_2S; [7783-06-4] 2. Tetrahydrothiophene, 1,1-dioxide,(®Sulfolane); $C_4H_8O_2S$; [126-33-0]	Rivas, O.R.; Prausnitz, J.M. *Am. Inst. Chem. Engnrs. J.* 1979, *25*, 975-984.
VARIABLES: Temperature	PREPARED BY: C.L. Young.

EXPERIMENTAL VALUES:

T/K	Henry's constant, H /MPa	Mole fraction of [+] hydrogen sulfide in liquid, x_{H_2S}
303.15	2.07	0.0489
323.15	3.07	0.0330
348.15	4.57	0.0222
373.15	6.33	0.0160

+ at a partial pressure of 101.3 kPa calculated by compiler assuming Henry's law applies at that pressure.

AUXILIARY INFORMATION

METHOD/APPARATUS/PROCEDURE:	SOURCE AND PURITY OF MATERIALS:
Volumetric apparatus with a fused quartz precision bourdon pressure gauge. Solubility apparatus carefully thermostatted. Solvent degassed *in situ*. Apparatus described in ref. (1). and modifications given in source.	1. and 2. Purity at least 99 mole per cent.
	ESTIMATED ERROR: $\delta T/K = \pm 0.05$; $\delta x_{H_2S} = \pm 1\%$.
	REFERENCES: 1. Cukor, P.M.; Prausnitz, J.M. *Ind. Eng. Chem. Fundam.* 1971, *10*, 638.

COMPONENTS:	ORIGINAL MEASUREMENTS:
1. Hydrogen sulfide; H_2S; [7783-06-4] 2. Tetrahydrothiophene, 1,1-dioxide, (®Sulfolane); $C_4H_8O_2S$; [126-33-0]	Byeseda, J.J.; Deetz, J.A.; Manning, W.P. Proc.Laurance Reid Gas Cond.Conf. 1985.

VARIABLES:	PREPARED BY:
	P.G.T. Fogg

EXPERIMENTAL VALUES:

T/K	P_{H_2S}/psia	P_{H_2S}/bar[*]	Ostwald coeff. L	Mole fraction in liquid[*] x_{H_2S}
297.1	14.73	1.016	15.2	0.0686

[*] calculated by compiler

AUXILIARY INFORMATION

METHOD/APPARATUS/PROCEDURE:	SOURCE AND PURITY OF MATERIALS:
The H_2S was contained in a thermostatted metal cylinder connected to a pressure gage, vacuum pump and supply of gas. A tight fitting internal piston sealed with an O-ring fitted into the cylinder so that the volume of gas could be changed by controlled movement of the piston. A measured volume of solvent was injected into the cylinder by a syringe. The absorption of gas was found from the movement of the piston which was necessary to maintain constant pressure.	
	ESTIMATED ERROR:
	REFERENCES:

COMPONENTS:	ORIGINAL MEASUREMENTS:
1. Hydrogen sulfide; H_2S; [7783-06-4] 2. Esters of phosphoric acid	Lenoir, J-Y.; Renault, P.; Renon, H. *J. Chem. Eng. Data*, <u>1971</u>, *16*, 340-2.

VARIABLES:	PREPARED BY:
Temperature	C. L. Young

EXPERIMENTAL VALUES:

T/K	Henry's constant H_{H_2S}/atm	Mole fraction at 1 atm* x_{H_2S}
Phosphoric acid, trimethyl ester; $C_3H_9O_4P$; [512-56-1]		
325.2	18.4	0.0543
Phosphoric acid, triethyl ester; $C_6H_{15}O_4P$; [78-40-0]		
325.2	28.6	0.0350
Phosphoric acid, tripropyl ester; $C_9H_{21}O_4P$; [513-08-6]		
298.2	4.19	0.2387
323.2	7.96	0.1256
343.2	12.6	0.0794
Phosphoric acid, tributyl ester; $C_{12}H_{27}O_4P$; [126-73-8]		
325.2	9.07	0.1103
Phosphoric acid, tris(2-methylpropyl) ester; $C_{12}H_{27}O_4P$; [126-71-6]		
325.2	8.89	0.1125

*   Calculated by compiler assuming a linear function of P_{H_2S} vs x_{H_2S}, i.e.,
    x_{H_2S}(1 atm) = $1/H_{H_2S}$

AUXILIARY INFORMATION

METHOD/APPARATUS/PROCEDURE:	SOURCE AND PURITY OF MATERIALS:
A conventional gas-liquid chromato-graphic unit fitted with a thermal conductivity detector was used. The carrier gas was helium. The value of Henry's law constant was calculated from the retention time. The value applies to very low partial pressures of gas and there may be a substantial difference from that measured at 1 atm. pressure. There is also considerable uncertainty in the value of Henry's constant since surface adsorption was not allowed for although its possible existence was noted.	(1) L'Air Liquide sample, minimum purity 99.9 mole per cent. (2) Touzart and Matignon or Serlabo sample, purity 99 mole per cent.
	ESTIMATED ERROR: $\delta T/K = \pm 0.1$; $\delta H/atm = \pm 6\%$ (estimated by compiler).
	REFERENCES:

COMPONENTS:	ORIGINAL MEASUREMENTS:
1. Hydrogen sulfide; H_2S; [7783-06-4] 2. Tributyl phosphate, (*phosphoric acid, tributyl ester*); $C_{12}H_{27}O_4P$; [126-73-8]	Sergienko, I.D.; Kosyakov, N.E.; Yushko, V.L.; Khokhlov, S.F.; Pushkin, A.G. *Vop.Khim.Tekhnol.* 1973, *29*, 57 - 60

VARIABLES:	PREPARED BY:
Temperature, pressure	P.G.T. Fogg

EXPERIMENTAL VALUES:

The authors have given a small (3.3 cm x 3.8 cm) graph showing mole fraction solubilities of H_2S as a function of partial pressure of H_2S. The compiler has calculated approximate values of mole fractions and the corresponding partial pressures from co-ordinates of experimental points.

T/K	P_{H_2S}/mmHg	Mole fraction solubility, x_{H_2S}	T/K	P_{H_2S}/mmHg	Mole fraction solubility, x_{H_2S}
313.15	121	0.029	253.15	56	0.063
	362	0.075		87	0.108
	495	0.101		169	0.186
	591	0.113		273	0.264
	644	0.128		321	0.296
293.15	106	0.041	233.15	48	0.118
	253	0.094		106	0.203
	386	0.130		169	0.304
	495	0.163		193	0.336
	543	0.175			
			223.15	72	0.211
273.15	84	0.051		145	0.310
	159	0.101		181	0.344
	275	0.162			
	398	0.219			
	442	0.240			

AUXILIARY INFORMATION

METHOD/APPARATUS/PROCEDURE:	SOURCE AND PURITY OF MATERIALS:
A static method described in ref.1 was used.	1. purity found by chromatographic analysis to be at least 99.9% 2. 'pure' grade material used.
	ESTIMATED ERROR:
	REFERENCES: 1. Braude, G.E.; Shakhova, S.F. *Khim. prom.* 1961, *3*, 29.

COMPONENTS:	ORIGINAL MEASUREMENTS:
1. Hydrogen sulfide; H_2S; [7783-06-4] 2. Tributyl phosphate, (*phosphoric acid, tributyl ester*); $C_{12}H_{27}O_4P$; [126-73-8]	Vei, D.; Furmer, I.E.; Sadilenko, A.S.; Efimova, N.M.; Stepanova, Z.G.; Gracheva, N.V. *Gazov.Prom-st.*1975, 7, (7), 47 - 49

VARIABLES:	PREPARED BY:
Temperature; pressure	P.G.T. Fogg

EXPERIMENTAL VALUES:

The authors obtained Henry's law constants, K, at infinite dilution from experimental data by use of the Krichevskii-Il'inskaya equation i.e.

$$\log_{10}\left[\frac{P_2/\text{mmHg}}{x_2}\right] = \log_{10}(K/\text{mmHg}) - \beta\,(1 - x_1^2)$$

where x_1 and x_2 are the mole fractions of solvent and gas respectively, P_2 is the partial pressure of the gas and β is related to a coefficient A by the equation:

$$\beta = A/(2.303\,RT)$$

T/K	Henry's Law constant K/mmHg	T/K	Henry's Law constant K/mmHg
298.15	3020	343.15	11176
313.15	4677	363.15	15510
328.15	7413	383.15	20850

The authors have given a small scale graph showing variations of α, the volume of gas absorbed per unit volume of solvent, corrected to 273.15 K and 1.013 bar, with partial pressure of H_2S. The compiler has calculated approximate values of α from the smooth curves which the authors drew through experimental points and has estimated corresponding mole fraction solubilities.

T/K	P/mmHg	α	Mole fraction solubility x_{H_2S} *	T/K	P/mmHg	α	Mole fraction solubility x_{H_2S} *
298.15	200	5.56	0.064	343.15	200	1.76	0.022
	400	9.80	0.108		400	3.01	0.037
	600	13.02	0.139		600	4.24	0.052
313.15	200	3.66	0.044		750	5.27	0.064
	400	6.56	0.061	363.15	200	1.26	0.016
	600	9.07	0.102		400	2.20	0.028
	750	10.68	0.118		600	3.04	0.038
328.15	200	2.75	0.034		750	3.66	0.046
	400	4.68	0.056	383.15	200	0.79	0.010
	600	6.41	0.075		400	1.46	0.019
	750	7.46	0.087		600	2.05	0.027
					750	2.63	0.034

* estimated by compiler using densities from 298 K to 338 K given in ref. 2.

AUXILIARY INFORMATION

METHOD/APPARATUS/PROCEDURE:	SOURCE AND PURITY OF MATERIALS:
A static method was used. The compiler was not able to consult the original Russian paper. Literature references could not be included in the English translation from which the compilation was prepared. (1)	No information.

ESTIMATED ERROR:

REFERENCES
1. *British Gas Corporation Translations*, T5408/BG/LRS/LRST492/80
2. J. Timmermans, "*Physico-Chemical Constants of Pure Organic Compounds*", Vol. 2, Elsevier, London. 1965.

COMPONENTS:	ORIGINAL MEASUREMENTS:
1. Hydrogen sulfide; H_2S; [7783-06-4] 2. Phosphoric acid, tributyl ester (*tributyl phosphate*); $C_{12}H_{27}PO_4$; [126-73-8]	Härtel, G.H. *J.Chem.Eng.Data* <u>1985</u>, *30*, 57-61.

VARIABLES:	PREPARED BY:
Temperature	P.G.T. Fogg

EXPERIMENTAL VALUES:

T/K	Henry's constant H_{H_2S}/bar	Mole fraction solubility x_{H_2S} (1.013 bar)*
293.15	3.540	0.286
313.15	5.429	0.187
333.15	8.571	0.118
353.15	10.084	0.100
373.15	16.304	0.062

Henry's constant is defined as :

$$H_{H_2S} = \frac{pressure}{mole\ fraction\ solubility}$$

Solubilities were measured at concentrations not greater than x_{H_2S} = 0.16. The variation of mole fraction solubility with pressure was found to be almost linear with a correlation coefficient of better than 0.99.

*
 Calculated by the compiler assuming that the variation of mole fraction solubility with variation of pressure was linear to 1.013 bar. (Values for 293.15 K and 313.15 K lie outside the range for which linearity was experimentally demonstrated by the author.)

AUXILIARY INFORMATION

METHOD/APPARATUS/PROCEDURE:	SOURCE AND PURITY OF MATERIALS:
Absorption was found from the decrease in pressure when a known volume of gas came into contact with a known mass of degassed solvent. Pressure changes were found by use of a transducer with a mercury manometer to provide a reference pressure.	1. From Matheson, Heusenstamm, FRG; minimum purity 99.5%.
	ESTIMATED ERROR: δP/mbar = ± 0.25 (author)
	REFERENCES:

COMPONENTS:	ORIGINAL MEASUREMENTS:
1. Hydrogen sulfide; H_2S; [7783-06-4] 2. Phosphoric triamide, hexamethyl; $C_6H_{18}NO_3P$; [680-31-9]	Lenoir, J-Y.; Renault, P.; Renon, H. *J. Chem. Eng. Data*, <u>1971</u>, *16*, 340-2

VARIABLES:	PREPARED BY:
	C. L. Young

EXPERIMENTAL VALUES:

T/K	Henry's constant H_{H_2S}/atm
298.2	1.61

Reliable extrapolation to a partial pressure of 1 atm is not possible for this system.

AUXILIARY INFORMATION

METHOD/APPARATUS/PROCEDURE:	SOURCE AND PURITY OF MATERIALS:
A conventional gas-liquid chromatographic unit fitted with a thermal conductivity detector was used. The carrier gas was helium. The value of Henry's law constant was calculated from the retention time. The value applies to very low partial pressures of gas and there may be a substantial difference from that measured at 1 atm. pressure. There is also considerable uncertainty in the value of Henry's constant since surface adsorption was not allowed for although its possible existence was noted.	(1) L'Air Liquide sample, minimum purity 99.9 mole per cent. (2) Touzart and Matignon or Serlabo sample, purity 99 mole per cent.
	ESTIMATED ERROR: $\delta T/K = \pm0.1$; δH/atm = $\pm6\%$ (estimated by compiler).
	REFERENCES:

COMPONENTS:	EVALUATOR:
1. Deuterium sulfide; D_2S; [13536-94-2] 2. Water - d_2; D_2O; [7789-20-0]	Peter G.T. Fogg, Department of Applied Chemistry and Life Sciences, Polytechnic of North London, Holloway, London N7 8DB, U.K. July 1987

CRITICAL EVALUATION:

The solubility of deuterium sulfide in water-d_2 (D_2O) has been measured by Clarke and Glew (1) over the temperature range 278 to 323 K and a total pressure range of 0.437 to 1.044 bar. These measurements are consistent with each other and appear to be reliable. No other data are available for this system and these data should therefore be accepted on a tentative basis.

The evaluator has calculated mole fraction solubilities for a partial pressure of 1.013 bar from the data which has been published. These mole fraction solubilities fit the following equation:

$$\ln x_{D_2S} = -54.487 + 5727.1/(T/K) + 3.6052 \ln(T/K) + 0.028224\ T/K$$

$$\text{Standard deviation for } x_{D_2S} = \pm 4.0 \times 10^{-6}$$

This equation is based upon measurements from 278 to 323 K.

The mole fraction solubilities for a partial pressure of 1.013 bar in this temperature range lie close to mole fraction solubilities of hydrogen sulfide in water (H_2O). The evaluator has also derived a similar smoothing equation for solubilities of hydrogen sulfide in water at a partial pressure of 1.013 bar. All the corresponding values for deuterium sulfide in D_2O lie within the standard deviation for this smoothing equation for hydrogen sulfide.

References

1. Clarke, E.C.W.; Glew, D.N. *Can. J. Chem.* 1971, *49*, 691-698.

COMPONENTS:	ORIGINAL MEASUREMENTS:
1. Deuterium sulfide; D_2S; [13536-94-2] 2. Water - d_2; D_2O; [7789-20-0]	Clarke, E.C.W.; Glew, D.N. *Can. J. Chem.* 1971, *49*, 691-698.

VARIABLES:	PREPARED BY:
Temperature, pressure	C.L. Young.

EXPERIMENTAL VALUES:

	Total pressure			Mole fractions	
T/K	P/atm	P/kPa	D_2S in gas y_{D_2S}	D_2S in liquid $10^3 x_{D_2S}$	DS^- in liquid* $10^6 x_{DS^-}$
278.104	0.46205	46.818	0.98441	1.4489	0.56
278.103	0.53869	54.583	0.98659	1.6916	0.61
278.104	0.65058	65.921	0.98886	2.0451	0.67
278.103	0.83037	84.138	0.99122	2.6113	0.76
278.119	0.48660	49.305	0.98514	1.5266	0.58
278.119	0.56686	57.438	0.98721	1.7806	0.63
278.121	0.68365	69.272	0.98935	2.1493	0.69
278.120	0.87050	88.204	0.99159	2.7373	0.78
283.162	0.49623	50.281	0.97917	1.3152	0.59
283.161	0.58177	58.948	0.98219	1.5453	0.64
283.161	0.70846	71.785	0.98531	1.8853	0.71
283.161	0.91626	92.840	0.98857	2.4409	0.81
283.166	0.49103	49.754	0.97895	1.3017	0.59
283.167	0.57614	58.378	0.98201	1.5308	0.64
283.167	0.70225	71.156	0.98519	1.8694	0.71
283.165	0.90982	92.188	0.98849	2.4248	0.80
288.146	0.49292	49.946	0.97051	1.1202	0.60
288.145	0.58044	58.814	0.97489	1.3239	0.65
288.144	0.71133	72.076	0.97943	1.6278	0.72
288.142	0.92970	94.203	0.98416	2.1331	0.83
288.148	0.50118	50.783	0.97098	1.1375	0.60
288.149	0.58946	59.728	0.97526	1.3425	0.66
288.148	0.72127	73.083	0.97970	1.6480	0.73
288.147	0.94014	95.261	0.98433	2.1535	0.83
293.165	0.47451	48.080	0.95748	0.9288	0.59

AUXILIARY INFORMATION

METHOD/APPARATUS/PROCEDURE:	SOURCE AND PURITY OF MATERIALS:
Static equilibrium cell described in source. Vapor pressure over a saturated aqueous solution were measured with a fused quartz precision Bourdon gauge. Great care was taken to make the necessary correction for water in the vapor phase.	1. Prepared by the action of D_2O on aluminium sulfide (pure grade K and K lab). 2. Contained 0.25 weight % protium oxide (water). Supplied by Atomic Energy of Canada.
	ESTIMATED ERROR: $\delta T/K = \pm 0.0015$; $\delta P/P = \pm 0.0002$; $\delta x_{D_2S}/x_{D_2S} = \pm 0.002$.
	REFERENCES:

COMPONENTS:	ORIGINAL MEASUREMENTS:
1. Deuterium sulfide; D_2S; [13536-94-2] 2. Water-d_2; D_2O; [7789-20-0]	Clarke, E.C.W.; Glew, D.N. *Can. J. Chem.* 1971, *49*, 691-698.

EXPERIMENTAL VALUES:

	Total Pressure			Mole fractions	
T/K	P/atm	P/kPa	D_2S in gas y_{D_2S}	D_2S in liquid $10^3 x_{D_2S}$	DS^- in liquid* $10^6 x_{DS^-}$
293.167	0.56008	56.751	0.96388	1.1027	0.65
293.168	0.68930	69.844	0.97054	1.3647	0.72
293.167	0.90742	91.945	0.97748	1.8056	0.83
298.152	0.53870	54.584	0.94869	0.9207	0.64
298.151	0.65900	66.774	0.95792	1.1360	0.71
298.150	0.85875	87.014	0.96752	1.4924	0.82
298.153	0.55844	56.584	0.95048	0.9577	0.65
298.153	0.68292	69.198	0.95936	1.1807	0.73
298.153	0.88929	90.108	0.96861	1.5494	0.83
298.154	0.46306	46.920	0.94042	0.7867	0.59
298.155	0.54848	55.575	0.94958	0.9401	0.65
298.156	0.67873	68.773	0.95911	1.1736	0.72
298.155	0.90170	91.366	0.96902	1.5721	0.84
303.141	0.55882	56.623	0.93309	0.8361	0.66
303.141	0.69356	70.276	0.94589	1.0506	0.74
303.141	0.92605	93.833	0.95922	1.4196	0.86
303.148	0.62235	63.060	0.93982	0.9373	0.70
303.147	0.77226	78.250	0.95130	1.1757	0.78
303.146	1.03080	104.447	0.96326	1.5853	0.91
313.134	0.43164	43.736	0.84752	0.4797	0.57
313.132	0.51421	52.103	0.87173	0.5874	0.63
313.133	0.64240	65.092	0.89699	0.7543	0.71
313.133	0.86778	87.929	0.92331	1.0469	0.84
323.141	0.54232	54.951	0.79389	0.4746	0.64
323.141	0.68071	68.974	0.83525	0.6260	0.73
323.140	0.92773	94.003	0.87840	0.8956	0.87

*Values of x_{DS^-} were calculated from the equation

$$\log_{10} K_{D_2S} = 54.50 - 3760 \, (T/K) - 20 \log (T/K)$$

COMPONENTS:	EVALUATOR:
1. Hydrogen selenide; H_2Se; [7783-07-5] 2. Aqueous and non-aqueous solvents.	Peter G.T. Fogg, Department of Applied Chemistry and Life Sciences, Polytechnic of North London, Holloway, London N7 8DB, U.K. July 1987

CRITICAL EVALUATION:

Few measurements of the solubility of hydrogen selenide have been reported in the literature. At 298.15 K liquid hydrogen selenide has a lower vapor pressure than liquid hydrogen sulfide. Raoult's law leads one to expect that hydrogen selenide is likely to have a higher mole fraction solubility in a given solvent than hydrogen sulfide, at the same temperature and partial pressure of gas. This appears to be the case for eight of the nine non-aqueous solvents for which data is available. Hydrogen sulfide is, however, the more soluble in water under the conditions which have been investigated.

Devyatykh *et al*. (1) measured distribution constants for certain non-aqueous solvents. These can only be equated with Ostwald coefficients if it can be established that surface effects made negligible contribution to the chromatographic process and that conditions were close to equilibrium. In the absence of other data these measurements are best considered as an indication of approximate relative solubilities in different solvents. The following solvents were studied:

2-Ethoxyethanol; $C_4H_{10}O_2$; [110-80-5]   1,1'-Oxybis(2-chloroethane);
Nitrobenzene; $C_6H_5NO_2$; [98-95-3] $C_4H_8Cl_2O$; [111-44-4]
2-Furancarboxaldehyde; $C_5H_4O_2$; [98-01-1]
1,2-Benzenedicarboxylic acid, didecyl ester; $C_{28}H_{46}O_4$; [84-77-5]
Triethoxysilane; $C_6H_{16}O_3Si$; [998-30-1]   Silicic acid, tetraethyl ester;
Liquid paraffin $C_8H_{20}O_4Si$; [78-10-4]

In all these solvents, with the exception of 1,2-benzenedicarboxylic acid, didecyl ester, hydrogen selenide has a higher distribution constant than hydrogen sulfide at 293.2 K.

Solubility in water at a partial pressure of 1.013 bar has been reported by McAmis & Felsing (2) for 287.7 to 308.2 K and by Dubeau *et al*. (3) for 298.2 to 343.2 K. There is, on the whole, good agreement between the two sets of measurements (see fig.7). Mole fraction solubilities for a partial pressure of 1.013 bar may be fitted to the following equation:

$$\ln x_{H_2Se} = 9.15 + 974/(T/K) - 3.542 \ln (T/K) + 0.00420 (T/K)$$

$$\delta x_{H_2Se} = \pm 2.3 \times 10^{-5}$$

This equation is valid in the temperature range 287.7 to 343.2 K.

The value given by McAmis and Felsing for 308.2 K seems to be out of line with the data at higher temperatures by Dubeau and has not been used by the evaluator in the calculation of this smoothing equation.

McAmis & Felsing also investigated the solubility at 298.15 K and a total pressure of 1.013 bar in solutions of hydrogen iodide of concentrations to 2.73 mol dm^{-3}. Solubilities increase with concentration of hydrogen iodide and follow a Sechenov relationship very closely. They may be fitted to the equation:
$$\log_{10}(\text{concn. } H_2S/\text{mol } dm^{-3}) = -1.0788 + 0.04011 (\text{concn. HI/mol } dm^{-3})$$

Standard deviation of $\log_{10}$(concn. of H_2S/mol dm^{-3}) = ± 0.0035

These data for solutions of hydrogen iodide appear to be reliable and may be accepted on a tentative basis.

References

1. Devyatykh, G.G.; Ezheleva, A.E.; Zorin, A.D.; Zueva, M.V.
    Zh. Neorgan. Khim. <u>1963</u>, 8, 1307-1313.
    Russ. J. Inorg. Chem. <u>1963</u>, 8, 678-682.

2. McAmis, A.J.; Felsing, W.A. *J. Am. Chem. Soc.* <u>1925</u>, 4, 2633-2637.

3. Dubeau, C.; Sisi, J.-C.; Ozanne, N. *J. Chem. Engng. Data* <u>1971</u>, 16, 78-79.

COMPONENTS:	EVALUATOR:
1. Hydrogen selenide; H_2Se; [7783-07-5] 2. Aqueous and non-aqueous solvents	Peter G.T. Fogg, Department of Applied Chemistry and Life Sciences, Polytechnic of North London, Holloway, London N7 8DB, U.K. July 1987

CRITICAL EVALUATION:

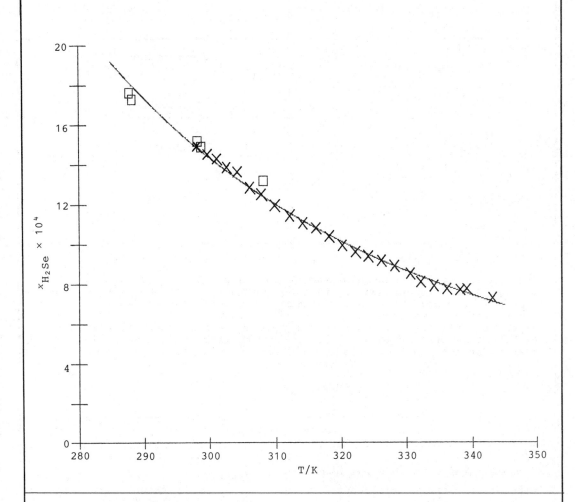

Fig. 7   Variation with temperature of the mole fraction solubility of hydrogen selenide in water at a partial pressure of 1.013 bar

Experimental points have been superimposed upon the curve corresponding to the equation:

$$\ln x_{H_2Se} = 9.15 + 974/(T/K) - 3.542 \ln (T/K) + 0.00420 (T/K)$$

☐ McAmis & Felsing (2)   ✕ Dubeau *et al.* (3)

COMPONENTS:	ORIGINAL MEASUREMENTS:
1. Hydrogen selenide; H_2Se; [7783-07-5] 2. Various liquids	Devyatykh, G. G.; Ezheleva, A. E.; Zorin, A. D.; Zueva, M. V. *Russ. J. Inorg. Chem.* <u>1963</u>, *8*, 678-682.

VARIABLES:	PREPARED BY:
Temperature, pressure	P. G. T. Fogg

EXPERIMENTAL VALUES:

Solvent	P_{H_2Se}/mmHg	Distribution constant vol_{H_2Se}/$vol_{solvent}$	Heat of solution /kcal mol^{-1}
2-Ethoxyethanol; $C_4H_{10}O_2$; [110-80-5]	0.3	14.42	- 1.6
1,1'-oxybis[2-chloroethane]; $C_4H_8Cl_2O$; [111-44-4]	0.2	12.5	- 3.4
Nitrobenzene; $C_6H_5NO_2$; [98-95-3]	0.2	22.0	- 2.45
Liquid paraffin	0.3	30.0	-

Temperature = 293.2 K.   760 mmHg = 1 atm = 1.013×10^5 Pa.

Distribution constants were measured between 278.2 K and 323.2 K with a total pressure of hydrogen selenide and carrier gas of about 760 mmHg. At a fixed temperature the distribution constants did not depart from the mean values by more than ±3%.   These mean values were reported at one temperature only but heats of solution, said to have been calculated from the variation of distribution constants with temperature, were given. If it is assumed that distribution constants vary with temperature according to equations of the form:

$$\ln K = (-\Delta H/RT) + A$$

(cont.)

AUXILIARY INFORMATION

METHOD/APPARATUS/PROCEDURE:	SOURCE AND PURITY OF MATERIALS:
A chromatographic method was used. Temperatures were controlled to ±0.5 K. The support phase consisted of Nichrome spirals. The carrier gas was either nitrogen or hydrogen. The volume, V_ℓ, of the liquid phase was calculated from the weight of the column before and after filling with liquid and allowing to drain. The free volume, V_g, was equated with the retention volume for hydrogen gas. The distribution constant, K, was calculated from the James and Martin equation: $$V_R = V_g + KV_\ell$$ where V_R is the retention volume for hydrogen selenide.	1. Prepared by hydrolysis of Al_2Se_3; purified by vacuum distillation; chromatographically pure. H_2 and N_2: passed through activated carbon and through molecular sieve.
	ESTIMATED ERROR:
	REFERENCES:

COMPONENTS:	ORIGINAL MEASUREMENTS:
1. Hydrogen selenide; H_2Se; [7783-07-5]	Devyatykh, G. G.; Ezheleva, A. E.; Zorin, A. D.; Zueva, M. V.
2. Various liquids	*Russ. J. Inorg. Chem.* 1963, *8*, 678-682.

EXPERIMENTAL VALUES:

where K is the distribution constant for a temperature T, ΔH is the heat of solution of hydrogen selenide in the solvent and A is a constant for the solvent, then distribution constants at any temperature in the range 278.2 K to 323.2 K may be estimated from a value at 293.2 K and the corresponding value of the heat of solution.   The equation for K may be written in the form:

$$K = \exp [A + (B/T)].$$

The following values of A and B have been calculated by the compiler:

Solvent	A	B/K
2-Ethoxyethanol	− 0.079	806
1,1'-oxybis[2-chloroethane]	− 3.313	1712
Nitrobenzene	− 1.116	1234

In the opinion of the compiler these distribution constants can not be equated with Ostwald coefficients unless the assumption is made that equilibrium was established between gas and liquid phases under the conditions of the experiment.

COMPONENTS:	ORIGINAL MEASUREMENTS:
1. Hydrogen selenide; H_2Se; [7783-07-5] 2. Various liquids	Devyatykh, G. G.; Ezheleva, A. E.; Zorin, A. D.; Zueva, M. V. *Russ. J. Inorg. Chem.* 1963, *8*, 678-682.
VARIABLES: Temperature	PREPARED BY: P. G. T. Fogg

EXPERIMENTAL VALUES:

Solvent	Distribution constant vol_{H_3P} /$vol_{solvent}$	Heat of solution /kcal mol^{-1}
2-Furancarboxaldehyde; $C_5H_4O_2$; [98-01-1]	22.5	-
1,2-Benzenedicarboxylic acid, didecyl ester; $C_{28}H_{46}O_4$; [84-77-5]	28.5	-
Triethoxysilane; $C_6H_{16}O_3Si$; [998-30-1]	5.21	- 5.0
Silicic acid, tetraethyl ester; $C_8H_{20}O_4Si$; [78-10-4]	11.52	- 5.4

Temperature = 293.2 K.

Distribution constants were measured between 278.2 K and 323.2 K.   The partial pressures of hydrogen selenide at which these measurements were made were not stated.   The total pressure of hydrogen selenide and carrier gas was about 760 mmHg in each case.   Distribution constants were reported at one temperature only but heats of solution, said to have been calculated from the variation of distribution constants with temperature, were given in two cases.

(cont.)

AUXILIARY INFORMATION

METHOD/APPARATUS/PROCEDURE:

A chromatographic method was used. Temperatures were controlled to ±0.5 K. The support phase consisted of Nichrome spirals.   The carrier gas was either nitrogen or hydrogen. The volume, V_ℓ, of the liquid phase was calculated from the weight of the column before and after filling with liquid and allowing to drain. The free volume, V_g, was equated with the retention volume for hydrogen gas.   The distribution constant, K, was calculated from the James and Martin equation:

$$V_R = V_g + KV_\ell$$

where V_R is the retention volume for hydrogen selenide.

SOURCE AND PURITY OF MATERIALS:

1. Prepared by hydrolysis of Al_2Se_3; purified by vacuum distillation; chromatographically pure.

H_2 and N_2: passed through activated carbon and through molecular sieve.

ESTIMATED ERROR:

REFERENCES:

COMPONENTS:	ORIGINAL MEASUREMENTS:
1. Hydrogen selenide; H_2Se; [7783-07-5] 2. Various liquids	Devyatykh, G. G.; Ezheleva, A. E.; Zorin, A. D.; Zueva, M. V. *Russ. J. Inorg. Chem.* <u>1963</u>, *8*, 678-682.

EXPERIMENTAL VALUES:

If it is assumed that distribution constants vary with temperature according to equations of the form:

$$\ln K = (-\Delta H/RT) + A$$

where K is the distribution constant for a temperature T, ΔH is the heat of solution in the solvent and A is a constant for the solvent, then distribution constants at any temperature in the range 278.2 K to 323.2 K may be estimated from a value at 293.2 K and the corresponding value of the heat of solution.   The equation for K may be written in the form:

$$K = \exp [A + (B/T)].$$

The following values of A and B have been calculated by the compiler:

Solvent	A	B/K
Triethoxysilane	- 6.936	2518
Silicic acid, tetraethyl ester	- 6.830	2719

In the opinion of the compiler these distribution constants can not be equated with Ostwald coefficients unless the assumption is made that equilibrium was established between gas and liquid phases under the conditions of the experiment.

COMPONENTS:	ORIGINAL MEASUREMENTS:
1. Hydrogen selenide; H_2Se; [7783-07-5] 2. Water; H_2O; [7732-18-5]	McAmis, A.J.; Felsing, W.A. *J. Am. Chem. Soc.* <u>1925</u>,*4*,2633-2637
VARIABLES:	PREPARED BY:
Temperature	C.L. Young

EXPERIMENTAL VALUES:
$$P_{H_2Se} = 760 \text{ mmHg} = 101.3 \text{ kPa}$$

T/°C	T/K	No. of determinations (Av. devn.%)	Moles H_2Se per dm^3 of soln. +	Mole fraction of H_2Se, x_{H_2Se}
14.6	287.8	5 (0.14)	0.09789	0.00176
15.0	288.2	5 (0.29)	0.09611	0.00173
25.0	298.2	6 (0.21)	0.08415	0.00152
25.6	298.8	4 (0.41)	0.08277	0.00149
35.0	308.2	4 (0.23)	0.07317	0.00132

+   it appears that this concentration is, as indicated, in units
    of mol dm^{-3} (soln) rather than mol dm^{-3} (water). It has been
    assumed that the molar volume of the solution is the same as
    that of water for the purpose of calculating the mole fraction.

AUXILIARY INFORMATION

METHOD/APPARATUS/PROCEDURE:	SOURCE AND PURITY OF MATERIALS:
Hydrogen selenide bubbled into water until saturated at the stated temperatures. Samples of saturated solutions removed and analysed either volumetrically or gravimetrically. The gravimetric method was based on the reaction between silver nitrate and hydrogen selenide. In the volumetric method saturated sample added to an excess of standard iodine solution, the selenium allowed to settle, and the excess iodine solution titrated against thiosulfate solution.	1. Prepared by action of water on aluminium selenide. 2. Boiled and distilled.
	ESTIMATED ERROR: $\delta T/K = \pm 0.1$; $\delta x_{H_2Se} = \pm 2\%$
	REFERENCES:

COMPONENTS:	ORIGINAL MEASUREMENTS:
1. Hydrogen selenide; H_2Se; [7783-07-5] 2. Water; H_2O; [7732-18-5]	Dubeau, C.; Sisi, J.-C.; Ozanne, N. *J. Chem. Engng. Data* <u>1971</u>, *16*, 78-79.

VARIABLES:	PREPARED BY:
Temperature	C. L. Young

EXPERIMENTAL VALUES:

Partial pressure = 1 atm = 101.3 kPa

T/°C	T/K	Solubility[#]	Mole fraction[§] of hydrogen selenide, x_{H_2Se}
25.1	298.25	0.08347	0.00150
		0.08289	0.00149
26.6	299.75	0.08062	0.00145
		0.08031	0.00145
28.0	301.15	0.07943	0.00143
		0.07909	0.00143
29.5	302.65	0.07709	0.00139
		0.07675	0.00138
31.1	304.25	0.0757	0.00136
		0.07533	0.00136
33.0	306.15	0.07091	0.00128
		0.07133	0.00129
34.7	307.85	0.06913	0.00125
		0.06915	0.00125
36.7	309.85	0.06647	0.00120
		0.06613	0.00119
39.0	312.15	0.06338	0.00114
		0.06331	0.00114
41.0	314.15	0.06097	0.00110
		0.06141	0.00111

(cont.)

AUXILIARY INFORMATION

METHOD/APPARATUS/PROCEDURE:	SOURCE AND PURITY OF MATERIALS:
Sample of water saturated with hydrogen selenide at lowest experimental temperature and sample withdrawn after 15 days. Temperature increased and after 4 hrs another sample withdrawn. Samples analyzed by gravimetric analysis using silver nitrate as precipitating agent. Details in source.	1. Matheson sample, purity 98 mole per cent. 2. Deoxygenated and demineralised.

ESTIMATED ERROR:

$\delta T/K = \pm 0.1$; $\delta S = \pm 1\%$ up to 50 °C and $\pm 2\%$ above 50 °C (estimated by compiler).

REFERENCES:

COMPONENTS:	ORIGINAL MEASUREMENTS:
1. Hydrogen selenide; H_2Se; [7783-07-5] 2. Water; H_2O; [7732-18-5]	Dubeau, C.; Sisi, J.-C.; Ozanne, N. *J. Chem. Engng. Data* 1971, *16*, 78-79.

EXPERIMENTAL VALUES:

$T/°C$	T/K	Solubility[#]	Mole fraction[§] of hydrogen selenide, x_{H_2Se}
43.0	316.15	0.05922	0.00107
		0.06033	0.00109
45.0	318.15	0.05833	0.00105
		0.05643	0.00102
47.0	320.15	0.05544	0.00100
		0.05504	0.000992
49.0	322.15	0.05361	0.000967
		0.05297	0.000955
51.0	324.15	0.05249	0.000947
		0.05161	0.000931
53.0	326.15	0.05079	0.000916
		0.05073	0.000915
55.0	328.15	0.05012	0.000904
		0.0487	0.000878
57.4	330.55	0.04672	0.000843
		0.04744	0.000856
59.0	332.15	0.04587	0.000827
		0.04384	0.000791
61.0	334.15	0.04465	0.000805
		0.04299	0.000775
63.0	336.15	0.04397	0.000793
		0.04197	0.000757
65.0	338.15	0.04377	0.000790
		0.04214	0.000760
		0.04255	0.000767
66.0	339.15	0.04228	0.000763
		0.04269	0.000770
		0.0440	0.000794
70.0	343.15	0.04017	0.000725

[#] Moles of H_2Se per dm^3 of solution.

[§] Calculated by compiler assuming that 1 dm^3 of solution contains 55.4 moles. It seems likely that the volume was measured at room temperature. The effect of the H_2Se on the density of water has been neglected.

COMPONENTS:	ORIGINAL MEASUREMENTS:
1. Hydrogen selenide; H$_2$Se; (7783-07-5]	McAmis, A.J. Felsing, W.A.
2. Hydrogen iodide; HI; [10034-85-2]	J. Am. Chem. Soc. 1925,4,2633-2637
3. Water; H$_2$O; [7732-18-5]	

VARIABLES:	PREPARED BY:
Concentration of comp. 2.	C.L. Young

EXPERIMENTAL VALUES:

T/°C	T/K	Conc. of HI/mol dm^{-3}	No. of determinations (Av.devn. %)	Conc. of H$_2$Se, S /mol dm^{-3} (soln)
25	298.2	0.20	4 (0.31)	0.08478
		0.40	4 (0.31)	0.08634
		2.73	4 (0.82)	0.11012

Total pressure = 760 mmHg = 1.013 bar

AUXILIARY INFORMATION

METHOD/APPARATUS/PROCEDURE:

Hydrogen selenide bubbled into hydriodic acid until saturated at the stated temperature. Samples of saturated solutions removed and analysed by volumetric method. A saturated sample was added to an excess of standard iodine solution, the selenium allowed to settle and the excess iodine solution titrated against thiosulfate solution.

SOURCE AND PURITY OF MATERIALS:

1. Prepared by action of water on aluminium selenide.

2.and 3.

Hydrogen iodide prepared by reaction of red phosphorus and iodine in the presence of water was dissolved in conductivity water.

ESTIMATED ERROR:

δT/K = ±0.1; δS = ±2%

REFERENCES:

SYSTEM INDEX

Page numbers preceded by E refer to evaluation texts whereas
page numbers not preceded by E refer to compiled tables
Compounds are named as in Chemical Abstracts indexes (toluene
appears under Benzene, methyl-).

REGISTRY NUMBER INDEX

Page numbers preceded by E refer to evalation texts whereas those not preceded by E refer to compiled tables.

694-48-4	E180-E182, 301
696-29-7	E166, E169, 221
779-02-2	E170, E171, 237
872-50-4	E180, 260, 305-313
929-06-6	E80, E82, 108, 155-159
998-30-1	E185, 277, 278, E330, 334, 335
1120-21-4	E166-E168, 189
1313-82-2	48
1321-94-4	E170, E171, 236
1333-74-0	75, 76, 237, 240, 276
1453-58-3	301
1559-34-8	E173-E175, 257
1678-91-7	E168, E169, 218
1678-92-8	E166, E169, 219, 220
2556-73-2	E180, E184, 314
6484-52-2	39
7447-40-7	39, 41
7601-89-0	E26, 38, 40
7601-90-3	E16, E26, 38
7631-99-4	39
7646-85-7	E26, 35
7647-01-0	E25, E26, 29, 30, 34, 35-37
7647-14-5	E25, E26, 36, 37, 39, 41, 42
7664-41-7	E54, E55, 56-78
7664-93-9	E26, 32, 33, 52, 53
7704-34-9	E185, 317
7727-37-9	75, 76
7732-18-5	E1-E3, 4-19, E20, 21-24, E25-E28, 29-50, E54, E55, 56-78, E79-E83, 84-165, 298, E330, E331, 336-339
7757-79-1	E25, 39
7757-82-6	E25, 39, 40, 43
7758-94-3	E26, 34
7778-80-5	39
7783-06-4	E1-E3, 4-19, E20, 21-24, E25-E28, 29-50, E51, 52-53, E54, E55, 56-78, E79-E83, 84-165, E166-E187, 188-326
7783-07-5	E330, E331, 332-339
7783-20-2	39, 67
7789-20-0	E327, 328, 329
7791-03-9	298
10034-85-2	E26, 31, E330, 339
10043-52-4	E25, E26, 40, 44
12125-02-9	39, 66
13536-94-2	E327, 328, 329
25265-71-8	E173, E175, 250
25322-68-3	E173-E175, 252, 253
26896-48-0	259
33100-27-5	E173-E176
63095-29-4	E174-E176, 261

AUTHOR INDEX

Page numbers preceded by E refer to evaluation texts whereas those not preceded by E refer to compilation tables.

SOLUBILITY DATA SERIES